Bioceramics: Status in Tissue Engineering and Regenerative Medicine

(Part 1)

Edited by

Saeid Kargozar
Department of Radiation Oncology
Simmons Comprehensive Cancer Center
UT Southwestern Medical Center
Harry Hines Blvd Dallas
TX75390, USA

&

Francesco Baino
Department of Applied Science and Technology (DISAT)
Institute of Materials Physics and Engineering
Politecnico di Torino
Torino, Italy

Bioceramics: Status in Tissue Engineering and Regenerative Medicine *(Part 1)*

Editors: Saeid Kargozar and Francesco Baino

ISBN (Online): 978-981-5238-39-6

ISBN (Print): 978-981-5238-40-2

ISBN (Paperback): 978-981-5238-41-9

need for a court order if at any point you breach any terms of this License Agreement. In no event will any delay or failure by Bentham Science Publishers in enforcing your compliance with this License Agreement constitute a waiver of any of its rights.

3. You acknowledge that you have read this License Agreement, and agree to be bound by its terms and conditions. To the extent that any other terms and conditions presented on any website of Bentham Science Publishers conflict with, or are inconsistent with, the terms and conditions set out in this License Agreement, you acknowledge that the terms and conditions set out in this License Agreement shall prevail.

Bentham Science Publishers Pte. Ltd.
80 Robinson Road #02-00
Singapore 068898
Singapore
Email: subscriptions@benthamscience.net

**BENTHAM
SCIENCE**

CONTENTS

David Bahati, Meriame Bricha and *Khalil El Mabrouk*

FOREWORD

The use of bioceramics for tissue engineering and regenerative medicine extends over two centuries. Dorozhkin provided a detailed review of the history of bioceramics [1]. He noted that Johan Gottlieb Gahn and Carl Wilhelm Scheele first described the presence of calcium and phosphorus in bone in the second half of the eighteenth century [1, 2]. The first use of bioceramics in medicine occurred in the late nineteenth century when Junius E. Cravens distributed a calcium orthophosphate powder called "Lacto-Phosphate of Lime" for capping the dental pulp during dental restorations [1, 3, 4]. Larry Hench's discovery in 1969 that a sodium-calcium-phosphorous—silicate glass possesses bone bonding functionality gave rise to the clinical use of "bioactive glass" materials for bone repair [5, 6]. The term "bioceramics" was first used shortly thereafter in 1971 [7]. The bioceramics field is now truly global in nature and includes research, pre-clinical, and clinical activities involving various types of bioactive and bioinert inorganic materials.

This is Part 1 by Saeid Kargozar, a research fellow in the Department of Radiation Oncology, Simmons Comprehensive Cancer Center, UT Southwestern Medical Center, and Francesco Baino, an associate professor in the Department of Applied Science and Technology at the Politecnico di Torino, provides a comprehensive overview of the use of bioceramics for tissue engineering and regenerative medicine. The first part of the book (Part 1) focuses on the fundamentals of biocompatible ceramics, bioactive glasses and composites, and collects 10 chapters. In Chapter 1, Kargozar and Baino provide a description of the status of bioceramics in tissue engineering and regenerative medicine. Chapter 2, by Moghanian *et al.*, provides an introduction to biocompatible glasses, ceramics, and glass ceramics. Batool *et al.* consider recent advances in bioactive glasses and glass ceramics in Chapter 3. Chapter 4, by Bahati *et al.*, describes the structure, properties, and processing of bioactive glasses. Kargozar *et al.* focus on the biocompatibility of bioactive glasses in Chapter 5. In Chapter 6, Moghanian and Nasiripour describe the use of bioinert ceramics for biomedical applications. Moghanian *et al.* review the processing and properties of bioresorbable ceramics in Chapter 7. Dorozhkin reviews the use of calcium orthophosphates in tissue engineering in Chapter 8. In Chapter 9, Hosseini *et al.* consider the use of carbon nanostructures for tissue engineering and cancer therapy. Benedini and Messina describe advances in polymer/ceramic composites for bone tissue engineering in Chapter 10. The second part of the book (Part 2) will be addressed to the applications of the bioceramic materials discussed in the present volume.

In this volume, Professors Kargozar and Baino as well as the chapter contributors have provided the bioceramics community with a comprehensive consideration of the bioceramics field. I anticipate that their volume will be beneficial to students as well as researchers in academia, government, and industry as they continue efforts to improve our understanding of the use of bioceramic materials for tissue engineering and regenerative medicine applications.

Prof. Roger Narayan
Joint Department of Biomedical Engineering
North Carolina and North Carolina State University
Raleigh, USA

REFERENCES

[1] Dorozhkin SV. A detailed history of calcium orthophosphates from 1770s till 1950. Mater Sci Eng C 2013; 33(6): 3085-110.
[http://dx.doi.org/10.1016/j.msec.2013.04.002] [PMID: 23706189]

[2] Dorozhkin SV. A history of calcium orthophosphates (CaPO$_4$) and their biomedical applications. Morphologie 2017; 101(334): 143-53.
[http://dx.doi.org/10.1016/j.morpho.2017.05.001] [PMID: 28595833]

[3] Dorozhkin SV. Calcium orthophosphates as a dental regenerative material 2019.
[http://dx.doi.org/10.1016/B978-0-08-102476-8.00016-5]

[4] Cravens JE. Lacto-phosphate of lime; Pathology and treatment of exposed dental pulps and sensitive dentine. Dent Cosmos 1876; 18: 463-9.

[5] Hench LL. The story of Bioglass®. J Mater Sci Mater Med 2006; 17(11): 967-78.
[http://dx.doi.org/10.1007/s10856-006-0432-z] [PMID: 17122907]

[6] Hench LL, Splinter RJ, Allen WC, Greenlee TK. Bonding mechanisms at the interface of ceramic prosthetic materials. J Biomed Mater Res 1971; 5(6): 117-41.
[http://dx.doi.org/10.1002/jbm.820050611]

[7] Blakeslee KC, Condrate RA, Sr. Vibrational spectra of hydrothermally prepared hydroxyapatites. J Am Ceram Soc 1971; 54(11): 559-63.
[http://dx.doi.org/10.1111/j.1151-2916.1971.tb12207.x]

List of Contributors

Amirhossein Moghanian — Department of Materials Engineering, Imam Khomeini International University, Qazvin 34149-16818, Iran

Amir K. Miri — Department of Biomedical Engineering, New Jersey Institute of Technology, Newark, New Jersey 07102, United States

Anuj Kumar — School of Materials Science and Technology, Indian Institute of Technology (BHU), Varanasi 221005, India

Danna Valentina Sánchez — Department of Biomedical Engineering, New Jersey Institute of Technology, Newark, New Jersey 07102, United States

David Bahati — Euromed University of Fes, UEMF, Fes, Morocco

Francesco Baino — Department of Applied Science and Technology (DISAT), Institute of Materials Physics and Engineering, Politecnico di Torino, 10129 Torino, Italy

Fabian Westhauser — Department of Orthopaedics, Heidelberg University Hospital, Schlierbacher Landstraße 200a, 69118 Heidelberg, Germany

Hae-Won Kim — Institute of Tissue Regeneration Engineering (ITREN), Dankook University, Cheonan 330-714, Republic of Korea
Department of Nanobiomedical Science & BK21 PLUS NBM Global Research Center for Regenerative Medicine, Dankook University, Cheonan 330-714, Republic of Korea
Department of Biomaterials Science, School of Dentistry, Dankook University, Cheonan 330-714, Republic of Korea

Khalil El Mabrouk — Euromed University of Fes, UEMF, Fes, Morocco

Luciano Benedini — INQUISUR-CONICET, Universidad Nacional del Sur, Bahía Blanca, Argentina
Department of Biology, Biochemistry and Pharmacy, Universidad Nacional del Sur, Bahía Blanca, Argentina

Memoona Akhtar — Department of Materials Science and Engineering, Institute of Space Technology Islamabad, Islamabad-44000, Pakistan
Department of Materials Science and Engineering, Institute of Biomaterials, University of Erlangen-Nuremberg, Erlangen-91058, Germany

Muhammad Rizwan — Department of Metallurgical Engineering, Faculty of Chemical and Process Engineering, NED University of Engineering and Technology, Karachi-75270, Pakistan

Muhammad Atiq Ur Rehman — Department of Materials Science and Engineering, Institute of Space Technology Islamabad, Islamabad-44000, Pakistan

Meriame Bricha — Euromed University of Fes, UEMF, Fes, Morocco

Niloofar Kolivand — Department of Materials Engineering, Imam Khomeini International University, Qazvin, 34149-16818, Iran

Paula Messina — NQUISUR-CONICET, Universidad Nacional del Sur, Bahía Blanca, Argentina
Department of Chemistry, Universidad Nacional del Sur, Bahía Blanca, Argentina

Saeid Kargozar Department of Radiation Oncology, Simmons Comprehensive Cancer Center, UT Southwestern Medical Center, 5323 Harry Hines Blvd, Dallas, TX75390, USA

Syeda Ammara Batool Department of Materials Science and Engineering, Institute of Space Technology Islamabad, Islamabad-44000, Pakistan

Saba Nasiripour School of Metallurgy and Materials Engineering, Faculty of Engineering, University of Tehran, Tehran, Iran

Sergey V. Dorozhkin Department of Physics, Moscow State University, Vorobievy Gory, Moscow, Russia

Seyede Atefe Hosseini Department of Medical Biotechnology and Nanotechnology, Faculty of Medicine, Mashhad University of Medical Sciences, Mashhad, Iran

Zahra Miri Department of Materials Engineering, Isfahan University of Technology, Isfahan 84156-83111, Iran

CHAPTER 1

Bioceramics: Status in Tissue Engineering and Regenerative Medicine

Saeid Kargozar[1,*] and Francesco Baino[2,*]

[1] *Department of Radiation Oncology, Simmons Comprehensive Cancer Center, UT Southwestern Medical Center, 5323 Harry Hines Blvd, Dallas, TX75390, USA*

[2] *Department of Applied Science and Technology (DISAT), Institute of Materials Physics and Engineering, Politecnico di Torino, 10129 Torino, Italy*

Abstract: Tissue engineering and regenerative medicine seek biomaterials with potent regenerative potential *in vivo*. The bioceramics superfamily represents versatile inorganic materials with exceptional compatibility with living cells and tissues. They can be classified into three distinctive groups including almost bioinert (*e.g.*, alumina and zirconia), bioactive (bioactive glasses (BGs)), and bioresorbable (*e.g.*, calcium phosphates (CaPs)) ceramics. Regarding their physicochemical and mechanical properties, bioceramics have been traditionally used for orthopedic and dental applications; however, they are now being utilized for soft tissue healing and cancer theranostics due to their tunable chemical composition and characteristics. From a biological perspective, bioceramics exhibit great opportunities for tissue repair and regeneration thanks to their capability of improving cell growth and proliferation, inducing neovascularization, and rendering antibacterial activity. Different formulations of bioceramics with diverse shapes (fine powder, particles, pastes, blocks, *etc.*) and sizes (micro/ nanoparticles) are now available on the market and used in the clinic. Moreover, bioceramics are routinely mixed into natural and synthetic biopolymers to extend their applications in tissue engineering and regenerative medicine approaches. Current research is now focusing on the fabrication of personalized bioceramic-based scaffolds using three-dimensional (3D) printing technology in order to support large-volume defect tissue regeneration. It is predicted that more commercialized products of bioceramics will be available for managing both hard and soft tissue injuries in the near future, either in bare or in combination with other biomaterials.

---

* Corresponding authors Saeid Kargozar and Francesco Baino: Department of Radiation Oncology, Simmons Comprehensive Cancer Center, UT Southwestern Medical Center, 5323 Harry Hines Blvd, Dallas, TX75390, USA; Department of Applied Science and Technology (DISAT), Institute of Materials Physics and Engineering, Politecnico di Torino, 10129 Torino, Italy; Tel: 214-648-3111; E-mails: Saeid.Kargozar@utsouthwestern.edu, francesco.baino@polito.it

Keywords: Additive manufacturing, Angiogenesis, Antibacterial activity, Anticancer activity, Bioactive glasses (BGs), Bioinert ceramics, Bioresorbable ceramics, Biofabrication, Bone regeneration, Calcium phosphates, Composite, Clinical trials, Drug delivery, Hydroxyapatite (HAp), Regenerative medicine, Scaffolds, Soft tissue healing, Tissue engineering, Three-dimensional (3D) printing, Wound healing.

INTRODUCTION

Tissue engineering is a multidisciplinary field that aims to regenerate damaged tissues by applying the principles of engineering, materials science, biology, and medicine. Pioneers in the field have introduced biomaterials, cells, and bioactive molecules as the three main building blocks of tissue engineering and regenerative medicine field [1]. Naturally, human tissues are formed from differentiated or undifferentiated cells located in an extracellular matrix (ECM) (mostly collagen) containing bioactive molecules (*e.g.*, growth factors). As a rule of thumb, the ECM of tissues is greatly destroyed following severe injuries and damages; therefore, various biocompatible materials can be utilized as three-dimensional (3D) scaffolds to restore the destroyed ECM. Up to now, many types of natural and synthetic materials have been successfully processed, developed, and used for managing different tissue damage and injuries [2, 3]. Naturally occurring substances suffer from critical restrictions including the risk of disease transmission, batch-to-batch variations, and limited availability [4, 5]. Accordingly, there is a great interest in the use of synthetic materials in tissue reconstruction approaches. Regarding the nature of hard tissues (*e.g.*, bone), bioceramics are recognized as the ideal implant materials for the replacement of degenerated or traumatized osseous tissues.

Bioceramics represent biocompatible ceramic materials that are being continuously developed for use as medical implants. In fact, they are inorganic biomaterials that comprise crystalline ceramics, amorphous glasses, and glass-ceramics. In other words, the bioceramics superfamily members can be classified into three distinct generations, *i.e.*, almost bioinert (*e.g.*, alumina and zirconia), bioactive (*e.g.*, bioactive glasses (BGs)), and bioresorbable (*e.g.*, most calcium phosphates (CaPs)). These substances are commonly synthesized in the laboratory using high temperatures and used in different formats, including fine powder, granules, and dense blocks. Furthermore, bioceramics can be fabricated into tissue-mimicking scaffolds through well-established techniques and protocols (*e.g.*, sponge replication method). In recent years, great efforts have been made to produce bioceramics-based constructs using 3D printing machines in order to fit the size and shape of the lost tissues. It should be mentioned that some types of bioceramics (BGs) are being utilized as coatings for other ceramics or metal implants.

The most fascinating feature of bioceramics for orthopedic and dental applications is related to their mechanical properties which are in the range of naïve hard tissues. In addition, bioceramics (*e.g.*, BGs and glass ceramics) exhibit excellent biological properties, including the ability to induce osteogenesis, osteoconduction, osteoinduction, and osteointegration. Moreover, bioceramics can be employed for the loading and delivery of various drugs, chemicals, and bioactive molecules to desired locations in the body. Although the first and foremost application of bioceramics is to restore hard tissue lesions, recent trends have also confirmed their suitability in soft tissue repair and regeneration (*e.g.*, skin wound healing). In this sense, they can be utilized as additives in polymeric substrates for improving particular biological events (*e.g.*, angiogenesis), and the reported data have been quite interesting. Still, some challenges remain to be solved regarding the widespread use of bioceramics in soft tissue healing strategies, including defining the most suitable composition and formulation. Since implantable materials must be compatible with living systems (*e.g.*, cells and tissues), bioceramics have been extensively examined for their potential adverse effects (toxicity) *in vitro* and *in vivo*. In general, bioceramics are known as safe substances for human beings; their main components (elements like silicon, calcium, phosphorus, *etc.*) are commonly found in low concentrations in the body and needed for the proper function of human cells [6]. However, some potentially toxic elements (*e.g.*, cobalt) can be incorporated into the basic composition of bioceramics for rendering particular activities, such as improving angiogenesis. In this case, caution should be taken to avoid any unwanted adverse effects on the human body at molecular and cellular levels. In addition, the positive potential effects of any new formulation of bioceramics may be of interest to researchers and scientists in the field.

In this chapter, we first introduce the structure, properties, and classifications of bioceramics and then highlight their possibilities in tissue engineering and regenerative medicine. The main challenges ahead will be discussed to shed light on their future applications for managing injured tissues.

BIOCERAMICS: STORY AND SIGNIFICANCE

The human body is a "marvelous machine" that efficiently incorporates different materials for different functions, such as structural support, filtration capacity, energy generation and storage, gas exchange, flexibility, and self-healing/regenerative ability, into one fascinating, integrated, and well-orchestrated bio-system. In other words, the human body is an exceptional "collection" of highly functional materials. Ideally, these materials should retain their functionality for many decades throughout the human lifespan, which is 73.2 years on average worldwide [7]. However, the biomaterials of the body are at risk of many harsh conditions and situations (*e.g.*, overuse, trauma, or different pathologies like osteoporosis) over years and sometimes fail. This is becoming more common as worldwide populations age owing to the overall increase in life expectancy. Hence, the development of "spare parts" for the restoration of injured and diseased tissues/organs/structures of the body is becoming of utmost importance.

The use of ceramic materials in biomedicine dates back to the 10th century AD when ancient Egyptians used calcium sulfate for the restoration of broken bones of cadavers and mummies [8]. The first official report about the implantation of calcium sulfate in living human patients to fill voids resulting from tuberculous osteomyelitis was published in 1892 [9]. After the Second World War, high-purity pellets of calcium sulfate (also called "plaster of Paris") started being routinely used as a bone substitute following the seminal work of Peltier in the early 1960s [10]. In the same years, other very important classes of bioceramics – *i.e.* alumina for joint prostheses, hydroxyapatite, and bioactive glasses for bone tissue regeneration in orthopedics and dentistry – began to attract the researchers' interest and to be systematically investigated for use in contact with bone [11].

At present, bone is globally the second tissue needing replacement after blood [12]. The bone restoration can be completed by using a variety of natural and synthetic substances including autografts from the patient, allografts from another donor/cadaver, xenografts from animals, as well as man-made bioceramics. Over the years, numerous compositions of bioceramics have been developed and explored for the replacement of injured bones. The common forms of bone grafts include monolithic devices (used in the reconstruction of middle ear small bones or orbital floor), fine particles, porous granules, rigid scaffolds for filling large bone defects, moldable pastes (*e.g.*, injectable blocks of cement for spine surgery), coatings on metallic prostheses and composites involving the dispersion of ceramic inclusions in a soft polymeric matrix. New emerging applications are mainly addressed to multifactorial tissue engineering and may involve special extra-functionalities, such as therapeutic actions in contact with soft tissues and controlled drug/ion delivery.

OVERVIEW OF BIOCERAMICS APPLICATIONS WITH FOCUS ON TISSUE ENGINEERING AND REGENERATIVE MEDICINE

Historically, bioceramics have been widely used for managing hard tissue lesions due to their appropriate physico-chemical, mechanical, and biological properties.

Almost all bioinert ceramics (*e.g.*, alumina (Al_2O_3), zirconia (ZrO_2), titania (TiO_2)) form the first generation of bioceramics and have been successfully used as musculoskeletal (hip and knee replacements) and dental implants. Indeed, this kind of ceramics exhibits excellent mechanical properties (*e.g.*, tensile, compressive, hardness, low wear, toughness) for long-term implantation in the body. Moreover, they show good corrosion resistance which makes them suitable devices for long-lasting implantation. Nonetheless, bioinert ceramics usually need to be coated with other types of bioceramics (*e.g.*, BGs and CaPs) in order to improve their biological characteristics (*e.g.*, osteoconduction, osteointegration, and osteogenesis) [13]. Interestingly, bioinert ceramics have been utilized for coating alloys (*e.g.*, Ti6Al4V) without causing any negative impacts on cell viability and proliferation [14].

As the second generation of bioceramics, BGs, and glass-ceramics offer great opportunities for tissue engineering and regenerative medicine strategies. These biocompatible substances can bind to the living tissues (both hard and soft) through a hydroxycarbonate apatite (HCA) layer which is formed on their surface upon contacting physiological fluids (*e.g.*, blood plasma). There are plentiful experimental studies in the literature that indicate the great suitability of BGs for the repair and regeneration of bone tissue [15]. BGs were first invented by Prof. Larry Hench in 1969 at the University of Florida; 45S5 Bioglass® is known as the parent of silicate-based BGs with the composition of $45SiO_2–24.5CaO–24.5NaO–6P_2O_5$ (wt%) [16]. Since then, two other subgroups of BGs have been successfully developed and named phosphate- and borate-based BGs. The primary application of BGs was to restore damaged bone and teeth due to their excellent inherent properties. BGs can induce the osteogenesis process (*i.e.*, supporting new bone growth) and thereby are known as osteoinductive materials [17]. Additionally, BGs can support human cell growth, proliferation, and differentiation, leading to accelerating bone tissue reconstruction [18]. Interestingly, BGs have been found as angiogenesis-inducing materials that can encourage new blood vessel formation *in vitro*, *ex vivo*, and *in vivo* [19, 20]. This potential can greatly accelerate bone regeneration since angiogenesis plays a pivotal role in all stages of the tissue healing process. As bacterial infections are a life-threatening issue in the clinic, particular types of BGs were developed and confirmed to act against both Gram-positive and Gram-negative bacteria [21]. Recent studies have revealed that specific formulations of BGs may modulate inflammatory responses through the stimulation of M1 to M2 phenotype switching of macrophages [22]. The main mechanism behind the mentioned biological activities is associated with the ion release process from BGs into the surrounding physiological environment [23]. Accordingly, several attempts have been made to incorporate metallic and non-metallic ions into the basic composition of BGs to enhance and extending their biological performance. For example, copper-doped BGs may show antibacterial and angiogenic effects [24]; while barium-containing BGs can elicit anti-inflammatory responses [25]. Focusing on tissue engineering and regenerative medicine, BGs have been selected in order to generate 3D scaffolds for bone tissue engineering applications [26, 27]. However, BG-based constructs suffer from low mechanical properties due to their brittle nature. Therefore, they are usually mixed with biocompatible natural and synthetic polymers to fabricate composite scaffolds having improved mechanical properties. On the other hand, BGs are added to otherwise-bioinert polymeric constructs for rendering specific biological features (*e.g.*, improving angiogenesis). The current research aims to utilize BGs for the fabrication of 3D printed scaffolds in the concept of personalized medicine [28].

Recently, BGs have been suggested as suitable additives for soft tissue healing (*e.g.*, skin wound healing) [29]. In fact, they were proposed for soft tissue engineering due to their outstanding biological features, including biocompatibility, angiogenesis-induction, and antibacterial activity.

The third generation of bioceramics is represented by bioresorbable ceramics that are prone to dissolution and degradation by the body cells. Bioresorbable bioceramics include amorphous calcium phosphates (CaPs), nano-sized HAp, α-/β-tricalcium phosphates (TCPs), and calcium sulfates (including plaster of Paris). Prior reports have shown that the resorption rate of different CaPs varies as the following trend of α-TCP > β-TCP > HAp. Apart from the mentioned compositions, other types of bioresorbable CaP-based materials have been developed and proven to be resorbed in the body, including dicalcium phosphate dihydrate (DCPD; $CaHPO_4 \cdot 2H_2O$), dicalcium phosphate (DCP; $CaHPO_4$), octocalcium phosphate (OCP; $Ca_8H_2(PO_4)_6 \cdot 5H_2O$). Compositionally, CaPs exhibit the greatest similarity to the minerals found in the natural bone tissue; therefore, they are extensively used for bone tissue engineering applications (*e.g.*, spinal surgery) [30]. The bioresorption process of these materials is determined by two main factors, including solubility kinetics and *in vivo* conversion [31]; hence, their degradation can be regulated by two main mechanisms of physico-chemical- and cell-mediated dissolution [32]. Many experimental studies have confirmed that by-products of bioresorbable ceramics are not toxic to human cells and tissues [32]. The possibility of generating nano-scaled bioresorbable ceramics has revolutionized their applications in orthopedic surgery. For tissue engineering and regenerative medicine, bioresorbable ceramics-based products in different forms and formulations (fine powders/particles, paste, *etc.*) have been successfully commercialized and utilized for clinical applications of hard tissue lesions [33]. These substances were proven to induce the osteogenic differentiation of bone-related cells (*i.e.*, osteogenesis) and stimulate neovessel formation, either in dopant-free or doped forms [34, 35]. The fabrication of 3D-printed scaffolds from

CaPs has opened up new horizons in the field regarding the next generation of patient-specified constructs [36]. Still, the low mechanical properties of CaPs have limited their use in load-bearing applications; they are commonly mixed into polymeric materials for generating composite constructs in different shapes (*e.g.*, 3D scaffolds, hydrogels, *etc.*) [37, 38]. It should be stated that CaPs can be used as coatings on metal alloys as well as bone cement [39]. Moreover, they are currently employed for drug delivery applications thanks to their suitable structure that enables loading and delivering various therapeutical drugs [40]. It is of interest that recent studies have elucidated some types of CaPs (*e.g.*, HAp) that can be used for managing soft tissue healing, like skin tissue repair and regeneration [41, 42].

CONCLUDING REMARKS

Bioceramics are routinely used for healthcare applications and the relevant market for these products is significant. The introduction of novel cost-effective therapeutics has been always welcome in the biomedical industry; thus, advancements in the field of biomaterials are progressing faster than just a few years ago due to the continuous introduction of new, smart materials options. In this context, bioceramics indeed play a pivotal role. Bioceramics make bonds to the human body and can provide great support for damaged and diseased tissues and organs. Regarding aging populations and the need for more sophisticated tissue replacements, a bright future with great opportunities can be forecast for ceramic-based technologies. In this regard, implantable biomaterials were previously estimated to have a global market of around $110 billion in 2019 [43].

In terms of the future of healthcare, regenerative medicine is a big business. Tissue engineering and regeneration-based technologies were previously estimated to have a global market of around $25 billion in 2018 and are predicted to reach $109.9 billion by 2023, representing an impressive growth rate [43]. While bone is a significant focus of this market, the attention is moving to soft tissues as well. On this matter, cardiovascular and gastrointestinal systems, muscle, neural, and skin tissues have been treated with some types of bioceramics with promising results *in vitro* and *in vivo* [29]. Also, there is potential for many different types of materials in this broad field. In the field of regenerative medicine and tissue engineering, no one material is going to tackle all the challenges. Many of the ceramic- and glass-based strategies to heal tissues often combine these bioactive materials with non-bioactive/resorbable organic phases, for example in polymer-matrix composites or hydrogels [44, 45]. For more progress in the field, it seems necessary to make more collaboration between different areas of science including materials science and engineering, biology, pharmacology, and medicine. In this regard, understanding genetic upregulation and activation by ionic stimuli released from bioactive ceramics and glasses offers the possibility of developing patient-specific therapies, which is a huge challenge for the aging population.

The next generation of biomaterials and scaffolds with the capability of simultaneous treatment of different tissues can meet the future of tissue engineering and regenerative medicine. In this regard, multifunctional stimuli-responsive biomaterials can be effective in facing coordinated and complex responses of the human body to any implanted substances. In this regard, additive manufacturing technologies [46] combined with biofabrication principles [47], involving the manipulation and printing of biomaterials (*e.g.* bioactive ceramics/glasses), biomolecules, and living cells, will be an exceptional resource.

REFERENCES

[1]   Hoffman T, Khademhosseini A, Langer R. Chasing the paradigm: clinical translation of 25 years of tissue engineering. Tissue Eng Part A 2019; 25(9-10): 679-87.
[http://dx.doi.org/10.1089/ten.tea.2019.0032] [PMID: 30727841]

[2]   Ashouri S, Hosseini SA, Hoseini SJ, *et al.* Decellularization of human amniotic membrane using detergent-free methods: Possibilities in tissue engineering. Tissue Cell 2022; 76: 101818.
[http://dx.doi.org/10.1016/j.tice.2022.101818] [PMID: 35580526]

[3]   Kelly CN, Miller AT, Hollister SJ, Guldberg RE, Gall K. Design and structure–function characterization of 3D printed synthetic porous biomaterials for tissue engineering. Adv Healthc Mater 2018; 7(7): 1701095.
[http://dx.doi.org/10.1002/adhm.201701095] [PMID: 29280325]

[4]   Brouki Milan P, Pazouki A, Joghataei MT, *et al.* Decellularization and preservation of human skin: A platform for tissue engineering and reconstructive surgery. Methods 2020; 171: 62-7.
[http://dx.doi.org/10.1016/j.ymeth.2019.07.005] [PMID: 31302179]

[5]   Milan PB, Amini N, Joghataei MT, *et al.* Decellularized human amniotic membrane: From animal models to clinical trials. Methods 2020; 171: 11-9.
[http://dx.doi.org/10.1016/j.ymeth.2019.07.018] [PMID: 31326597]

[6]   Kargozar S, Hamzehlou S, Baino F. Effects of the biological environment on ceramics. 2018.

[7]   WorldoMeter. 2022. Available from: https://www.worldometers.info/demographics/life-expectancy

[8]   Dorozhkin SV. A detailed history of calcium orthophosphates from 1770s till 1950. Mater Sci Eng C 2013; 33(6): 3085-110.
[http://dx.doi.org/10.1016/j.msec.2013.04.002] [PMID: 23706189]

[9]   Dressman H. Ueber knochenplombierung bei hohlenformigen defekten des knochens. Beitr Klin Chir 1892; 9: 804-10.

[10]   Peltier LF. The use of plaster of Paris to fill defects in bone. Clin Orthop 1961; 21(21): 1-31.
[PMID: 14485018]

[11]   Hench LL. Bioceramics: From concept to clinic. J Am Ceram Soc 1991; 74(7): 1487-510.
[http://dx.doi.org/10.1111/j.1151-2916.1991.tb07132.x]

[12]   Shegarfi H, Reikeras O. Review article: Bone transplantation and immune response. J Orthop Surg 2009; 17(2): 206-11.
[http://dx.doi.org/10.1177/230949900901700218] [PMID: 19721154]

[13]   Vernè E, Bosetti M, Brovarone CV, *et al.* Fluoroapatite glass-ceramic coatings on alumina: Structural,

mechanical and biological characterisation. Biomaterials 2002; 23(16): 3395-403.
[http://dx.doi.org/10.1016/S0142-9612(02)00040-6] [PMID: 12099282]

[14]   Chen T, Deng Z, Liu D, Zhu X, Xiong Y. Bioinert TiC ceramic coating prepared by laser cladding: Microstructures, wear resistance, and cytocompatibility of the coating. Surf Coat Tech 2021; 423: 127635.
[http://dx.doi.org/10.1016/j.surfcoat.2021.127635]

[15]   Kermani F, Mollazadeh Beidokhti S, Baino F, Gholamzadeh-Virany Z, Mozafari M, Kargozar S. Strontium- and cobalt-doped multicomponent mesoporous bioactive glasses (MBGs) for potential use in bone tissue engineering applications. Materials 2020; 13(6): 1348.
[http://dx.doi.org/10.3390/ma13061348] [PMID: 32188165]

[16]   Hench LL. The story of Bioglass®. J Mater Sci Mater Med 2006; 17(11): 967-78.
[http://dx.doi.org/10.1007/s10856-006-0432-z] [PMID: 17122907]

[17]   Schmitz SI, Widholz B, Essers C, *et al.* Superior biocompatibility and comparable osteoinductive properties: Sodium-reduced fluoride-containing bioactive glass belonging to the CaO–MgO–SiO$_2$ system as a promising alternative to 45S5 bioactive glass. Bioact Mater 2020; 5(1): 55-65.
[http://dx.doi.org/10.1016/j.bioactmat.2019.12.005] [PMID: 31956736]

[18]   Ojansivu M, Hyväri L, Kellomäki M, Hupa L, Vanhatupa S, Miettinen S. Bioactive glass induced osteogenic differentiation of human adipose stem cells is dependent on cell attachment mechanism and mitogen-activated protein kinases. Eur Cell Mater 2018; 35: 54-72.
[http://dx.doi.org/10.22203/eCM.v035a05]

[19]   Kargozar S, Lotfibakhshaiesh N, Ai J, *et al.* Strontium- and cobalt-substituted bioactive glasses seeded with human umbilical cord perivascular cells to promote bone regeneration *via* enhanced osteogenic and angiogenic activities. Acta Biomater 2017; 58: 502-14.
[http://dx.doi.org/10.1016/j.actbio.2017.06.021] [PMID: 28624656]

[20]   Kargozar S, Baino F, Hamzehlou S, Hill RG, Mozafari M. Bioactive glasses: Sprouting angiogenesis in tissue engineering. Trends Biotechnol 2018; 36(4): 430-44.
[http://dx.doi.org/10.1016/j.tibtech.2017.12.003] [PMID: 29397989]

[21]   Zhang D, Leppäranta O, Munukka E, *et al.* Antibacterial effects and dissolution behavior of six bioactive glasses. J Biomed Mater Res A 2010; 93A(2): 475-83.
[http://dx.doi.org/10.1002/jbm.a.32564] [PMID: 19582832]

[22]   Zhang W, Zhao F, Huang D, Fu X, Li X, Chen X. Strontium-substituted submicrometer bioactive glasses modulate macrophage responses for improved bone regeneration. ACS Appl Mater Interfaces 2016; 8(45): 30747-58.
[http://dx.doi.org/10.1021/acsami.6b10378] [PMID: 27779382]

[23]   Kargozar S, Baino F, Hamzehlou S, Hill RG, Mozafari M. Bioactive glasses entering the mainstream. Drug Discov Today 2018; 23(10): 1700-4.
[http://dx.doi.org/10.1016/j.drudis.2018.05.027] [PMID: 29803626]

[24]   Kargozar S, Mozafari M, Ghodrat S, Fiume E, Baino F. Copper-containing bioactive glasses and glass-ceramics: From tissue regeneration to cancer therapeutic strategies. Mater Sci Eng C 2021; 121: 111741.
[http://dx.doi.org/10.1016/j.msec.2020.111741] [PMID: 33579436]

[25]   Majumdar S, Hira SK, Tripathi H, *et al.* Synthesis and characterization of barium-doped bioactive glass with potential anti-inflammatory activity. Ceram Int 2021; 47(5): 7143-58.
[http://dx.doi.org/10.1016/j.ceramint.2020.11.068]

[26]   Johari B, Kadivar M, Lak S, *et al.* Osteoblast-seeded bioglass/gelatin nanocomposite: a promising bone substitute in critical-size calvarial defect repair in rat. Int J Artif Organs 2016; 39(10): 524-33.
[http://dx.doi.org/10.5301/ijao.5000533] [PMID: 27901555]

[27]   Kargozar S, Mozafari M, Hashemian SJ, *et al.* Osteogenic potential of stem cells-seeded bioactive

nanocomposite scaffolds: A comparative study between human mesenchymal stem cells derived from bone, umbilical cord Wharton's jelly, and adipose tissue. J Biomed Mater Res B Appl Biomater 2018; 106(1): 61-72.
[http://dx.doi.org/10.1002/jbm.b.33814] [PMID: 27862947]

[28] Fathi A, Kermani F, Behnamghader A, *et al.* Three-dimensionally printed polycaprolactone/multicomponent bioactive glass scaffolds for potential application in bone tissue engineering. Biomedical Glasses 2020; 6(1): 57-69.
[http://dx.doi.org/10.1515/bglass-2020-0006]

[29] Kargozar S, Singh RK, Kim HW, Baino F. "Hard" ceramics for "Soft" tissue engineering: Paradox or opportunity? Acta Biomater 2020; 115: 1-28.
[http://dx.doi.org/10.1016/j.actbio.2020.08.014] [PMID: 32818612]

[30] Schröter L, Kaiser F, Stein S, Gbureck U, Ignatius A. Biological and mechanical performance and degradation characteristics of calcium phosphate cements in large animals and humans. Acta Biomater 2020; 117: 1-20.
[http://dx.doi.org/10.1016/j.actbio.2020.09.031] [PMID: 32979583]

[31] Bohner M. 5 - Bioresorbable ceramics. In: Buchanan F, Ed. Degradation Rate of Bioresorbable Materials. Woodhead Publishing 2008; pp. 95-114.
[http://dx.doi.org/10.1533/9781845695033.2.95]

[32] Sheikh Z, Abdallah MN, Hanafi A, Misbahuddin S, Rashid H, Glogauer M. Mechanisms of *in vivo* degradation and resorption of calcium phosphate based biomaterials. Materials 2015; 8(11): 7913-25.
[http://dx.doi.org/10.3390/ma8115430] [PMID: 28793687]

[33] Habraken W, Habibovic P, Epple M, Bohner M. Calcium phosphates in biomedical applications: Materials for the future? Mater Today 2016; 19(2): 69-87.
[http://dx.doi.org/10.1016/j.mattod.2015.10.008]

[34] Malhotra A, Habibovic P. Calcium phosphates and angiogenesis: Implications and advances for bone regeneration. Trends Biotechnol 2016; 34(12): 983-92.
[http://dx.doi.org/10.1016/j.tibtech.2016.07.005] [PMID: 27481474]

[35] Kermani F, Mollazadeh S, Kargozar S, Vahdati Khakhi J. Improved osteogenesis and angiogenesis of theranostic ions doped calcium phosphates (CaPs) by a simple surface treatment process: A state-of-the-art study. Mater Sci Eng C 2021; 124: 112082.
[http://dx.doi.org/10.1016/j.msec.2021.112082] [PMID: 33947573]

[36] Kumar A, Kargozar S, Baino F, Han SS. Additive manufacturing methods for producing hydroxyapatite and hydroxyapatite-based composite scaffolds: A review. Front Mater 2019; 6: 313.
[http://dx.doi.org/10.3389/fmats.2019.00313]

[37] Backes EH, Fernandes EM, Diogo GS, *et al.* Engineering 3D printed bioactive composite scaffolds based on the combination of aliphatic polyester and calcium phosphates for bone tissue regeneration. Mater Sci Eng C 2021; 122: 111928.
[http://dx.doi.org/10.1016/j.msec.2021.111928] [PMID: 33641921]

[38] Choi JB, Kim YK, Byeon SM, *et al.* Fabrication and characterization of biodegradable gelatin methacrylate/biphasic calcium phosphate composite hydrogel for bone tissue engineering. Nanomaterials 2021; 11(3): 617.
[http://dx.doi.org/10.3390/nano11030617] [PMID: 33801249]

[39] Gao J, Su Y, Qin YX. Calcium phosphate coatings enhance biocompatibility and degradation resistance of magnesium alloy: Correlating *in vitro* and *in vivo* studies. Bioact Mater 2021; 6(5): 1223-9.
[http://dx.doi.org/10.1016/j.bioactmat.2020.10.024] [PMID: 33210020]

[40] Fosca M, Rau JV, Uskoković V. Factors influencing the drug release from calcium phosphate cements. Bioact Mater 2022; 7: 341-63.
[http://dx.doi.org/10.1016/j.bioactmat.2021.05.032] [PMID: 34466737]

[41]   Cunha CS, Castro PJ, Sousa SC, *et al.* Films of chitosan and natural modified hydroxyapatite as effective UV-protecting, biocompatible and antibacterial wound dressings. Int J Biol Macromol 2020; 159: 1177-85.
[http://dx.doi.org/10.1016/j.ijbiomac.2020.05.077] [PMID: 32416293]

[42]   Xu X, Liu X, Tan L, *et al.* Controlled-temperature photothermal and oxidative bacteria killing and acceleration of wound healing by polydopamine-assisted Au-hydroxyapatite nanorods. Acta Biomater 2018; 77: 352-64.
[http://dx.doi.org/10.1016/j.actbio.2018.07.030] [PMID: 30030176]

[43]   Gocha A, McDonald L. Better bodies with biomaterials: How ceramic and glass contribute to the $110 B global market for implantable biomaterials. Am Ceram Soc Bull 2020; 99(9): 17-31.

[44]   Kargozar S, Mozafari M, Hill RG, *et al.* Synergistic combination of bioactive glasses and polymers for enhanced bone tissue regeneration. Mater Today Proc 2018; 5(7): 15532-9.
[http://dx.doi.org/10.1016/j.matpr.2018.04.160]

[45]   Zeimaran E, Pourshahrestani S, Fathi A, *et al.* Advances in bioactive glass-containing injectable hydrogel biomaterials for tissue regeneration. Acta Biomater 2021; 136: 1-36.
[http://dx.doi.org/10.1016/j.actbio.2021.09.034] [PMID: 34562661]

[46]   Baino F, Fiume E. 3D Printing of hierarchical scaffolds based on mesoporous bioactive glasses (MBGs)—fundamentals and applications. Materials 2020; 13(7): 1688.
[http://dx.doi.org/10.3390/ma13071688] [PMID: 32260374]

[47]   Mironov V, Trusk T, Kasyanov V, Little S, Swaja R, Markwald R. Biofabrication: A 21st century manufacturing paradigm. Biofabrication 2009; 1(2): 022001.
[http://dx.doi.org/10.1088/1758-5082/1/2/022001] [PMID: 20811099]

CHAPTER 2

Introduction to Biocompatible Glasses, Ceramics, and Glass-Ceramics

Amirhossein Moghanian[1,*], **Zahra Miri**[2], **Danna Valentina Sánchez**[3] and **Amir K. Miri**[3]

[1] *Department of Materials Engineering, Imam Khomeini International University, Qazvin 34149-16818, Iran*

[2] *Department of Materials Engineering, Isfahan University of Technology, Isfahan 84156-83111, Iran*

[3] *Department of Biomedical Engineering, New Jersey Institute of Technology, Newark, New Jersey 07102, United States*

Abstract: Glass ceramics and ceramics have a vast range of applications in tissue engineering and regenerative medicine. Biocompatible glasses and ceramics, including bioinert ceramics, bioactive glasses (BGs), and calcium phosphate have been reviewed in this chapter detailing the history, properties, structure, and application. Ceramics and glasses with bioactivity and biocompatibility properties are pioneer solutions for a variety of clinical needs. The capacity of ceramics in hydroxyapatite formation (HA) has also been explained in this section. This chapter includes the invention of the first generation of ceramics and an explanation of how significant are their clinical applications.

Keywords: Hydroxyapatite, Tissue engineering, Bioceramics, Glass-ceramics, Ceramics, Bioactivity, Biocompatibility, Hydroxyapatite (HA), Calcium phosphate (CP), Bioactive glasses, Sol-gel, Bone tissue engineering, Bioinert, Dentistry, Melt-quench, Arthrodesis application, Bone-fillers applications, Scaffolds, Implantation, Dentin hypersensitivity.

INTRODUCTION

Biomaterials appeared 2000 years ago when applied for prosthetics and similar cases [1]. Biomaterials are selected to mimic both the physical and chemical properties of human organs and tissues [2]. Forming a bond with the host tissue, and defining the fidelity of an appropriate environment for cell and bone growth

---

* **Corresponding author Amirhossein Moghanian:** Department of Materials Engineering, Imam Khomeini International University, Qazvin, 34149-16818, Iran; Tel: +989123816103; E-mail: moghanian@eng.ikiu.ac.ir

Saeid Kargozar and Francesco Baino (Eds.)

[1, 3]. Among current biomaterials, ceramics such as cement, porcelain, and glass are used in energy, environment, health, and transportation sectors because of their corrosion resistance, osteoconductivity, brittleness, and stiffness [5]. In clinics, ceramics have been used for bone reconstruction and implantations (known as bioceramics) [4].

Dental regeneration is a recent application because of its 3D scaffold structure [6 - 8]. To illustrate, teeth composition contains dentine and enamel, and teeth cannot self-repair like bones when injured. Bioceramics have been recognized as materials that meet the significant demands for different dental repairs and treatments. Bioceramics are categorized based on their composition, solid structure, non-metallic or inorganic substrate content, and response to the host tissue [4]. Bioinert ceramics are known for corrosion resistance without inflicting on the tissue. Bioactive ceramics, including glass and BGs, have excellent bioactivity properties and interact with the targeted tissue for other processes. Bioresorbable ceramics involve calcium carbonates, calcium phosphates, and calcium silicates. Several glass ceramics can have magnetic properties for different clinical applications. Glass ceramics have shown thermal, chemical, biological, and dielectric properties leading to significant recognition of glass ceramics for clinical treatments [9].

CERAMICS

Structure

Ceramic materials have different atoms arrangements, which depend on the size of atoms and the bonding in the structure [10]. The bonding between atoms in ceramics is covalent or ionic and can be a combination of both, affecting their chemical and physical system [11].

Classifications

Bioinert Ceramics

Bioinert ceramics are characterized by their hardness, excellent mechanical behavior, corrosion resistance, and durability. Zirconia (ZrO_2) and alumina (Al_2O_3) are two famous bioinert ceramics in this field [12, 13] being promising materials for orthopedic applications because of their compressive strength [13, 14]. The first generation of Al_2O_3 was introduced in the 1970s, not only being applied in dentistry [15], but also used to replace corneal and bone, dental implants, and maxillofacial regeneration [16]. On the other hand, ZrO_2 has a different crystalline structure depending on the temperature: below 1170 °C, it has a monoclinic system, at 1170 °C, it is tetragonal, and lastly, at 2370 °C, it is cubic.

The structural transformation is visible on the ceramics' surface when placed in body fluid, improving the implant's durability [19]. ZrO_2 enhances differentiation and cell proliferation for osteogenic and osseointegration applications [20]. Alumina-toughened zirconia (ATZ) or zirconia-toughened alumina (ZTA) is a mixed composition of Al_2O_3 with ZrO_2 to increase the toughness degree and versatility [17, 18]. Bioactive materials differ from inactive materials because of the chemical reaction when placed in the biological fluid [21]. Both Al_2O_3 and ZrO_2 are biocompatible while they are passive without a direct bond with the bone and tissue.

Glass-Ceramics and Bioactive Glasses (BGs)

Glass-ceramics and BGs are superior materials in tissue engineering. When glass is heated, it crystallizes and improves its toughness and strength. Glasses contain main ions, including silica (Si), sodium (Na), calcium (Ca), and phosphate (P), which are released when BGs dissolve. By releasing ions, BGs can promote various biological events such as angiogenesis and vascularization [22 - 24]. The first generation of biomaterials, invented by Larry Hench, was called 45S5 bioactive glass. It contained 45%SiO_2, 24.5%Na_2O, 24.5%CaO and 6%P_2O5 (mol.%). Studies showed that 45S5 has good osteoconduction and biocompatibility, playing an important role in bone regeneration [25]. The *in vivo* and *in vitro* evaluations on 45S5 highlight properties such as bioactivity and its capacity to interact with the host tissue by forming hydroxyapatite (HA) particles. 45S5 is a silica-based BG, with Si particles playing an important role in bone regeneration by improving osteogenesis [26]. Phosphates (PO_4) are found in a tetrahedral shape and are asymmetric in nature; consequently, it has a high level of solubility when placed in biological fluid [29]. Similar BGs involve a network containing SiO_4 tetrahedrons and oxygen surrounded by two numbers of silicon; this open structure breaks into a solution [27]. The first applicable glass based on borosilicate was discovered by Brink in 1997. This glass had reactive properties with a lower level of chemical durability.

In 1987, BGs were first defined as materials with specific biological responses [30]. BGs became essential for bone applications because of their ability to form an HA layer on a bone surface and provide a substrate for the generation of injured tissue [31 - 33]. It is also recognized because of their high bioactivity property by placing these materials in simulated body fluid (SBF) solutions with a similar composition to the human body plasma. The formation of HA on the surface is essential and determining in some processes such as regeneration, treatment of injured tissue, and osteoblast stimulation [33].

BGs have applications in dentistry, implantation, and drug delivery [28, 29]. For instance, the first BG 45S5 was mainly used as an implant for hearing loss patients [34, 35]. Other glass ceramics have also been applied to treat dental roots if inert [36]. The ion-releasing property of BGs opened a new understanding of its applications. Ionic dissolution helps the stimulation of angiogenesis critical for healing in the body [37]. BGs that contain magnetic properties cause cytocompatibility during the HA formation between the bioactive material and mineralization of cells and can be used for cancer therapy by the induction of magnetic hyperemia. These BGs can influence the death of tumor cells because of heat generation. During this process, tumor cells go under apoptosis while the other cells remain viable [38]. Biological improvements of BGs have been recently suggested by adding fluoride (F), P, and borate (B) [39]. The process of bioactivity happens through the corrosion of glass and different mechanisms and chemical operations [40], shown schematically in Fig. (**1**).

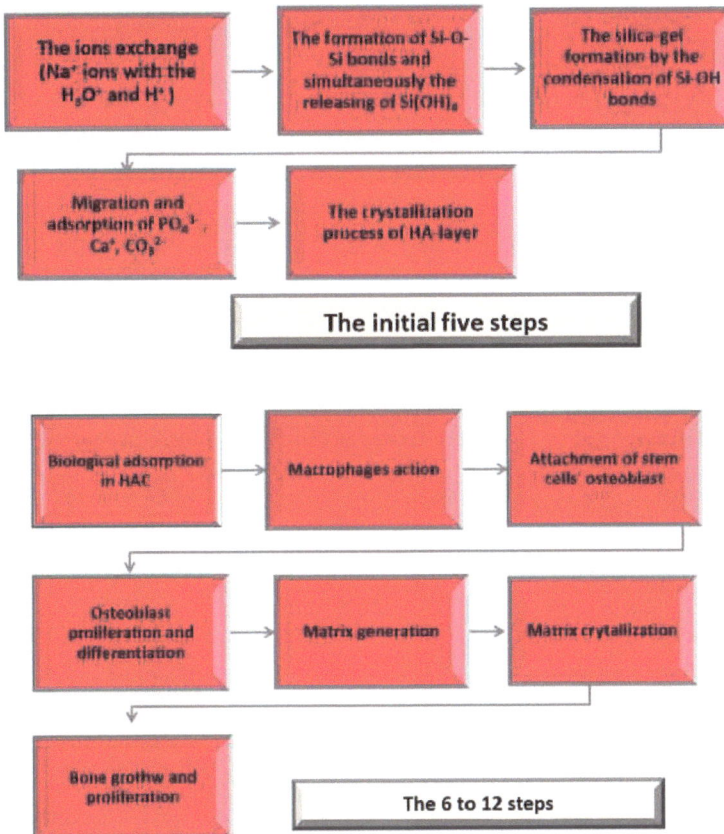

Fig. (1). The bioactivity process of BGs; 1 to 5 steps is a chemical process and from 6 to 12 is a biochemical process.

The Formation of HA Layer on BGs' Surface

The formation of HA is caused by the exchange of Na^+ and Ca^{2+} ions, which increases the pH's solution. Na^+ ions exchange with the H_3O^+ and H^+ from the fluid. Si-O-Si bonds break, and Si-O bonds form between the tissue and glass. The silica-gel is formed by the silanols' condensation and ions such as PO_4^{3-} and Ca^+ migrate from the fluid to the glass surface. A layer containing phosphorous and calcium oxides overlaps with the silica gel. Calcium phosphate then converts to the crystallized structure called hydroxyapatite (HA). This formation happens by the CO_3^{2-} and OH^- incorporation from the fluid. These five steps can occur in the *in vitro* environment with SBF solution, whereas an *in vivo* environment is needed to continue the process. When the HA layer forms, it is not recognized as a foreign or harmful material since its composition is similar to bone tissue minerals. The HA layer stimulates stem cells to attach while cell differentiation can create diverse cells. This process helps the bone to generate a matrix and crystallize so that bone can be generated [41, 42].

Factors on the Formation of Apatite

Many ions can inhibit or promote the formation of HA on the BGs' surface. It has been reported that P ions have a leading role in HA formation [43]. Different ions such as strontium (Sr) can be substituted instead of Ca for a greater impact on HA formation [44]. By contrast, HA formation can be inhibited by magnesium (Mg) ions [45]. Moreover, for some applications, faster HA formation is required, while for others slower; therefore, the composition should be designed accordingly [44].

Different BGs with Diverse Compositions

Many ranges of glasses have been introduced so far. Each of them has a specific composition and unique properties. Some different BGs are displayed in Table 1 [39].

Table 1. Chemical composition of BGs.

Bioactive Glass Type	CaO	SiO₂	P₂O₅	Na₂O	K₂O	MgO	B₂O₃
45S5	24.5	45.0	6.0	24.5	0	0	0
58S	32.6	58.2	9.2	0	0	0	0
13-93B1	19.5	34.4	3.8	5.8	11.7	4.9	19.9
6P53B	18.0	52.7	6.0	10.3	2.8	10.2	0
70S30C	28.6	71.4	0	0	0	0	0

Synthesis of BGs

Two different methods have been recognized for synthesizing BGs: melt quench and sol-gel. The melt quench method consists of high temperature; however, for sol-gel, the room temperature is enough. Additionally, the structure of sol-gels obtained is purer than the melt quench [46].

Melt-Quench Method

The first generation of 45S5 was produced by the melt quench method [47]. The targeted glass is produced by the melt and fusion process of different components. The raw materials and precursors are mixed to a prominent point to the highest pure degree to prevent contamination. The final powder should be dried in the air and then the melting process can start. The melting process is done in crucibles and at a varied temperature between 1200°C and 1500°C. The temperature depends on the kind of BGs. The mentioned temperature can be chosen for BGs based on borate and silicate. Lower temperatures between 1000°C and 1200°C can be designed for phosphorous-based BGs [48].

Sol-Gel Method

Graham proposed the sol-gel method approximately 150 years ago [49]. He reported that it is possible to form a glass based on SiO_2 if the tetraethyl orthosilicate (TEOS) is hydrolyzed. The name sol-gel is based on the colloidal suspension. When sol has been produced, a rigid network of gel is produced. The gel has a porous structure and contains chains of silicate. The TEOS is the primary material and precursor for the sol-gel process for silica [50-51]. The sol-gel method has two important processes: poly-condensation and hydrolysis [38]. The sol-gel method has been recognized as a more effective process for producing BGs with better properties in tissue engineering than the melt-quench method [52]. The highlighted advantage of sol-gel is its lower temperature capacity [53].

Calcium Phosphates

There is a similarity between calcium phosphate and the mineral part of the tissue. Therefore, it can be used in various dental and orthopedic applications [54 - 56]. Calcium phosphate has been recognized as a biocompatible, bioresorbable, and osteoconductive material. After an ionic release, there is an absorption of protein and the formation of a physical layer [57]. Calcium phosphates have different dissolution rates due to diverse chemical compositions, pore sizes, and porosity [58]. Furthermore, calcium phosphates have poor strength being applied as filler or coating [52, 59]. The common calcium phosphates are CDHA, biphasic calcium phosphates (combination of TCP and HAp), β-TCP, HAp, and α-TCP

[41, 46]. HAp is stable and has a crystalline structure. It can be produced by different processes such as hydrothermal, precipitation, and solid-state reactions with temperatures above 1200 °C [60, 61]. β-TCP can be obtained at a lower temperature than HAp above 800°C. Three materials can introduce TCP: β-TCP can stay stable at a temperature under 1120 °C, α-TCP stays from 1120 °C to 1470 °C, and finally, α'-TCP at a temperature more than 1470°C. β-TCP is considered a biodegradable material that can extensively be used as a substitute for bone [55]. α-TCP can be produced by β-TCP. α-TCP also is a biodegradable and biocompatible material [62]. Although HAp and β-TCP have a similar chemical composition, HAp has a slower resorption rate than β-TCP [63, 64]. CDHA can be gained by precipitation in a solution that has a pH higher than 7 [54]. It has poor crystallinity, and its solubility can increase by decreasing the molar ratio of Ca-P, size, and crystallinity and can be decomposed if heated to more than 700 °C [65]. Calcium phosphates have properties such as bioactivity, biocompatibility, resorbability, and enough compressive strength for different applications [41, 57, 63, 66].

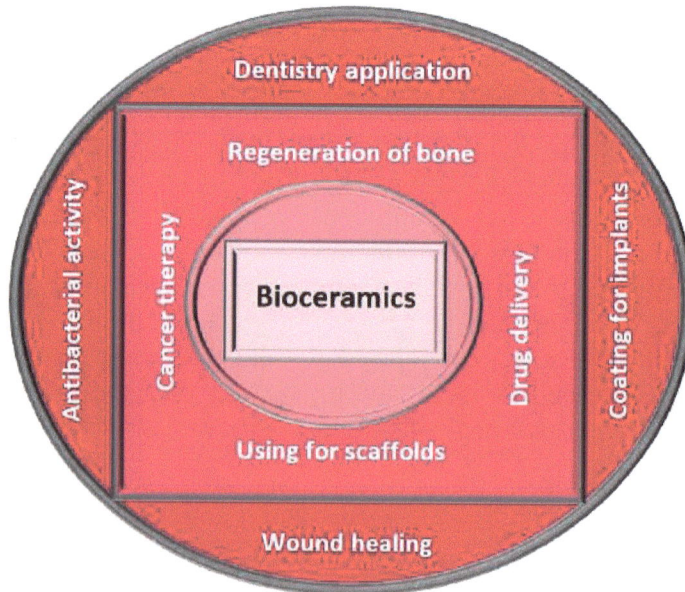

Fig. (2). Different bioactive glasses' applications.

Bioceramics Applications

Bioceramics have been recognized as materials applicable in numerous fields such as dental [67], treatment and replacement of the knee, ligaments, hips [68], implantation [69], reconstruction of maxillofacial [70], and so on. The use of ceramics in bone tissue engineering has been widely suggested recently [71], shown schematically in Fig. (**2**).

Arthrodesis Application

Ceramics based on silicon nitride are recognized as ceramics with high mechanical properties and are very influential as an implant. Many clinical experiments were conducted on this ceramic to prove its effectiveness spine of the lumbar as an arthrodesis piece [72]. Ceramics were successfully used as an implant in the improvement of the backbone of the thoracolumbar and cervical. Silicon-nitride ceramic can be more applicable in the articulation fields [72]. Bioceramics can be applied as a bone graft for the fusion of the lumbar spine because of its advantages compared to other bone graft materials, including flexibility, safe profile, inertness, and easier sterilization. By contrast, the lower tensile strength and sensitivity to fracture of bioceramics can limit their usage [73].

Bioceramics Coating

The materials used as a coat for different implants are osteoconductive, biocompatible, and have appropriate mechanical properties based on the targeted application. The used skin has antibacterial activity, avoiding infections [74, 75]. Bioceramics are more applicable and practical for articulating prostheses because of their high resistance against wear, therefore, the wear rate can be minimized. This coat can be a barrier between the bones or tissue and implants, minimizing the allergy in the human body [75]. Also, when bioceramics involve an electrospray for the surface coating, it can be influential for drug delivery applications and cell behavior control affecting the interaction of protein [76 - 78].

Hydroxyapatite (HA) Coating

In orthopedic implants, the use of HA as a coat is essential. HA contains calcium phosphate, which can potentially improve osteoconductivity, adhesion, and osteoblast differentiation through the exchange of ions between biological fluid and HA coating in clinical experiments and *in vitro* evaluations [38]. Adding different ions to the HA enhances and promotes its ability. For instance, fluoride (F) substitution in the HA layer stimulates cell proliferation and differentiation. Recently, a new process of HA coating has been reported for obtaining better results, such as plasma-spraying, thermal spraying, pulsed-laser deposition, sputter coating, dip coating, hot isotactic pressing, and electrophoretic deposition, among others [79-81].

Calcium Phosphate Coating

TCP and HA are two famous members of the calcium phosphate group for orthopedic implantation applications. For coating on femoral prostheses, HA has been suggested to prevent the complications of using PMMA. Some properties of HA, such as low tensile strength, caused fewer compositions for bone grafts [82]. In another study, however, it has been reported that there is no specific evidence about the effect of HA on the improvement of prostheses [82]. It has been illustrated that many factors are involved in the fixation process, such as the thickness of the coat, the composition of HA, the roughness of the surface, and the kind of substrate.

BG Coating

All BGs are biocompatible, osteoconductive, and antimicrobial, therefore applied for different clinical applications. One of the recent applications of BGs is a coating for other implants. The majority of implants do not have bioactivity properties for the formation of HA, so they lack the shape of a silica-layer rich pivotal in connection with tissue. BGs provide a solution for diverse implants, using them as a coat for metal implants to protect them from corrosion and avoid toxic *in vivo* environments by releasing cations [83]. BGs application for coating is still limited as these glasses are highly bioactive, dissolve quickly, and can put implants in direct contact with the tissue [38]. Different methods have been suggested for applying the BGs for coating implants, including dipping or deposition by sintering the particles of glasses or thermal spraying [38].

Nitride Coating

Since 1980, titanium-based implants have been coated with titanium nitride. The coating is applied from 1 to 1.5 (μm) thickness and used for resurfacing hip implants for arthroplasty of the whole hip or as a component of the femoral knee [72, 84].

Bone Tissue Engineering (BTE)

Treating injured and damaged bone has been the main concern for surgeries. For bone-tumor surgeries or spinal fixation, BTE has been investigated for the use of stem cell mesenchymal instead of autografts [85]. Introducing different scaffolds with the potential to carry stem cells and growth factors plays a vital role in the treatment of injured bone [85]. These diverse scaffolds can include HA, TCPs, or BGs. The materials can also be used as 3D scaffolds for different bone applications: a filler for bone space, augmentation of the alveolar ridge, and reconstruction of maxillofacial and periodontal reduction [86].

Bioceramics as Bone Filler

Bioceramics composition containing HA with collagen acts as an osteoconductive and biodegradable barrier for bone-morphogenetic protein [87]. The design based on HA, collagen, and alginate as filler illustrated osteogenesis after implantation without any deformation in the collagen sponge [85]. In dentistry and orthopedics, bioceramics products are placed on the bone or teeth defect for resorption, and lastly, new bone grows [86]. Bioglass is used as a filler on the defect easily compared to other products during surgeries. When introduced to dentistry, bioglass fillers are prominent in the field such as PerioGlas, BonAlive, and Biogran [88].

Bioceramic Scaffolds

Combining bioglass with other materials such as graphene can introduce an effective and influential composition for regeneration and bone repair. Previous studies showed that this composition had perfect biocompatibility and bioactivity in addition to a potential application in BTE [89]. BTE scaffolds have a porous structure with properties including bioactivity and appropriate mechanical strength. Bioceramics, in combination with other polymers and materials, can be considered a candidate for this field [90].

Bioceramics in Dentistry

Prothesis Application

Bioceramics such as ZrO_2 and Al_2O_3 have been reinforced with titanium-disilicate and spinel, providing the need for prosthesis [91].These kinds of ceramics are recognized only for a limited application, including restoration of the anterior crown or bridge for showing resistance against stress [92, 93]. ZrO_2 and Al_2O_3 have been identified as materials that have long-term effects [94, 95]. Therefore, one of the reasons for applying ZrO_2 in dentistry is its aesthetic ability and similar appearance in terms of color and opacity to human teeth [96]. Al_2O_3 can also work for transparency applications. With this in mind, methods were designed by the computer for the creation of a framework for applying ZrO_2 in prostheses [97].

Veneers for Laminates

Ceramics advancement helps dentists have various choices for producing and making veneers applicable and aesthetic. The ceramics that are translucent and functional are the best for shells. Ceramics containing glassier nanocrystalline structures have been recognized as translucent compared to those with only crystalline structures [98]. The traditional ceramics used in dentistry are feldspar-porcelain, kaolin, and quartz which resist temperatures of more than 870 °C. Felspar-based ceramics are brittle when used, so other ceramics have been enhanced and improved with zirconia and alumina. Ceramics with crystalline structures are applicable for treating cores in teeth [99].

Bioceramics for Implantation

Zirconia-based implants are suggested instead of titanium-based ones because of their bioinert properties [100]. Zirconia is an influential material for implants given its biocompatibility, suitable strength, color-similarity to teeth, and high toughness against fracture and bone resorption [100, 102]. Zirconia implants have limitations for application because of the problematic surface modification, which can cause insufficient osseointegration [100].

Restorative Materials

Used restorative materials should be bioactive. This property prevents the growth of bacteria and new caries [103, 104]. When restorative materials are strengthened, they become threatened by bacterial growth and dentin mineralization. Creating a strong bond between used components, tissue and BGs is a solution because of the formation of the HA layer [105]. BGs are modified with different ions for this purpose. The incorporation of F ions protects the collagen network together with remineralizing properties [106]. Additionally, substituted Ag in BGs has antibacterial properties and prevents caries [107].

Enamel Remineralization

The effective BGs on remineralization include Novamin compositions, which are Ca–P–Na–Si glass [108, 109]. Other studies suggested that incorporating F ion in BGs promotes this property. However, due to the formation of fluorapatite (FAP), this composition shows better remineralization properties compared with other BGs free F ions [110, 111].

Maxillofacial Surgery

BGs such as 45S5 are used in maxillofacial surgeries because of the bone formation stimulation [112]. Different 45S5-BGs have been introduced, including Biogran, PerioGlas [113], and NovaBone [114]. The presence of strontium (Sr) ions in BGs could decrease the resorption of bone [115].

Air Abrasion

Teeth are in danger of hypersensitivity because of enamel wear. As a result, used materials are less abrasive, particularly for the outer parts of enamel [116]. In orthodontic therapy, major damage to the enamel is done by removing adhesive materials. So, studies have shown that BGs with less hardness have a better function [117].

Dentin Hypersensitivity

Hypersensitivity (DH) occurs when dentin is exposed to abstraction, periodontal disease, and erosion [118, 119]. BGs-containing F ions are suggested to prevent this complication because of FAP formation [120]. Additionally, BGs (Borosilicate) is considered another solution for the treatment of DH [103], given that the HA layer can overcome DH [118].

Periodontics

Periodontics is an inflammatory response [121], that challenge dentin implants [122, 123]. Therefore, BGs are suggested to treat the defects created by periodontics [118, 124]. Some types of 45S5-BG, such as PerioGlas, are reported to have an effective impact in this matter [124 - 126].

Bioceramics in Endodontic

Bioceramics are prominent in endodontics because of biocompatibility, osteoconductivity, radiopacity, and host tissue bond formation [127]. Calcium phosphate materials are used for pulp capping, stimulation of complex tissue information, apexification, and apical barrier. Sealers that are calcium phosphate-based led to high osteogenic response and fewer cytotoxic complications compared to zinc oxide-based [127]. Clinical trials proved a positive effect on pulp capping, defects of the periapical, improvement of the barrier of apical, and regeneration of endodontic [127]. Bioceramics HA has also shown effective in endodontics by repairing the dentin without necrosis [128] and root stimulation compared to the TCP [128].

CONCLUDING REMARKS

This chapter explained ceramics and glass ceramics together with their applications in various fields like TE and dentistry. As mentioned precisely in this chapter, ceramics contain different groups such as BGs, bioinert ceramics, and CPs. The history, structure, and methods for synthesis have been reviewed in this chapter. The great and important happening in this field was the invention of BGs. These kinds of materials showed excellent properties during many investigations and reports that have been prepared so far. This is due to the fact that they can provide an initial and prominent requirement for repairing and regeneration of injured bone which is the formation of HA on the specimen's surface. In addition to the BGs, CPs, and HA also have been used in TE applications for many reasons. Because they also showed perfect biocompatibility and bioactivity properties for this application. Recently, the use of bioceramics has attracted more attention in the field of dentistry that has been described in this chapter like endodontics, periodontics, hypersensitivity, air abrasion, maxillofacial surgery, enamel remineralization, veneers for laminate, and implantation. The applicability of bioceramics for clinical and medical applications is more and varied and research and investigation on these materials can produce many innovations to help humans to have a healthier life in the modern world. These materials have always been a hot topic for scientists to develop and improve clinical needs.

REFERENCES

[1]   Hench LL, Splinter RJ, Allen WC, Greenlee TK. Bonding mechanisms at the interface of ceramic prosthetic materials. J Biomed Mater Res 1971; 5(6): 117-41.
[http://dx.doi.org/10.1002/jbm.820050611]

[2]   Cao W, Hench LL. Bioactive materials. J Ceram 1996; 22: 493-507.

[3]   Kundu S. Silk biomaterials for tissue engineering and regenerative medicine 2014; 74: 592-5.

[4]   Rahmati M, Mozafari M. Biocompatibility of alumina-based biomaterials–A review. J Cell Physiol 2019; 234(4): 3321-35.
[http://dx.doi.org/10.1002/jcp.27292] [PMID: 30187477]

[5]   Hasan MS, Ahmed I, Parsons AJ, Rudd CD, Walker GS, Scotchford CA. Investigating the use of coupling agents to improve the interfacial properties between a resorbable phosphate glass and polylactic acid matrix. J Biomater Appl 2013; 28(3): 354-66.
[http://dx.doi.org/10.1177/0885328212453634] [PMID: 22781920]

[6]   Pina S, Oliveira JM, Reis RL. Natural-based nanocomposites for bone tissue engineering and regenerative medicine: A review. Adv Mater 2015; 27(7): 1143-69.
[http://dx.doi.org/10.1002/adma.201403354] [PMID: 25580589]

[7]   Pina S, Rebelo R, Correlo VM, Oliveira JM, Reis RL. Bioceramics for osteochondral tissue engineering and regeneration. Adv Exp Med Biol 2018; 1058: 53-75.
[http://dx.doi.org/10.1007/978-3-319-76711-6_3] [PMID: 29691817]

[8]   Ghaemi MH, Reichert S, Krupa A, *et al.* Zirconia ceramics with additions of Alumina for advanced tribological and biomedical applications. Ceram Int 2017; 43(13): 9746-52.
[http://dx.doi.org/10.1016/j.ceramint.2017.04.150]

[9]   Banerjee S, Tyagi AK. Functional materials : Preparation, processing and applications. Elsevier 2012; pp. 730-10.

[10]   Kundu S. Silk biomaterials for tissue engineering and regenerative medicine. Woodhead Publishing 2014.

[11]   David Kingery W, Bowen HK, Donald R. Uhlmann. Introduction to Ceramics 1976.

[12]   Maccauro G, Rossi Iommetti P, Raffaelli L, Manicone PF. Alumina and zirconia ceramic for orthopaedic and dental devices. Intech 2011; p. 470.
[http://dx.doi.org/10.5772/23917]

[13]   Webster TJ, Siegel RW, Biios R. Design and evaluation of nanophase alumina for orthopaedic/dental applications. J Nanomater 1999; 12: 983-6.

[14]   Sinn Aw M, Losic D. 6 Nanoporous alumina as an intelligent nanomaterial for biomedical applications 2017; 592.

[15]   Madfa AA, Al-Sanabani FA, Al-Qudaimi NH, Al-Qudaimi Alumina NH. Ceramic for dental applications. Am J Mater Sci 2014; 1: 26-34.

[16]   Greenspan DC. Glass and medicine: The larry hench story. Int J Appl Glass Sci 2016; 7(2): 134-8.
        [http://dx.doi.org/10.1111/ijag.12204]

[17]   Kurtz SM, Kocagöz S, Arnholt C, Huet R, Ueno M, Walter WL. Advances in zirconia toughened alumina biomaterials for total joint replacement. J Mech Behav Biomed Mater 2014; 31: 107-16.
        [http://dx.doi.org/10.1016/j.jmbbm.2013.03.022] [PMID: 23746930]

[18]   Biamino S, Fino P, Pavese M, Badini C. Alumina–zirconia–yttria nanocomposites prepared by solution combustion synthesis. Ceram Int 2006; 32(5): 509-13.
        [http://dx.doi.org/10.1016/j.ceramint.2005.04.004]

[19]   Pieralli S, Kohal RJ, Jung RE, Vach K, Spies BC. Clinical outcomes of zirconia dental implants: A systematic review. J Dent Res 2017; 96(1): 38-46.
        [http://dx.doi.org/10.1177/0022034516664043] [PMID: 27625355]

[20]   Afzal A. Implantable zirconia bioceramics for bone repair and replacement: A chronological review. Mater Express 2014; 4(1): 1-12.
        [http://dx.doi.org/10.1166/mex.2014.1148]

[21]   Huang J, Best S. Ceramic biomaterials. Woodhead Publ Ser Biomater 2007; 3-31.

[22]   Xynos ID, Edgar AJ, Buttery LDK, Hench LL, Polak JM. Ionic products of bioactive glass dissolution increase proliferation of human osteoblasts and induce insulin-like growth factor II mRNA expression and protein synthesis. Biochem Biophys Res Commun 2000; 276(2): 461-5.
        [http://dx.doi.org/10.1006/bbrc.2000.3503] [PMID: 11027497]

[23]   Lobel KD, Hench LL. *In-vitro* protein interactions with a bioactive gel-glass. J Sol-Gel Sci Technol 1996; 7(1-2): 69-76.
        [http://dx.doi.org/10.1007/BF00401885]

[24]   Gorustovich AA, Roether JA, Boccaccini AR. Effect of bioactive glasses on angiogenesis 2009; 16: 199-207.

[25]   Wilson J, Pigott GH, Schoen FJ, Hench LL. Toxicology and biocompatibility of bioglasses. J Biomed Mater Res 1981; 15(6): 805-17.
        [http://dx.doi.org/10.1002/jbm.820150605] [PMID: 7309763]

[26]   Xynos ID, Hukkanen MVJ, Batten JJ, Buttery LD, Hench LL, Polak JM. Bioglass 45S5 stimulates osteoblast turnover and enhances bone formation *In vitro*: implications and applications for bone tissue engineering. Calcif Tissue Int 2000; 67(4): 321-9.
        [http://dx.doi.org/10.1007/s002230001134] [PMID: 11000347]

[27]   Brink M, Turunen T, Happonen RP, Yli-Urpo A. Compositional dependence of bioactivity of glasses in the system Na2O-K2O-MgO-CaO-B2O3-P2O5-SiO2. J Biomed Mater Res 1997; 37(1): 114-21.
        [http://dx.doi.org/10.1002/(SICI)1097-4636(199710)37:1<114::AID-JBM14>3.0.CO;2-G] [PMID: 9335356]

[28]   Andersson H, Liu G, Karlsson KH, Niemi L, Miettinen J, Juhanoja J. *In vivo* behaviour of glasses in the SiO2-Na2O-CaO-P2O5-Al2O3-B2O3 system. J Mater Sci Mater Med 1990; 1(4): 219-27.
        [http://dx.doi.org/10.1007/BF00701080]

[29]   Knowles JC. Phosphate based glasses for biomedical applications. J Mater Chem 2003; 13(10): 2395-401.
        [http://dx.doi.org/10.1039/b307119g]

[30]   Williams DF. On the mechanisms of biocompatibility. Biomaterials 2008; 29(20): 2941-53.
        [http://dx.doi.org/10.1016/j.biomaterials.2008.04.023] [PMID: 18440630]

[31]   Hench LL. Bioactive ceramics: Theory and clinical applications. Bioceramics 1994; pp. 3-14.

[32]   Farooq I, Tylkowski M, Müller S, Janicki T, Brauer DS, Hill RG. Influence of sodium content on the properties of bioactive glasses for use in air abrasion. Biomed Mater 2013; 8(6): 065008.
[http://dx.doi.org/10.1088/1748-6041/8/6/065008] [PMID: 24287337]

[33]   Jones JR. Review of bioactive glass: From Hench to hybrids. Acta Biomater 2013; 9(1): 4457-86.
[http://dx.doi.org/10.1016/j.actbio.2012.08.023] [PMID: 22922331]

[34]   Jones JR, Brauer DS, Hupa L, Greenspan DC. Bioglass and bioactive glasses and their impact on healthcare. Int J Appl Glass Sci 2016; 7(4): 423-34.
[http://dx.doi.org/10.1111/ijag.12252]

[35]   Merwin GE. Bioglass middle ear prosthesis: Preliminary report. Ann Otol Rhinol Laryngol 1986; 95(1): 78-82.
[http://dx.doi.org/10.1177/000348948609500117] [PMID: 3947007]

[36]   Montazerian M, Zanotto ED. Bioactive and inert dental glass-ceramics. J Biomed Mater Res A 2017; 105(2): 619-39.
[http://dx.doi.org/10.1002/jbm.a.35923] [PMID: 27701809]

[37]   Kargozar S, Baino F, Hamzehlou S, Hill RG, Mozafari M. Bioactive glasses: Sprouting angiogenesis in tissue engineering. Trends Biotechnol 2018; 36(4): 430-44.
[http://dx.doi.org/10.1016/j.tibtech.2017.12.003] [PMID: 29397989]

[38]   Ghione G, Ferrero R. Ceramics, glass and glass-ceramics from early manufacturing steps towards modern frontiers. Springer 2021; p. 351.

[39]   O'Donnell MD, Watts SJ, Hill RG, Law RV. The effect of phosphate content on the bioactivity of soda-lime-phosphosilicate glasses. J Mater Sci Mater Med 2009; 20(8): 1611-8.
[http://dx.doi.org/10.1007/s10856-009-3732-2] [PMID: 19330429]

[40]   Pantano CG Jr, Clark AE Jr, Hench LL. Multilayer corrosion films on bioglass surfaces. J Am Ceram Soc 1974; 57(9): 412-3.
[http://dx.doi.org/10.1111/j.1151-2916.1974.tb11429.x]

[41]   Nejatian T. "Dental biocomposites," in Biomaterials for Oral and Dental Tissue Engineering. Elsevier Inc. 2017; pp. 65-84.
[http://dx.doi.org/10.1016/B978-0-08-100961-1.00005-0]

[42]   Hench LL, Roki N, Fenn MB. Bioactive glasses: Importance of structure and properties in bone regeneration. J Mol Struct 2014; 1073: 24-30.
[http://dx.doi.org/10.1016/j.molstruc.2014.03.066]

[43]   Farooq I, Ali S, Husain S, Khan E, Hill RG. "Bioactive glasses—structure and applications," in Advanced Dental Biomaterials. Elsevier 2019; pp. 453-76.
[http://dx.doi.org/10.1016/B978-0-08-102476-8.00017-7]

[44]   Diba M, Tapia F, Boccaccini AR, Strobel LA. Magnesium-containing bioactive glasses for biomedical applications. Int J Appl Glass Sci 2012; 3(3): 221-53.
[http://dx.doi.org/10.1111/j.2041-1294.2012.00095.x]

[45]   O'donnell M D. 2 Melt-Derived Bioactive Glass 2012.
[http://dx.doi.org/10.1002/9781118346457.ch2]

[46]   Wren AW, Coughlan A, Hassanzadeh P, Towler MR. Silver coated bioactive glass particles for wound healing applications. J Mater Sci Mater Med 2012; 23(5): 1331-41.
[http://dx.doi.org/10.1007/s10856-012-4604-8] [PMID: 22426653]

[47]   E. El-Meliegy, R. van Noort Glasses and Glass Ceramics for Medical Applications." Springer-Verlag New York 2012:244-269.

[48]   Vallet-Regi M. Bioceramics with Clinical Applications 2014.

[49]   Hench LL. Io! bioceramics and the future. J Am Ceram Soc 1995; 20.

[50]   Chakraborty J, Basu D. Bioceramics—A new era. Trans Indian Ceram Soc 2005; 64(4): 171-92.
[http://dx.doi.org/10.1080/0371750X.2005.11012210]

[51]   Wilson J, Pigott GH, Schoen FJ, Hench LL. Toxicology and biocompatibility of bioglasses. J Biomed Mater Res 1981; 15(6): 805-17.
[http://dx.doi.org/10.1002/jbm.820150605] [PMID: 7309763]

[52]   Xynos ID, Hukkanen MVJ, Batten JJ, Buttery LD, Hench LL, Polak JM. Bioglass 45S5 stimulates osteoblast turnover and enhances bone formation *In vitro*: Implications and applications for bone tissue engineering. Calcif Tissue Int 2000; 67(4): 321-9.
[http://dx.doi.org/10.1007/s002230001134] [PMID: 11000347]

[53]   Brink M, Turunen T, Happonen RP, Yli-Urpo A. Compositional dependence of bioactivity of glasses in the system Na2O-K2O-MgO-CaO-B2O3-P2O5-SiO2. J Biomed Mater Res 1997; 37(1): 114-21.
[http://dx.doi.org/10.1002/(SICI)1097-4636(199710)37:1<114::AID-JBM14>3.0.CO;2-G]   [PMID: 9335356]

[54]   Dorozhkin S. Calcium orthophosphates in nature, biology and medicine. Materials 2009; 2(2): 399-498.
[http://dx.doi.org/10.3390/ma2020399]

[55]   Dorozhkin SV. Calcium orthophosphates. J Mater Sci 2007; 42(4): 1061-95.
[http://dx.doi.org/10.1007/s10853-006-1467-8]

[56]   LeGeros RZ, LeGeros JP. Calcium phosphate bioceramics: Past, present and future. Key Eng Mater 2003; 240-242: 3-10.
[http://dx.doi.org/10.4028/www.scientific.net/KEM.240-242.3]

[57]   Yuan H, Fernandes H, Habibovic P, *et al.* Osteoinductive ceramics as a synthetic alternative to autologous bone grafting. Proc Natl Acad Sci USA 2010; 107(31): 13614-9.
[http://dx.doi.org/10.1073/pnas.1003600107] [PMID: 20643969]

[58]   Bohner M. Physical and chemical aspects of calcium phosphates used in spinal surgery. Eur Spine J 2001; 10(Suppl 2) (Suppl. 2): S114-21.
[PMID: 11716008]

[59]   Daculsi G, Laboux O, Malard O, Weiss P. Current state of the art of biphasic calcium phosphate bioceramics. J Mater Sci Mater Med 2003; 14(3): 195-200.
[http://dx.doi.org/10.1023/A:1022842404495] [PMID: 15348464]

[60]   Kannan S, Lemos AF, Ferreira JMF. Synthesis and mechanical performance of biological-like hydroxyapatites. Chem Mater 2006; 18(8): 2181-6.
[http://dx.doi.org/10.1021/cm052567q]

[61]   Yin X, Stott MJ, Rubio A. α- and β-tricalcium phosphate: A density functional study. Phys Rev B Condens Matter 2003; 68: 205.
[http://dx.doi.org/10.1103/PhysRevB.68.205205]

[62]   Takahashi Y, Yamamoto M, Tabata Y. Osteogenic differentiation of mesenchymal stem cells in biodegradable sponges composed of gelatin and β-tricalcium phosphate. Biomaterials 2005; 26(17): 3587-96.
[http://dx.doi.org/10.1016/j.biomaterials.2004.09.046] [PMID: 15621249]

[63]   Ginebra MP, Traykova T, Planell JA. Calcium phosphate cements: Competitive drug carriers for the musculoskeletal system? Biomaterials 2006; 27(10): 2171-7.
[http://dx.doi.org/10.1016/j.biomaterials.2005.11.023] [PMID: 16332349]

[64]   Dorozhkin SV. Calcium orthophosphate cements for biomedical application. J Mater Sci 2008; 43(9): 3028-57.
[http://dx.doi.org/10.1007/s10853-008-2527-z]

[65]   Kannan S, Goetz-Neunhoeffer F, Neubauer J, Ferreira JMF. Ionic substitutions in biphasic

hydroxyapatite and β-tricalcium phosphate mixtures: Structural analysis by rietveld refinement. J Am Ceram Soc 2008; 91(1): 1-12.
[http://dx.doi.org/10.1111/j.1551-2916.2007.02117.x]

[66]   Khalid H, Suhaib F, Zahid S, *et al.* Microwave-assisted synthesis and *in vitro* osteogenic analysis of novel bioactive glass fibers for biomedical and dental applications. Biomed Mater 2018; 14(1): 015005.
[http://dx.doi.org/10.1088/1748-605X/aae3f0] [PMID: 30251708]

[67]   Bunpetch V, Zhang X, Li T, *et al.* Silicate-based bioceramic scaffolds for dual-lineage regeneration of osteochondral defect. Biomaterials 2019; 192: 323-33.
[http://dx.doi.org/10.1016/j.biomaterials.2018.11.025] [PMID: 30468999]

[68]   Furko M, Havasi V, Kónya Z, *et al.* Development and characterization of multi-element doped hydroxyapatite bioceramic coatings on metallic implants for orthopedic applications. Bol Soc Esp Ceram Vidr 2018; 57(2): 55-65.
[http://dx.doi.org/10.1016/j.bsecv.2017.09.003]

[69]   Pina S, Rebelo R, Correlo VM, Oliveira JM, Reis RL. Bioceramics for osteochondral tissue engineering and regeneration. Adv Exp Med Biol 2018; 1058: 53-75.
[http://dx.doi.org/10.1007/978-3-319-76711-6_3] [PMID: 29691817]

[70]   Ma H, Feng C, Chang J, Wu C. 3D-printed bioceramic scaffolds: From bone tissue engineering to tumor therapy. Acta Biomater 2018; 79: 37-59.
[http://dx.doi.org/10.1016/j.actbio.2018.08.026] [PMID: 30165201]

[71]   McEntire BJ, Bal BS, Rahaman MN, Chevalier J, Pezzotti G. Ceramics and ceramic coatings in orthopaedics. J Eur Ceram Soc 2015; 35(16): 4327-69.
[http://dx.doi.org/10.1016/j.jeurceramsoc.2015.07.034]

[72]   Nickoli MS, Hsu WK. Ceramic-based bone grafts as a bone grafts extender for lumbar spine arthrodesis: A systematic review. Global Spine J 2014; 4(3): 211-6.
[http://dx.doi.org/10.1055/s-0034-1378141] [PMID: 25083364]

[73]   Tobin EJ. Recent coating developments for combination devices in orthopedic and dental applications: A literature review. Adv Drug Deliv Rev 2017; 112: 88-100.
[http://dx.doi.org/10.1016/j.addr.2017.01.007] [PMID: 28159606]

[74]   Zhang B, Myers D, Wallace G, Brandt M, Choong P. Bioactive coatings for orthopaedic implants-recent trends in development of implant coatings. Int J Mol Sci 2014; 15(7): 11878-921.
[http://dx.doi.org/10.3390/ijms150711878] [PMID: 25000263]

[75]   Siebers MC, Walboomers XF, Leeuwenburgh SCG, Wolke JGC, Jansen JA. Electrostatic spray deposition (ESD) of calcium phosphate coatings, an *in vitro* study with osteoblast-like cells. Biomaterials 2004; 25(11): 2019-27.
[http://dx.doi.org/10.1016/j.biomaterials.2003.08.050] [PMID: 14741616]

[76]   Praveena J, Chinnasamy G, Jayarama RV, David LB, Seeram R, Dinesh KS. Controlled release of drugs in electrosprayed nanoparticles for bone tissue engineering. J Advanced Drug Delivery Reviews 2015; 94: 77-95.
[http://dx.doi.org/10.1016/j.addr.2015.09.007] [PMID: S0169409X15002100]

[77]   Siebers MC, Walboomers XF, Leeuwenburgh SCG, Wolke JGC, Jansen JA. The influence of the crystallinity of electrostatic spray deposition-derived coatings on osteoblast-like cell behavior, *in vitro*. J Biomed Mater Res A 2006; 78A(2): 258-67.
[http://dx.doi.org/10.1002/jbm.a.30700] [PMID: 16628711]

[78]   Wang J, de Groot K, van Blitterswijk C, de Boer J. Electrolytic deposition of lithium into calcium phosphate coatings. Dent Mater 2009; 25(3): 353-9.
[http://dx.doi.org/10.1016/j.dental.2008.07.013] [PMID: 18804857]

[79]   De Groot K, Geesink R, Klein CPAT, Serekian P, Serekian P. Plasma sprayed coatings of

hydroxylapatite. J Biomed Mater Res 1987; 21(12): 1375-81.
[http://dx.doi.org/10.1002/jbm.820211203] [PMID: 3429472]

[80]   Mohseni E, Zalnezhad E, Bushroa AR. Comparative investigation on the adhesion of hydroxyapatite
coating on Ti–6Al–4V implant: A review paper. Int J Adhes Adhes 2014; 48: 238-57.
[http://dx.doi.org/10.1016/j.ijadhadh.2013.09.030]

[81]   Prakash L. Ceramics in arthroplasty, arthritis and orthopaedics. Arthritis Res Ther 2018; 1: 4.

[82]   Bohner M. Calcium orthophosphates in medicine: From ceramics to calcium phosphate cements.
Injury 2000; 31: D37-47.
[http://dx.doi.org/10.1016/S0020-1383(00)80022-4] [PMID: 11270080]

[83]   Subramanian B, Muraleedharan CV, Ananthakumar R, Jayachandran M. A comparative study of
titanium nitride (TiN), titanium oxy nitride (TiON) and titanium aluminum nitride (TiAlN), as surface
coatings for bio implants. Surf Coat Tech 2011; 205(21-22): 5014-20.
[http://dx.doi.org/10.1016/j.surfcoat.2011.05.004]

[84]   Sotome S, Uemura T, Kikuchi M, *et al.* Synthesis and *in vivo* evaluation of a novel
hydroxyapatite/collagen–alginate as a bone filler and a drug delivery carrier of bone morphogenetic
protein. Mater Sci Eng C 2004; 24(3): 341-7.
[http://dx.doi.org/10.1016/j.msec.2003.12.003]

[85]   Hashmat G, Shahreen Z, Muhammad K, Shahab UD. bioglass, a new trend towards clinical bone
tissue engineering. J Pak Dent Assoc 2015; 35: 706-12.

[86]   Kikuchi M, Itoh S, Ichinose S, Shinomiya K, Tanaka J. Self-organization mechanism in a bone-like
hydroxyapatite/collagen nanocomposite synthesized *in vitro* and its biological reaction *in vivo*.
Biomaterials 2001; 22(13): 1705-11.
[http://dx.doi.org/10.1016/S0142-9612(00)00305-7] [PMID: 11396873]

[87]   Bagheri R. Bioactive glasses in dentistry: A review. J Dent 2015; 2: 1-9.

[88]   Gao C, Liu T, Shuai C, Peng S. Enhancement mechanisms of graphene in nano-58S bioactive glass
scaffold: Mechanical and biological performance. Sci Rep 2014; 4(1): 4712.
[http://dx.doi.org/10.1038/srep04712] [PMID: 24736662]

[89]   Hasanzadeh M, Pournaghi-Azar MH, Shadjou N, Jouyban A. A new mechanistic approach to elucidate
furosemide electrooxidation on magnetic nanoparticles loaded on graphene oxide modified glassy
carbon electrode. RSC Advances 2014; 4(13): 6580-90.
[http://dx.doi.org/10.1039/c3ra46973e]

[90]   Raigrodski AJ. Contemporary materials and technologies for all-ceramic fixed partial dentures: A
review of the literature. J Prosthet Dent 2004; 92(6): 557-62.
[http://dx.doi.org/10.1016/j.prosdent.2004.09.015] [PMID: 15583562]

[91]   Lüthy H, Filser F, Loeffel O, Schumacher M, Gauckler L, Hammerle C. Strength and reliability of
four-unit all-ceramic posterior bridges. Dent Mater 2005; 21(10): 930-7.
[http://dx.doi.org/10.1016/j.dental.2004.11.012] [PMID: 15923031]

[92]   Özcan M, Vallittu PK. Effect of surface conditioning methods on the bond strength of luting cement to
ceramics. Dent Mater 2003; 19(8): 725-31.
[http://dx.doi.org/10.1016/S0109-5641(03)00019-8] [PMID: 14511730]

[93]   Heffernan MJ, Aquilino SA, Diaz-Arnold AM, Haselton DR, Stanford CM, Vargas MA. Relative
translucency of six all-ceramic systems. Part I: Core materials. J Prosthet Dent 2002; 88(1): 4-9.
[http://dx.doi.org/10.1067/mpr.2002.126794] [PMID: 12239472]

[94]   Tinschert J, Natt G, Mohrbotter N, Spiekermann H, Schulze KA. Lifetime of alumina- and zirconia
ceramics used for crown and bridge restorations. J Biomed Mater Res B Appl Biomater 2007; 80B(2):
317-21.
[http://dx.doi.org/10.1002/jbm.b.30599] [PMID: 16838354]

[95]   Mclaren E A, Ii R A G. Zirconia-based ceramics: Material properties, esthetics, and layering techniques of a new veneering porcelain, VM9. Quintessence Dent Technol 2005; 28: 0362-913.

[96]   Manicone PF, Rossi Iommetti P, Raffaelli L. An overview of zirconia ceramics: Basic properties and clinical applications. J Dent 2007; 35(11): 819-26.
[http://dx.doi.org/10.1016/j.jdent.2007.07.008] [PMID: 17825465]

[97]   Pascotto R, Pini N, Aguiar FHB, Lima DANL, Lovadino JR, Terada RSS. Advances in dental veneers: Materials, applications, and techniques. Clin Cosmet Investig Dent 2012; 4: 9-16.
[http://dx.doi.org/10.2147/CCIDE.S7837] [PMID: 23674920]

[98]   Ho GW, Matinlinna JP. Insights on ceramics as dental materials. Part I: Ceramic material types in dentistry. Silicon 2011; 3(3): 109-15.
[http://dx.doi.org/10.1007/s12633-011-9078-7]

[99]   Özkurt Z, Kazazoğlu E. Zirconia dental implants: A literature review. J Oral Implantol 2011; 37(3): 367-76.
[http://dx.doi.org/10.1563/AAID-JOI-D-09-00079] [PMID: 20545529]

[100]   Piconi C, Maccauro G, Maccauro G. Zirconia as a ceramic biomaterial. Biomaterials 1999; 20(1): 1-25.
[http://dx.doi.org/10.1016/S0142-9612(98)00010-6] [PMID: 9916767]

[101]   Depprich R, Zipprich H, Ommerborn M, *et al.* Osseointegration of zirconia implants compared with titanium: an *in vivo* study. Head Face Med 2008; 4(1): 30.
[http://dx.doi.org/10.1186/1746-160X-4-30] [PMID: 19077228]

[102]   A. Langalia MDS PGDHHM. A. Buch MDS, M. Khamar, and P. Patel BDS, "Polymerization shrinkage of composite resins: A review,". J Med Dent Sci 2015; 2: 23-7.

[103]   Esteves CM, Ota-Tsuzuki C, Reis AF, Rodrigues JA. Antibacterial activity of various self-etching adhesive systems against oral streptococci. Oper Dent 2010; 35(4): 448-53.
[http://dx.doi.org/10.2341/09-297-L] [PMID: 20672730]

[104]   Profeta AC. Dentine bonding agents comprising calcium-silicates to support proactive dental care: Origins, development and future. Dent Mater J 2014; 33(4): 443-52.
[http://dx.doi.org/10.4012/dmj.2013-267] [PMID: 24500368]

[105]   Tezvergil-Mutluay A, Seseogullari-Dirihan R, Feitosa VP, Cama G, Brauer DS, Sauro S. Effects of composites containing bioactive glasses on demineralized dentin. J Dent Res 2017; 96(9): 999-1005.
[http://dx.doi.org/10.1177/0022034517709464] [PMID: 28535357]

[106]   Chatzistavrou X, Lefkelidou A, Papadopoulou L, *et al.* Bactericidal and bioactive dental composites. Front Physiol 2018; 9: 103.
[http://dx.doi.org/10.3389/fphys.2018.00103] [PMID: 29503619]

[107]   Gjorgievska E, Nicholson JW. Prevention of enamel demineralization after tooth bleaching by bioactive glass incorporated into toothpaste. Aust Dent J 2011; 56(2): 193-200.
[http://dx.doi.org/10.1111/j.1834-7819.2011.01323.x] [PMID: 21623812]

[108]   Burwell AK, Litkowski LJ, Greenspan DC. Calcium sodium phosphosilicate (NovaMin): Remineralization potential. Adv Dent Res 2009; 21(1): 35-9.
[http://dx.doi.org/10.1177/0895937409335621] [PMID: 19710080]

[109]   Brauer DS, Karpukhina N, O'Donnell MD, Law RV, Hill RG. Fluoride-containing bioactive glasses: Effect of glass design and structure on degradation, pH and apatite formation in simulated body fluid. Acta Biomater 2010; 6(8): 3275-82.
[http://dx.doi.org/10.1016/j.actbio.2010.01.043] [PMID: 20132911]

[110]   Farooq I, Majeed A, Alshwaimi E, Almas K. Efficacy of a novel fluoride containing bioactive glass based dentifrice in remineralizing artificially induced demineralization in human enamel. Fluoride 2019; 53: 447-55.

[111] Peltola MJ, Aitasalo KMJ, Suonpää JTK, Yli-Urpo A, Laippala PJ, Forsback AP. Frontal sinus and skull bone defect obliteration with three synthetic bioactive materials. A comparative study. J Biomed Mater Res B Appl Biomater 2003; 66B(1): 364-72.
[http://dx.doi.org/10.1002/jbm.b.10023] [PMID: 12808596]

[112] De Lange GL, Lyaruu DM, Kuiper L, Burger EH, Tadjoedin E. Ette S. Tadjoedin High concentrations of bioactive glass material (BioGranA) vs. autogenous bone for sinus floor elevation Histomorphometrical observations on three split mouth clinical cases. Clin Oral Implants Res 2002; 13: 428-36.
[http://dx.doi.org/10.1034/j.1600-0501.2002.130412.x]

[113] Hasanzadeh M, Pournaghi-Azar MH, Shadjou N, Jouyban A. A new mechanistic approach to elucidate furosemide electrooxidation on magnetic nanoparticles loaded on graphene oxide modified glassy carbon electrode. RSC Advances 2014; 4(13): 6580.
[http://dx.doi.org/10.1039/c3ra46973e]

[114] Fujikura K, Karpukhina N, Kasuga T, Brauer DS, Hill RG, Law RV. Influence of strontium substitution on structure and crystallisation of Bioglass® 45S5. J Mater Chem 2012; 22(15): 7395-402.
[http://dx.doi.org/10.1039/c2jm14674f]

[115] Banerjee A, Hajatdoost-Sani M, Farrell S, Thompson I. A clinical evaluation and comparison of bioactive glass and sodium bicarbonate air-polishing powders. J Dent 2010; 38(6): 475-9.
[http://dx.doi.org/10.1016/j.jdent.2010.03.001] [PMID: 20223272]

[116] Taha AA, Hill RG, Fleming PS, Patel MP. Development of a novel bioactive glass for air-abrasion to selectively remove orthodontic adhesives. Clin Oral Investig 2018; 22(4): 1839-49.
[http://dx.doi.org/10.1007/s00784-017-2279-8] [PMID: 29185145]

[117] Skallevold HE, Rokaya D, Khurshid Z, Zafar MS. Bioactive glass applications in dentistry. Int J Mol Sci 2019; 20(23): 5960.
[http://dx.doi.org/10.3390/ijms20235960] [PMID: 31783484]

[118] West NX, Lussi A, Seong J, Hellwig E. Dentin hypersensitivity: Pain mechanisms and aetiology of exposed cervical dentin. Clin Oral Investig 2013; 17(S1): 9-19.
[http://dx.doi.org/10.1007/s00784-012-0887-x] [PMID: 23224116]

[119] Brauer D S, Karpukhina N, O'donnell M D, v Law R, Hill R G. Fluoride-containing bioactive glasses: Effect of glass design and structure on degradation, pH and apatite formation in simulated body fluid 2010; 49: 225-37.
[http://dx.doi.org/10.1016/j.actbio.2010.01.043]

[120] Hirschfeld L, Wasserman B. A long-term survey of tooth loss in 600 treated periodontal patients. J Periodontol 1978; 49(5): 225-37.
[http://dx.doi.org/10.1902/jop.1978.49.5.225] [PMID: 277674]

[121] Heitz-Mayfield LJA. Peri-implant diseases: Diagnosis and risk indicators. J Clin Periodontol 2008; 35(s8) (Suppl.): 292-304.
[http://dx.doi.org/10.1111/j.1600-051X.2008.01275.x] [PMID: 18724857]

[122] Renvert S, Persson GR. Periodontitis as a potential risk factor for peri-implantitis. J Clin Periodontol 2009; 36(s10) (Suppl. 10): 9-14.
[http://dx.doi.org/10.1111/j.1600-051X.2009.01416.x] [PMID: 19432626]

[123] Lovelace TB, Mellonig JT, Meffert RM, Jones AA, Nummikoski PV, Cochran DL. Clinical evaluation of bioactive glass in the treatment of periodontal osseous defects in humans. J Periodontol 1998; 69(9): 1027-35.
[http://dx.doi.org/10.1902/jop.1998.69.9.1027] [PMID: 9776031]

[124] Profeta AC, Prucher GM. Bioactive-glass in periodontal surgery and implant dentistry. Dent Mater J 2015; 34(5): 559-71.

[http://dx.doi.org/10.4012/dmj.2014-233] [PMID: 26438980]

[125] Profeta A, Huppa C. Bioactive-glass in oral and maxillofacial surgery. Craniomaxillofac Trauma Reconstr 2016; 9(1): 1-14.
[http://dx.doi.org/10.1055/s-0035-1551543]

[126] Utneja S, Nawal RR, Talwar S, Verma M. Current perspectives of bio-ceramic technology in endodontics: Calcium enriched mixture cement - review of its composition, properties and applications. Restor Dent Endod 2015; 40(1): 1-13.
[http://dx.doi.org/10.5395/rde.2015.40.1.1] [PMID: 25671207]

[127] Oguntebi B, Clark A, Wilson J. Pulp capping with Bioglass and autologous demineralized dentin in miniature swine. J Dent Res 1993; 72(2): 484-9.
[http://dx.doi.org/10.1177/00220345930720020301] [PMID: 8423245]

[128] Sawsan T. Reparative activity of bioglass and tricalcium phosphate to induce apical closure compared to normal apexogensis. Histological and Histochemical Study). Cairo Dental Journal 1999; 15(3): 1151-1160.

Recent Advances in Bioactive Glasses and Glass Ceramics

Syeda Ammara Batool[1], **Memoona Akhtar**[1,2], **Muhammad Rizwan**[3] and **Muhammad Atiq Ur Rehman**[1,*]

[1] *Department of Materials Science and Engineering, Institute of Space Technology Islamabad, Islamabad-44000, Pakistan*

[2] *Department of Materials Science and Engineering, Institute of Biomaterials, University of Erlangen-Nuremberg, Erlangen-91058, Germany*

[3] *Department of Metallurgical Engineering, Faculty of Chemical and Process Engineering, NED University of Engineering and Technology, Karachi-75270, Pakistan*

Abstract: Bone is a self-healing part of the body, which if damaged, repairs itself in the natural course of events. However, this healing process is deficient if the defect is too large or malignant to mend naturally. Bone regeneration is an age-dependent phenomenon where the older generation is at a disadvantage as compared to the younger generation due to the compromised biological performance as a result of aging. Therefore, it is crucial to create novel and effective ways to treat bone-related troubles. Bioactive glasses (BGs) and glass ceramics (GCs) belong to the third-generation bioactive materials. They not only have the potential to survive in the harsh physiological environment but can also renovate the defects present around them. They also come with the advantage of tunable chemical, physical, and biological properties. Designing an implant or scaffold while playing with distinct characteristics of metals, polymers, and ceramics, bestows a large selection pane in front of humankind for customized and patient-specific products. In this chapter, an overview of the recent advances in the BGs and GCs application in coatings and hydrogels for bone tissue engineering (BTE) is presented. BGs and GCs incorporated coatings and hydrogels loaded with metallic ions, growth factors, and biomolecules provide a complete bundle of features essential for bone repair and growth. Although many BGs and CGs-based products have made it into the market, some inherent challenges like high brittleness and low fracture toughness persist to overcome to date.

Keywords: 3D printed, Antibacterial, Bioactivity, Bioceramics, Biomaterials, Biopolymers, Bone, Coatings, Degradation, Electrospinning, Glasses, Hydrogels, Hydroxyapatite, Metallic ions, Osteoconductive, Osteointegration, Regeneration, Resorbable, Scaffolds, Tissue engineering.

---
* **Corresponding author Muhammad Atiq-Ur-Rehman:** Department of Materials Science and Engineering, Institute of Space Technology Islamabad, Islamabad-44000, Pakistan; Tel: +92-51-99075869; E-mail: atique1.1@hotmail.com

Saeid Kargozar and Francesco Baino (Eds.)

INTRODUCTION

Bioactive glasses (BGs), since the invention of the original 45S5 Bioactive glass (Bioglass®) in 1969 have revolutionized the spectrum of biomaterials [1]. Professor Larry Hench developed Bioglass® in search of materials capable of bonding with natural tissues [2]. Hench's research idea centered around the hypothesis that good bonding ability can be obtained in implant materials using elements/ ions already available abundantly in physiological environment (such as Ca & P) [3]. This invention introduced a new dimension of bioactive materials that are potentially able to develop bonds with tissues in the physiological environments [4]. The developed Bioglass® could bind to the natural bone so strongly that the detachment was not possible without breaking the bone [5]. Originally, BGs were prepared for bone substitute [6], but later they have found applications in dentistry [7] and soft tissue engineering (TE) [8]. BGs belong to class A bioactive materials which apart from employing already differentiated bone cells (osteoconduction), can also stimulate primitive undifferentiated cells to yield bone-related cells (osteoinduction) for enhanced osteoblasts' proliferation [9]. Moreover, BGs also upregulate genes for revascularization, enhance enzyme activity, exhibit antibacterial character, and deliver drugs [10, 11]. This set of unique properties have made them a biomaterial of significantly high research interest for almost half a century [11, 12].

Despite an extremely attractive set of biological properties, the major limiting factor for BGs is their inadequate mechanical performance [13]. Numerous efforts have been made to address this shortcoming, centered around doping, architectural designing, structure control and synthesis techniques [14, 15]. The glass ceramics (GCs) could exhibit similar biological performance along with good mechanical stability [16]. GCs are potentially similar to BGs but crystalline phases are yielded in the glassy matrix using a special heat treatment [17]. Kokubo *et al.* [18] developed Apatite and β-Wollastonite (A-W) BGC in $3Cao.P_2O_5-CaO.SiO_2-CaO. MgO.2SiO_2$ system. When the melt-quenched glass of above mentioned system was exposed to 1050 °C (slow heating rate 5 °C/ min), the formation of fibrous wollastonite and fine crystals of oxyapatite took place [19]. The crystallization of wollastonite and oxyapatite yield bending strength (215 MPa) higher than that of cortical bone (*i.e.* 160 MPa) and enhanced fracture toughness (*i.e.* 2 MPa · $m^{1/2}$) [20, 21]. However, in comparison to the fracture toughness of the cortical bone (*i.e.* 2–12 MPa · $m^{1/2}$), this enhancement is negligible and a marked improvement is needed to compete with the natural bone [22]. Currently, the size of crystallites in GCs is in the micrometer range and it is believed that if the crystallite size can be decreased to nanometer range, a marked increase in mechanical performance can be achieved [23, 24]. Several

BGC products have been developed commercially for bone and dental applications including: Bioverit®, Biosilicate®, Cerabone®, and Ceravital® [25].

During this journey of more than half a century, BGs and glass ceramics have witnessed several important milestones. They have been employed for a wide range of applications from hard and soft TE to theranostics [26]. After the discovery of Bioglass® in 1969, its first clinical trial was reported in 1977 for the replacement of middle ear (small) bones and later in 1978, as an ocular implant. For the liver cancer treatment, BG was employed in 1987 and later in 2004, lungs treatment was also carried out using it. More recently in 2018, TheraSphere® (a radioactive BG) has been employed to treat colorectal liver cancer. On the other hand, A-W GC found its application as a prosthesis for the reconstruction of iliac crest in 1987. Bioverit®, which is a common name for two types of mica-apatite GCs, had been implanted in more than 850 patients as middle ear implant or bone spacers till 1992 [27]. A relatively new GC (Biosilicate®), which exhibited bioactivity as high as that of Bioglass® 45S5 and mechanical performance similar to that of A-W was patented in 2007 for the dental ailments [28]. Fig. (**1**) shows the list of publications on BGs per year for the last ten years.

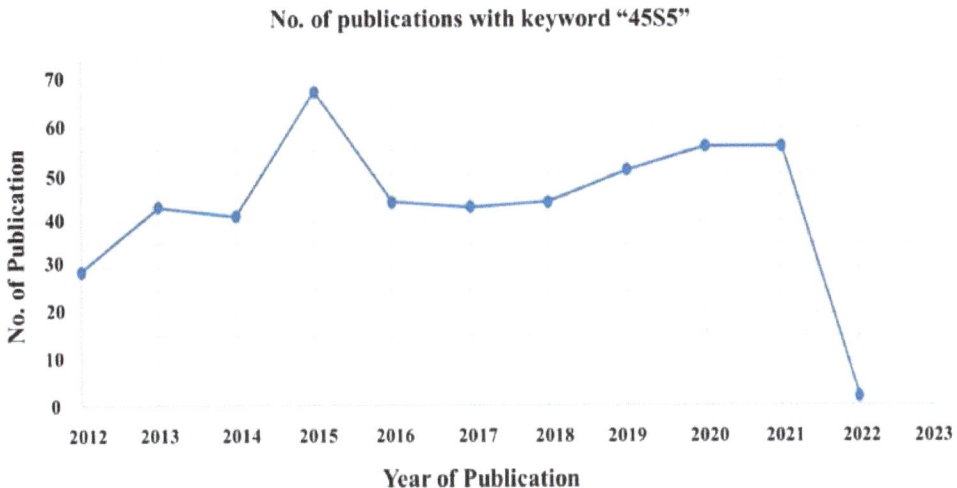

Fig. (1). List of publications on BGs.

Recent Developments in GCs and BGs Incorporated Coatings

GCs and BGs are used in medical applications due to their unique interaction with human body. Following the discovery of BGs in 1971, clinical uses of BGs were not achieved until the 1980s. The successful implantation of middle ear replacement prosthesis (MEP) [29] and endosseous ridge maintenance implants

(ERMI) [30] in clinical trials paved the door for wider use of BGs and GCs. As the research progressed throughout the 1990s, it produced sufficient evidence for the Food and Drug Administration (FDA) in the United States to approve the safe use of BGs and GCs for bone regeneration [31, 32].

Biomaterials have progressed from bio-inert materials to the creation of bioactive and bioresorbable materials. Metallic implant materials offer sufficient mechanical properties, such as strength, elastic modulus, and ductility, as compared to the polymers and ceramics. In last few decades, traditional metallic biomaterials such as cobalt-chromium (Co-Cr) alloys, stainless steels (SS), and titanium (Ti) and its alloys were considered gold standards for load-bearing implants in bone tissue regeneration. However, these implants had poor osteointegration due to their intrinsic inertness and became loose over time as a result of fibrous encapsulation [33]. Thus, the need to tailor the surface of inert materials arose to induce an interaction with the nearby hard and soft tissues. Metals such as iron (Fe), magnesium (Mg), and zinc (Zn) were also developed as new biodegradable implant materials with varying *in vivo* decomposition rates [34 - 36]. Similar to above mentioned metallic implants, they do not have good bioactivity. Controlling the rate of degradation in these materials is also a difficult task. As a result, metallic substrates must be coated with compounds that increase bioactivity and control degradation of implanted material. Surface coating is one of the most commonly used ways for improving the implant materials' overall life by manipulating surface biocompatibility, bioactivity and ion-release rate [37].

GCs and BGs coatings have the potential to make an inert surface bioactive. GCs include silicate (SiO_2), borate (B_2O_3), and phosphate (P_2O_5) type materials. BGs also belong to the family of bioceramics, mainly consisting of silica, calcium oxide, and phosphorous. The nature of the GCs and BGs' dissociation products contributes to their ability to bind with soft and hard tissues, tissue repair and regeneration [38]. Bone apatite (inorganic mineral composed of calcium phosphate) is the major component of bones, imparting strength to the bones. GCs and BGs upon degradation make a layer of inorganic phase 'hydroxyapatite (HA)' which is a marker of its bioactivity [39] and very much similar to the natural bone apatite in terms of chemical composition. The intrinsic ability to develop this HA layer renders GCs and BGs best substitute materials for bone [40]. Degradation of these substitute materials is accompanied by the repair/regeneration of tissues.

Calcium phosphate (CaP)-based bone regenerative substance is another well-known material from GCs family. Alpha and beta-tri calcium phosphate (α and β-TCP), biphasic calcium phosphate, and HA *etc.* all are differentiated on the basis of Ca to P ratio. The variable concentration of both phases provide them different features. Similar to CaP based materials, BGs also differ on the basis of their chemical composition. Table **1** gives an overview of different CaP-based GCs and BGs used for biomedical applications.

Table 1. CaP-based GCs and different BG compositions for biomedical applications. Adapted from [41, 42].

CaP Based GCs	Formula	Ca/P
Monocalcium phosphate monohydrate (MCPM)	$Ca(H_2PO_4)_2 \cdot H_2O$	0.5
Monocalcium phosphate anhydrous (MCPA)	$Ca(H_2PO_4)_2$	0.5
Dicalcium phosphate dihydrate (DCPD)	$CaHPO_4 \cdot 2H_2O$	1.0
Dicalcium phosphate anhydrous (DCPA)	$CaHPO_4$	1.0
Octacalcium phosphate (OCP)	$Ca_8(HPO_4)_2(PO_4)_4 \cdot 5H_2O$	1.33
α-Tricalcium phosphate (α-TCP)	$\alpha\text{-}Ca_3(PO_4)_2$	1.5
β-Tricalcium phosphate (β-TCP)	$\beta\text{-}Ca_3(PO_4)_2$	1.5
Amorphous calcium phosphates (ACP)	$Ca_xH_y(PO_4)_z \cdot nH_2O$ $n = 3\text{--}4.5$; 15–20% H_2O	1.2–2.2
Calcium-deficient hydroxyapatite (CDHA or Ca-def HA)	$Ca_{10-x}(HPO_4)_x(PO_4)_{6-x}(OH)_{2-x}$ $(0 < x < 1)$	1.5–1.67
Hydroxyapatite (HA, or HAp)	$Ca_{10}(PO_4)_6(OH)_2$	1.67
Tetracalcium phosphate (TTCP, or TetcP)	$Ca_4(PO_4)_2O$	2
Type of BG	**Composition**	
45S5	45 SiO_2 -24.5 Na_2O-24.5 CaO-6 P_2O_5	
42S5	42.1 SiO_2 -26.3 Na_2O-29 CaO-2.6 P_2O_5	
S53P4	53 SiO_2 -23 Na_2O-20 CaO-4 P_2O_5	
55S4	52.1 SiO_2 -21.5 Na_2O-23.8 CaO-2.6 P_2O_5	
58S	60 SiO_2 -36 CaO-4 P_2O_5	
70S30C	70 SiO_2-30 Na_2O	
45S5F	45 SiO_2 -24.5 Na_2O-12.25 CaO-6 P_2O_5-12.5 CaF_2	
40S5B5	40 SiO_2 -24.5 Na_2O-24.5 CaO-6 P_2O_5-5 B_2O_3	

GCs and BGs have demonstrated good biocompatibility and enhanced biological performance when applied as coatings on metallic implants. Co-Cr alloys, SS and Ti based alloys are the most commonly used metallic implants. However, each one of them has its own pros and cons. Co-Cr alloys are a material of choice for metal-on-metal (MoM) hip implants due to their good mechanical properties, low wear and corrosion rates, still studies show that the nano-sized wear particles can cause toxicity due to the Co ion leaching in blood stream and immune system impairment in patients with MoM implants [43, 44]. 316L SS (medical grade SS) show good ductility and fatigue properties. On the other hand, lack of bioactivity, low blood compatibility and susceptibility to corrosion limit its use for tissue regeneration [45]. Pure Ti and its alloy Ti-6Al-4V are used for light weight and strong implants with excellent corrosion resistance and biocompatibility, however, if corroded they can prove highly toxic in the body. Also, their relatively lower hardening coefficient render them difficult to process and tailor the mechanical properties [46]. In case of biodegradable metals, Fe is an essential element for all living beings, responsible for oxygen transport in the body [47]. Mg is an important cofactor for several enzyme systems in the body [48]. Zn is an important factor in growth and synthesis of proteins and DNA [49]. The presence of these ions in suitable quantity may contribute to various functional roles in physiological environment. However, their over-exposure has adverse effects on human health. Surface modification can not only improve the bio-functionality of permanent metallic implants, but also control the degradation rate of bioresorbable implant materials like Fe, Mg and Zn.

GCs and BGs Based Coatings

Coatings consisting of BGs along with other GCs can provide the best solution to most metal-related problems. A composite of BG and HA with varying compositions of both was deposited over commercially pure Ti (CP-Ti) *via* electrophoretic deposition (EPD) [50]. Results demonstrated that BG coating with 50 wt% HA had the highest bioactivity after immersion in simulated body fluid (SBF) and bonding strength. Another study was conducted to improve the corrosion resistance of CP-Ti by introducing a CaP-containing titanium oxide interlayer [51]. Subsequently, the EPD of nano-HA/45S5 coating was carried out which showed good bonding strength and enhanced corrosion resistance. Li *et al.* [52] successfully decreased the degradation rate of Mg alloy AZ31 by an interior layer of mesoporous 45S5 and an outer layer of phytic acid. The degradation rate was decreased to 0.62 mg per cm^2/day from the uncoated sample of 2.93 mg per cm^2/day on the 16th day. The bioactivity of the Mg alloy was also improved due to the presence of phytic acid. Singh *et al.* [53] deposited a novel composition of BG on laser-textured 316L SS. The corrosion current density for the coated substrate dropped to 35 nA/cm^2 as compared to the uncoated substrate for which the corrosion current density was 353 nA/cm^2. The formation of apatite layer confirmed the improved bioactivity of 316L SS.

Glass matrix has the capacity to accommodate various doping elements that can induce additional functions in the body. For example, the incorporation of strontium (Sr) ions stimulates bone growth [54], while silver (Ag) and Zn ions impart antibacterial effect [55, 56], and copper (Cu) ions improve angiogenesis [57], and manganese (Mn) ions help osteogenic differentiation [58], *etc.* Luo *et al.* [59] deposited multi-metal ion (Zn, Ag, and Cu) doped HA on Ti surface by pulse electrochemical method. The coatings proved effective against *Escherichia coli (E. coli)* and *Staphylococcus aureus (S. aureus)* and demonstrated good osteogenic potential.

It is well-known that *in-vitro* results are not enough to ensure biocompatibility and bioactivity of the materials. Materials can behave very differently when placed in the real physiological conditions. Various *in-vivo* trials are performed to analyze the true behavior of GCs-BGs in different conditions [60 - 62]. One such study was carried out by Souza *et al.* [63]. 45S5 (control group) and 45S5 substituted with 1.3 and 2.6 mol% of Nb (BGPN1.3 and BGPN2.6, respectively) were implanted in rat tibia to evaluate their biocompatibility and osteogenic potential. Images of histological sections of the rat tibia were taken on day 14 and the subperiosteal bone area was quantified on day 14 and 28. All the test samples were highly biocompatible and showed good osteostimulation at day 14 of implantation. However, quantitative analysis of the subperiosteal bone area at days 14 and 28 showed that the bone formation rate for BGPN samples was significantly higher than the control group. Overall, the replacement of Nb with Ca enhanced the bioactivity and osteostimulative properties of the implant.

Polymer-based Composite Coatings

Polymers have emerged as intriguing and flexible candidates for biomedical coatings. Natural or synthetic polymers-based coatings have the potential to overcome the intrinsic brittleness of GCs and BGs. The use of composite coatings with GCs and BGs as dispersed phases in the polymer matrix provides value by simultaneously improving biocompatibility, bioactivity, and bonding strength of metallic substrates. Furthermore, biodegradable polymer coatings can be used as corrosion resistant coatings on implants [64]. Examples of polymer-based composite coatings are listed below to give the readers an idea of their potential uses in TE.

Chitosan

Chitosan (CS) is a natural polymer with good biocompatibility. The structural similarity of CS with glycosaminoglycan (GAG); a component of extra cellular matrix (ECM) of bone, makes it suitable for cell adhesion and proliferation [65]. An important characteristic of CS is its biodegradability. Rehman *et al.* [66] have deposited a composite coating consisting of CS/gelatin (GEL)/BGs on 316L SS to improve its surface properties by controlling EPD parameters. The scanning electron microscopy (SEM) of CS/GEL/BGs coating displayed homogeneous dispersion of BGs in the CS/GEL matrix which promoted adequate bonding strength and surface wettability for cell adhesion.

Pawlik *et al.* [67] electrophoretically deposited CS/HA coatings on anodic titanium dioxide layer to improve the surface roughness parameter to achieve a more suitable bone-contacting surface. Another CS-based composite coating system was studied by Avcu *et al.* [68]. CS/45S5 micro-BG (mBG), CS/nano-BG (nBG), and CS/mBG/nBG coatings were deposited on pretreated Ti-6Al-4 V alloy by EPD to improve surface properties for enhanced bioactivity.

Zein

Zein is derived from maize endosperm, which is made up of prolamins. It is a biocompatible polymer that can be used for TE and drug delivery [69]. Zein has low mechanical strength due to which different composites with GCs and BGs are produced for coating purposes. Ahmed *et al.* [70] studied zein/HA based coatings on 316L SS for improved corrosion resistance and bioactivity. Zein/BGs coating was also used to control the degradation rate of pre-treated Mg [35]. Electrochemical analyses confirmed the protective nature of zein/BGs coating against corrosion. Pretreatment of Mg is also considered effective in controlling its degradation rate in physiological conditions [71, 72].

Polyetheretherketone

Polyetheretherketone (PEEK) is a synthetic polymer adopted for biological coatings. PEEK is a bio-inert and bio-stable polymer for coating permanent metallic implants. Mechanical properties of PEEK, such as elastic modulus can be easily tailored to match the elastic modulus of cortical bone [73]. However, the problem lies with the bio-inertness of PEEK which makes it in-effective for cellular bonding. Therefore, the addition of bioactive substances including GCs and BGs facilitate PEEK in tissue bonding. Rehman *et al.* [74] optimized EPD parameters for PEEK/45S5 coating on 316L SS using the design of experiment (DoE) approach. A similar study was carried out by Bastan *et al.* [75]. He optimized EPD parameters for PEEK/HA coatings on 316L SS to improve adhesion strength and bioactivity of substrate material. However, it was concluded that adhesion strength and bioactivity had a trade-off in between and improving one deteriorated the other. The bioactivity enhanced with the HA content. Hence, there is a need to optimize the composition of the coating.

GEL

GEL is another biopolymer used for surface modification of various substrate materials. GEL is nontoxic, water-soluble, and produces gel at low concentrations. Similar to other biopolymers, GEL also lacks mechanical strength. Therefore, a composite systems was developed comprising GEL and BGs along with some other biopolymers [76].

A majority of natural and synthetic polymers are being used for coatings in combination with GCs and BGs on different metal substrates. Synthetic polymers include poly lactic acid (PLA), poly (lactide-co-glycolic) acid (PLGA), polypropylene (PP), polycaprolactone (PCL), polymethyl methacrylate (PMMA), polyethylene (PE), and polyurethane (PU) *etc* [64]. However, it is not possible to include all the coating materials combined with GCs and BGs in this chapter. Different coating combinations can be chosen depending on the polymer and glass qualities, metal substrates, and specialized applications to improve the chances of creating a coating with high adherence while retaining bioactivity.

Challenges

GCs and BGs possess excellent features that can be easily manipulated by adding polymeric and metallic inclusion. Regrettably, their relatively poor mechanical properties, such as high brittleness, inadequate bending strength and fracture toughness, limit their applications to non-load-bearing implants. Another challenge is to minimize the difference in the elastic modulus of implant materials and surrounding tissues which is generally overcome by synthesizing porous base material, however, using highly porous materials for bioactive/resorbable implants render them mechanically weak. Controlling the dissolution rate of resorable materials is also a major concern. Dissolution rate of the coating must closely match with the growth rate of new tissue, maintaining a state of dynamic equilibrium. Another drawback is the mismatch of the thermal coefficient of GCs and BGs to that of metals, ultimately getting down to stress generation in coating and implant failure.

The challenge for materials scientists and researchers is to provide strong evidence proving the optimal performance of designed composite coatings in physiological settings. It is vital to provide trials-based proof before their safe implementation in clinical setting is allowed.

Hydrogels for Bone Regeneration

Trauma, genetic disease, surgery, or degenerative conditions like osteoarthritis can all cause bone abnormalities, which represent a serious health concern by reducing a patient's quality of life and mobility [77]. Osseous regeneration is limited, despite the fact that bone has the real regenerative capacity and is restored by the formation of new, healthy bone cells termed 'osteoblasts'. Bone flaws that are too large to be repaired naturally result in non-union and loss of function [78]. Although autologous bone grafting remains the cornerstone to heal major bone-related issues, it is hampered by donor site morbidity, high infection risk, and a limited ability to fill complex defects [79]. As a result, the orthopaedic community is becoming increasingly interested in TE options for bone healing.

Bone tissue engineering (BTE) is a unique approach to bone repair and regeneration that uses scaffolds seeded with cells or integrated with bioactive growth factors [80]. The scaffolds employed in BTE are designed to provide structural support and create an environment conducive to cell adhesion, migration, proliferation, as well as differentiation [81, 82]. Inorganic materials such as GCs have good mechanical and osteoconductive qualities that have been effectively used in alveolar bone replacement [83]. GCs and BGs' scaffolds in combination with natural and synthetic polymers are being researched in BTE during the last few decades. Natural materials like collagen (Col) and CS are degradable and bioresorbable, while synthetic polymeric materials like PLA and PLGA have configurable degradation behavior; yet, in the design of polymeric scaffolds, the structure and attributes of cells and native tissues are rarely addressed [84]. As a result, the scaffolds frequently show poor integration with the underlying bone structures. To fulfil medical needs, new materials and solutions are always being created that highlighted the compatibility of scaffolds and native tissues, specifically the pivotal function of the natural ECM in bone regeneration [85]. Polymeric materials and composites proved to interact well with surrounding bone tissues due to their biocompatible and biodegradable nature, in comparison with ceramics, allowing for a stable anchoring of implants and reducing the risk of an immune response [79].

Hydrogels, a form of polymeric material, provide a number of potential benefits in BTE [86]. Hydrogels are 3D polymeric chains with outstanding mechanical resistance, excellent hydrophilicity and the ability to produce nutritional conditions conducive to endogenous cell growth. They can resemble the bone's native ECM, which means they might be used to encapsulate bioactive chemicals/cells [87].

The entrapped proteins or cells are restricted in the meshes due to the hydrogels' 3D network, and the release of the materials can be controlled as needed [88]. Furthermore, hydrogels are resorbable and integrate well with surrounding tissues, decreasing the need for surgical removal and lowering the risk of an inflammatory response [89]. Furthermore, basic materials for making hydrogels are plentiful and easy to come by, and they may be adjusted to provide the necessary geometry for implantation or injection, with the degradation rate, porosity, and release profile easily regulated by changing the crosslinking method and degree [90, 91].

Hydrogel-based cell distribution and medication delivery have emerged as possible options in TE and regenerative medicine to enhance bone repair [92, 93]. They can support new bone development by providing a natural hydrophilic environment suitable for cell survival [81, 94].

Selection Criteria for Hydrogels

Optimized hydrogel formulations for bone regeneration should ideally meet the following criteria: 1) non-cytotoxic and non-immunogenic in order to inhibit an inflammatory response [95]; 2) improved bone regeneration by being osteoinductive, osteoconductive, osteogenic [96], and osteocompatible; 3) to the best extent possible, imitate the natural ECM to aid cell adherence, proliferation, and ultimately osteogenic differentiation at the implantation site [77]; 4) degradation under endogenous enzymes/ hydrolysis, synchronized with new bone ingrowth to provide sufficient space for new bone formation [97]; 5) structurally and mechanically stable that can be employed to cure load-bearing defects as well as avoid denaturation during sterilization [98]; 6) appropriate pore size and interconnected porosity, which can be optimized by varying the concentration and type of polymer and cross linkers to improve cell interaction, control the release of encapsulated bioactive factors, and allow the exchange of nutrients, oxygen, and metabolic waste within the hydrogels [99]; and 7) injectable capability with patient compliance to reduce pain and simplify the administration process [84].

Recent Advancements in GCs and BGs Incorporated Hydrogels

Natural materials like proteins (silk fibroin (SF), Col, and GEL) [100 - 102] and polysaccharides can be used to make hydrogels (CS, hyaluronan (HY) and alginate (ALG)) [80, 103, 104]. Natural polymers are biocompatible, less immunogenic, and cytotoxic, and can stimulate cell adhesion, proliferation, as well as tissue regeneration [89]. Natural polymers have a structure similar to the ECM, giving bone tissues mechanical integrity and structural stability while also avoiding inflammatory and immunological reactions [105]. They can be absorbed through enzyme-controlled or metabolic breakdown [89, 106]. Degradable polymeric compounds, like polyethylene glycol (PEG), polyvinyl alcohol (PVA), polyacrylamide (PAA), methyl carbonate, PLA, and its copolymers, can be utilized to make hydrogels for bone repair/regeneration. Synthetic polymers, which have basic structural units, allow polymer features (porosity, degradation rate, and mechanical properties) to be tailored for distinct purposes. Synthetic polymers offer consistent material sources and lengthy shelf lives, allowing them to be manufactured in huge quantities without fear of immunogenicity [107].

Gönen *et al.* [97] constructed a multifunctional scaffold of Sr or Cu-doped BGs and GEL/PCL blends *via* electrospinning. The ability of fiber mats to generate hydroxyapatite provides information about their bioactivity, which is important for bone regeneration as shown in Fig. (**2A**). However, the addition of >2% SrO and CuO to the BG composition can enhance osteogenesis, angiogenesis, and antimicrobial properties of nanocomposite fiber mat used for bone scaffold. In another study conducted on GEL-based scaffold, Sarker *et al.* [108] produced freeze-dried scaffolds of an oxidized alginate (ADA)-GEL hydrogel matrix reinforced with BG 45S5 to manage degradation and improve mechanical properties. It was discovered that BG 45S5 accelerates gelation, increases degree of crosslinking in ADA-GEL, and stimulates HA formation as shown in Fig. (**2B**). When compared to the ADA-GEL matrix alone, the compressive stress and modulus of scaffolds containing different concentrations of BGs rose from 54 ± 12 kPa to 638 ± 81 kPa and from 5 ± 1 kPa to 250 ± 42 kPa, respectively. The compressive behavior was investigated at day 28 of immersion in SBF. In addition, porosity lowered to 40 and 30% for 1 and 5 wt% addition of BGs into ADA-GEL hydrogel, respectively.

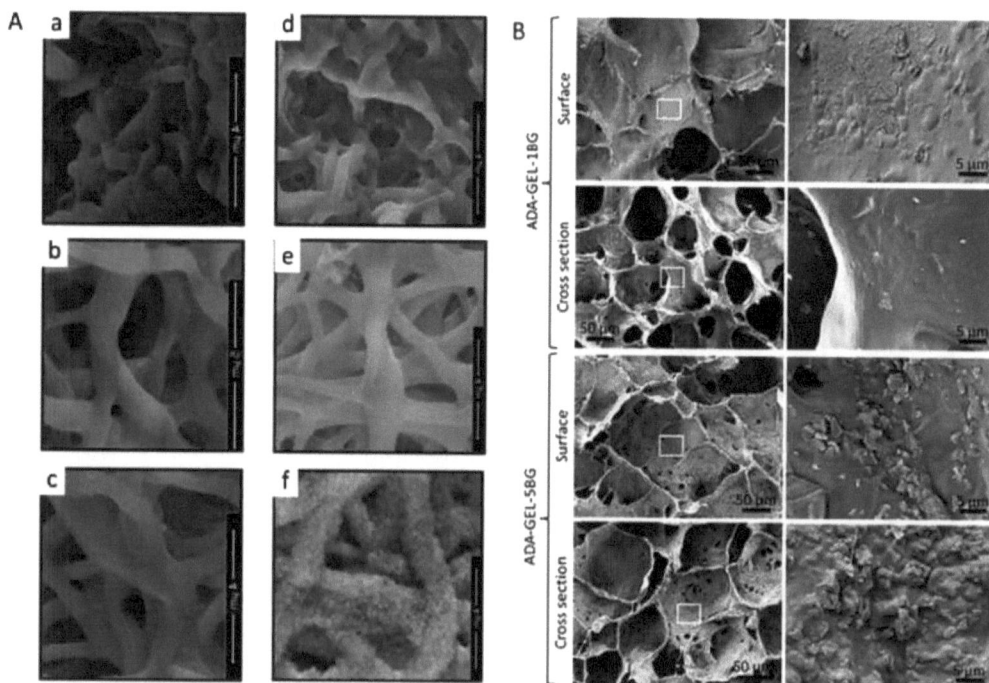

Fig. (2). (A) SEM images of fiber mats after being soaked in SBF for **(a-c)** 1 day and **(d-f)** 28 days: **(a, d)** GEL/PCL, **(b, e)** GEL/PCL/7.5Sr-BG, and **(c, f)** GEL/PCL/7.5Cu-BG fiber mats. Adapted from [97]. Reproduced with the permission from Elsevier™. **(B)** SEM images of the ADA-GEL/BG scaffolds after 28 days of immersion in SBF. Adapted from [108]. Reproduced with the permission from ACS™.

Bai *et al.* [109] developed a triple crosslinked injectable hydrogel by incorporating doubly modified ALG and BGs into chondroitin sulfate - PEG hydrogels paralleled to different crosslinker combinations and tested *in-vivo* for cranial bone regeneration. *In-vitro* studies included swelling, degradability, and mechanical characteristics. The results showed that triple crosslinked hydrogels had a storage modulus of up to 4 kPa and loaded with BGs had a storage modulus of over 4.5 kPa due to the release of Ca^{+2} ions from the BG.

Fernandes *et al.* [110] created PLLA electrospun membranes for BTE using Sr-Borosilicate BG. The elasticity increased from 14.6 ± 3.8 MPa (PLLA) to 24.7 ± 5.3 MPa after BG-Sr microparticles were added (PLLA-BGs-Sr). When compared to PLLA membranes (0.55 ± 0.6MPa), PLLA-BGs-Sr had the maximum tensile strength (0.75 ± 0.7 MPa). Rescignano *et al.* [111] created ALG hydrogels crosslinked with calcium carbonate ($CaCO_3$) and d-glucono-δ-lactone (GDL) that were effective against *E. coli* and *Pseudomonas aeruginosa* after 24 hours of incubation, with inhibitory zones between 5 and 7 cm. The hydrogels containing 2.5 wt% AgNPs, on the other hand, increased the Young's modulus of the ALG hydrogels from 0.293 to 0.530 kPa, showing a contact between the ALG and AgNPs.

Łańcucka *et al.* [112] used ALG and methacrylated functionalized GEL (Me-GEL) and methacrylamide (MAA) to create photo-crosslinked hydrogels for bone regeneration that were loaded with submicron silica (SiO_2) particles. With effective mineralization produced by SiO_2 particles, the materials supported MEFs and MG-63 mitochondrial activity. Furthermore, the storage modulus of GEL/ALG was found to be greater (6.88 - 9.03 kPa) than that of ALG based materials (0.178 - 0.238 kPa). Qiu *et al.* [113] developed highly viscous ALG hydrogel incorporated with Sr-containing mesoporous calcium silicate ($MCaSiO_3$) NPs. The inclusion of GDL in Na-ALG hydrogels incorporated with $MCaSiO_3$ NPs, reduced viscous nature of hydrogels, and increased injectability more than 90%. The Na-ALG composites were cytocompatible, promoted osteogenic differentiation and apatite formation. Several studies have shown that problematic locations lack blood supply, indicating that angiogenic agents could be included into hydrogels to improve bone repair as well as nerve regeneration. In the aforementioned study, Sr is a strong angiogenic material that is often employed to augment vascularization at the injury site.

Wu *et al.* [77] produced highly porous (80%) BG/Na-ALG/GEL scaffolds *via* 3D printing, demonstrating that scaffolds with good sustained release capabilities, similar to human cancellous bone, have potential for individualised bone defect repair. Mao *et al.* [78] used a modified melt moulding approach to create a natural-synthetic hybrid scaffold of Na-ALG/GEL composite mixed with PCL/58S BG with a compressive strength of 1.44 ± 0.02 MPa, near to that of cancellous bone (1 - 10 MPa).

The encapsulation of stem cells and growth factors in a hydrogel composites will, in theory, speed up the production of new ECM. Prompt hydrogel breakdown prior to the creation of fresh ECM could be one of the constraints, compromising mechanical integrity as well as disturbing mechanism of bone regeneration. Furthermore, appropriate stem cell differentiation is essential for ensuring optimal cell lineage differentiation and avoiding harmful impact when implanting stem cells to the problem site. Singh *et al.* [100] created a BG/SF/CMC composite used to create bone tissue constructs with human mesenchymal stem cells (hMSCs). SF/CMC composite exhibited excellent tensile properties, which declined when BGs were added. Cheng *et al.* [114] injected vancomycin into PLGA/BG to create a localized antibiotic delivery system *via* freeze-drying fabrication of bone scaffold. The addition of vancomycin-loaded BG improves both antibacterial and *in-vitro* bioactivity simultaneously, thus boosting bone ingrowth while limiting biofilm formation.

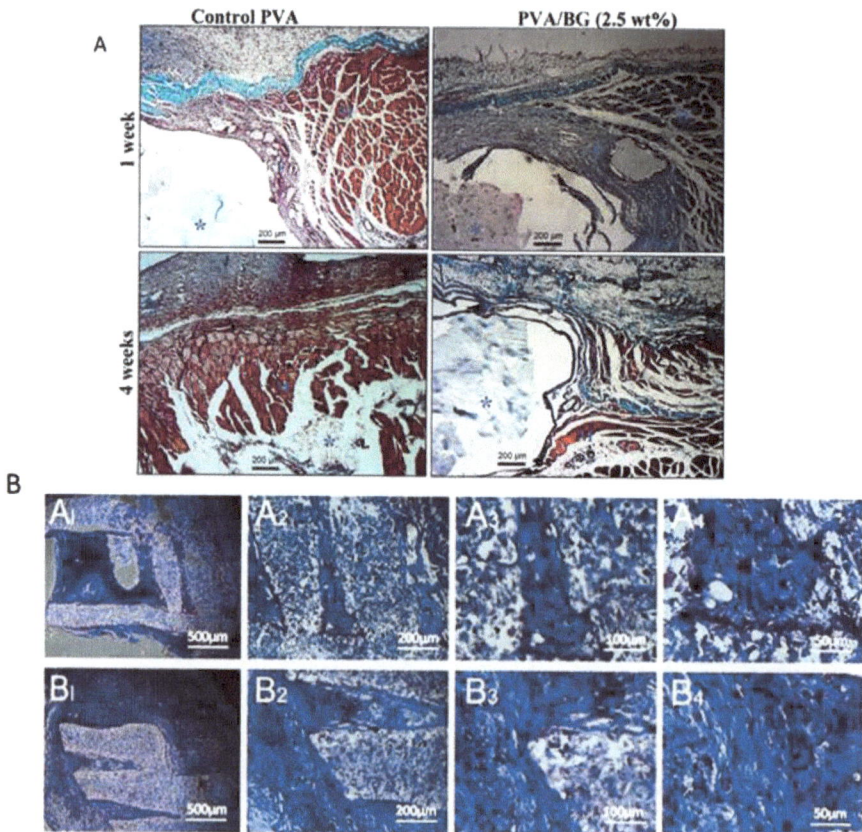

Fig. (3). (A) Light micrographs stained with Masson's Trichrome showing the histological outcomes of PVA and PVA/BG (2.5 wt%) explants after intramuscular implantation in rats for 1 and 4 week(s). Adapted from [116]. Reproduced with the permission from Taylor & Francis™. **(B)** *In-vivo* bone formation assay of BGs/SF (A$_1$-A$_4$) and BGs/PCL (B$_1$-B$_4$) scaffolds. Reproduced from [96].

Bonetti *et al.* [115] produced CS/GEL sheets, incorporated with Nb modified silicate BG *via* one-pot EPD. CS/GEL/BGNb composite appeared to be particularly promising scaffold design for BTE applications due to their intrinsic low cytotoxicity, antibacterial effect and apatite formation in SBF.

Abd-El-Fattah *et al.* [116] used a physical crosslinking technique to create an injectable PVA/2.5 wt% BGs hydrogel, which increased static compressive strength and Young's modulus by 325% and 150%, respectively. *In-vivo* subcutaneous implantation in rats verified the composite hydrogel's biocompatibility and biodegradability, as well as its ability to repair a non-loa--bearing bone defect as shown in Fig. (**3A**). Du *et al.* 2019 [96] created BG/SF composite and BG/PCL scaffolds using 3D printing technology and post-processing, which showcased remarkable mechanical properties and favored apatite formation. Furthermore, hMSCs were loaded into BG/SF scaffolds, which exhibited good bioactivity. Designed scaffolds were subcutaneously implanted on backs of rats for bone regeneration. 3D printed BG/SF scaffolds significantly promoted bone regeneration at the implanted site (Fig. **3B**) and proved their worth for effective BTE.

Dhinasekaran *et al.* [79] developed biocompatible, bone and nerve regenerative 3D membranes of Col and 45S5 BG nanofibers *via* electrospinning. Two morphological types of BG nanofibers were investigated *i.e.* hollow and solid termed BGH and BGS, respectively. Researchers measured the elastic modulus (MPa) of all membranes (Col, Col-BGS, and Col-BGH) while placing BGH containing Col membranes under two loading directions *i.e.* parallel and perpendicular. Elastic moduli of 213, 169, 173, and 126 were obtained for Col, Col- BGS, Col-BGH, (parallel), and Col-BGH (perpendicular), respectively. The study concluded that Col-BGH may be used to stimulate the directional growth of cells and mimic the bone ECM with an anisotropic alignment. Ensoylu *et al.* [85] used the foam replication approach to make tungsten disulfide (WS$_2$) NPs containing (0.1 to 2 wt%) PCL and PCL/PLGA-coated, borate-based porous BG scaffolds. The WS$_2$ NPs were biocompatible with MC3T3- E1 cells, and showed no cytotoxic response after 72 hours of growth.

Kumar *et al.* [117] fabricated *in-situ* double-network PVA/polyacrylamide (PAM)/BGs/ halloysite nanotubes (HNTs) nanocomposite hydrogels *via* freeze-thaw process. The synthesized hydrogels exhibited significantly good mechanical integrity (102.1 kPa strain and 3115.0 N/m stiffness). Valanezhad *et al.* 2021 [118] integrated a sol-gel-derived PLGA/ BG foam scaffolds to increase the mechanical integrity of scaffolds, which might be used to regenerate bone tissues.

Smart hydrogel systems with on-demand delivery competency have recently been a hot topic in the field of BTE. Hydrogels that are stimulus-responsive may detect stimuli in their surroundings and behave accordingly. Temperature, light, and pH are among the environmental stimuli that have been explored. Hydrogels that respond to biological stimuli such as enzymes, antigens, as well as ligands are also being investigated. The ability of a stimuli-responsive hydrogel to provide platform for bone defects caused by accidents, cancer, or ageing is being demonstrated, investigated and optimised to overcome the release of bioactive payload in response to changes in the body's surrounding physiological environment.

Challenges

A coordinated chemistry of cells, growth factors, and hydrogels is required for effective bone repair. Despite the fact that hydrogels possess key benefits in osteogenesis, there are still a few issues to be handled. Biocompatible nature should be considered while constructing hydrogels to avoid inflammation. Natural polymers such as Col, GEL, and CS are usually regarded as biocompatible, however their mechanical strength and structural stability are restricted. The scope of bone regeneration *via* hydrogels is limited by this significant problem involving the burst release of loaded bioactive growth factors and stem cells, which is directly connected to the hydrogel degradation rate. Because the main purpose of hydrogels is to improve *de novo* bone production, they must have a strong interface with the surrounding tissues (osteointegration), which might be achieved by incorporating bioactive GCs and BGs in the hydrogels.

To some extent, synthetic polymers can overcome the aforementioned challenges, but they come with drawbacks such as unexpected immunological responses, poor breakdown, and cell adhesion. As a result, further research is needed to optimize the polymeric composition, GCs/BGs concentration, and crosslinking approaches to enhance osteogenic potential while controlling the degradation rate of hydrogels. Regardless of significant challenges, GCs/BGs based-hydrogel exhibits immense potential for the treatment of bone-related problems.

CONCLUDING REMARKS

It is widely accepted that different combinations of GCs and BGs composite coatings and hydrogels can address multiple issues regarding biocompatibility, bioactivity and regenerative capability of implants. The pre-treatment of substrate and applied coating technique plays a major role in the final characteristics of metallic implants. Likewise, in hydrogels, the polymer content, degree of crystallization, crosslinking, and the type of incorporated GCs and BGs define the end-product. However, there is no set rule or principal for designing a coating or hydrogel for BTE which will deliver 100% of the expected outcomes. Despite the marvelous out turns of GCs and BGs based coatings and hydrogels obtained in *in-vitro* and *in-vivo* experiments, their realization in clinical setups is still unapproachable. There is a lot to consider while selecting a suitable biomaterial for a specified biomedical application. The underlying process of cell reaction and regeneration in response to an implant in actual humans is still a mystery to us, which demands for aggressive research and a cautious step towards human trials.

REFERENCES

[1]   Jones JR. Review of bioactive glass: From hench to hybrids. Acta Biomater 2013; 9(1): 4457-86.
        [http://dx.doi.org/10.1016/j.actbio.2012.08.023] [PMID: 22922331]

[2]   Hench LL. The story of Bioglass®. J Mater Sci Mater Med 2006; 17(11): 967-78.
        [http://dx.doi.org/10.1007/s10856-006-0432-z] [PMID: 17122907]

[3]   Hench LL, Splinter RJ, Allen WC, Greenlee TK. Bonding mechanisms at the interface of ceramic prosthetic materials. J Biomed Mater Res 1971; 5(6): 117-41.
        [http://dx.doi.org/10.1002/jbm.820050611]

[4]   Cao W, Hench LL. Bioactive materials. Ceram Int 1996; 22(6): 493-507.
        [http://dx.doi.org/10.1016/0272-8842(95)00126-3]

[5]   Baino F, Hamzehlou S, Kargozar S. Bioactive glasses: Where are we and where are we going? J Funct Biomater 2018; 9(1): 25.
        [http://dx.doi.org/10.3390/jfb9010025] [PMID: 29562680]

[6]   Hench L.L. The story of bioglass: From concept to clinic. Imperial College Inaugural Lectures In Materials Science And Materials Engineering. World Scientific 2001; pp. 203-29.
        [http://dx.doi.org/10.1142/9781848161740_0006]

[7]   Tiskaya M, Shahid S, Gillam D, Hill R. The use of bioactive glass (BAG) in dental composites: A critical review. Dent Mater 2021; 37(2): 296-310.
        [http://dx.doi.org/10.1016/j.dental.2020.11.015] [PMID: 33441250]

[8]   Miguez-Pacheco V, Hench LL, Boccaccini AR. Bioactive glasses beyond bone and teeth: Emerging applications in contact with soft tissues. Acta Biomater 2015; 13: 1-15.
        [http://dx.doi.org/10.1016/j.actbio.2014.11.004] [PMID: 25462853]

[9]   Rizwan M, Hamdi M, Basirun WJ. Bioglass® 45S5-based composites for bone tissue engineering and functional applications. J Biomed Mater Res A 2017; 105(11): 3197-223.
        [http://dx.doi.org/10.1002/jbm.a.36156] [PMID: 28686004]

[10]   Kaur G, Pandey OP, Singh K, Homa D, Scott B, Pickrell G. A review of bioactive glasses: Their structure, properties, fabrication and apatite formation. J Biomed Mater Res A 2014; 102(1): 254-74.
        [http://dx.doi.org/10.1002/jbm.a.34690] [PMID: 23468256]

[11]   Rizwan M, Alias R, Zaidi UZ, Mahmoodian R, Hamdi M. Surface modification of valve metals using plasma electrolytic oxidation for antibacterial applications: A review. J Biomed Mater Res A 2018; 106(2): 590-605.
        [http://dx.doi.org/10.1002/jbm.a.36259] [PMID: 28975693]

[12]   Kumar P, Dehiya BS, Sindhu A. Bioceramics for hard tissue engineering applications: A review. Int J Appl Eng Res 2018; 13: 2744-52.

[13]   Hench LL, Kokubo T. Properties of bioactive glasses and glass-ceramics.Handbook of biomaterial properties. Springer 1998; pp. 355-63.
        [http://dx.doi.org/10.1007/978-1-4615-5801-9_22]

[14] Brauer DS. Bioactive glasses—structure and properties. Angew Chem Int Ed 2015; 54(14): 4160-81.
[http://dx.doi.org/10.1002/anie.201405310] [PMID: 25765017]

[15] Liu X, Rahaman MN, Hilmas GE, Bal BS. Mechanical properties of bioactive glass (13-93) scaffolds fabricated by robotic deposition for structural bone repair. Acta Biomater 2013; 9(6): 7025-34.
[http://dx.doi.org/10.1016/j.actbio.2013.02.026] [PMID: 23438862]

[16] Kokubo T. Bioactive glass ceramics: Properties and applications. Biomaterials 1991; 12(2): 155-63.
[http://dx.doi.org/10.1016/0142-9612(91)90194-F] [PMID: 1878450]

[17] Zanotto ED. Bright future for glass-ceramics. Am Ceram Soc Bull 2010; 89: 19-27.

[18] Kokubo T, Shigematsu M, Nagashima Y, *et al.* Apatite-and wollastonite-containg glass-ceramics for prosthetic application. Bull Inst Chem Res Kyoto Univ 1982; 60: 260-8.

[19] Kokubo T, Ito S, Sakka S, Yamamuro T. Formation of a high-strength bioactive glass-ceramic in the system MgO-CaO-SiO2-P2O5. J Mater Sci 1986; 21(2): 536-40.
[http://dx.doi.org/10.1007/BF01145520]

[20] Yamamuro T. A/W glass-ceramic: Clinical applications. An Introduction to Bioceramics. World Scientific 2013; pp. 189-207.
[http://dx.doi.org/10.1142/9781908977168_0014]

[21] Kokubo T. Bioceramics and Their Clinical Applications. Elsevier 2008.

[22] Montazerian M, Dutra Zanotto E. History and trends of bioactive glass-ceramics. J Biomed Mater Res A 2016; 104(5): 1231-49.
[http://dx.doi.org/10.1002/jbm.a.35639] [PMID: 26707951]

[23] Fu L, Engqvist H, Xia W. Glass–ceramics in dentistry: A review. Materials 2020; 13(5): 1049.
[http://dx.doi.org/10.3390/ma13051049] [PMID: 32110874]

[24] Salinas AJ, Vallet-Regí M. Bioactive ceramics: From bone grafts to tissue engineering. RSC Advances 2013; 3(28): 11116-31.
[http://dx.doi.org/10.1039/c3ra00166k]

[25] Izquierdo-Barba I, Salinas AJ, Vallet-Regí M. Bioactive glasses: From macro to nano. Int J Appl Glass Sci 2013; 4(2): 149-61.
[http://dx.doi.org/10.1111/ijag.12028]

[26] Asano S, Kaneda K, Satoh S, Abumi K, Hashimoto T, Fujiya M. Reconstruction of an iliac crest defect with a bioactive ceramic prosthesis. Eur Spine J 1994; 3(1): 39-44.
[http://dx.doi.org/10.1007/BF02428315] [PMID: 7874540]

[27] Crovace MC, Souza MT, Chinaglia CR, Peitl O, Zanotto ED. Biosilicate® — A multipurpose, highly bioactive glass-ceramic. *In vitro, in vivo* and clinical trials. J Non-Cryst Solids 2016; 432: 90-110.
[http://dx.doi.org/10.1016/j.jnoncrysol.2015.03.022]

[28] Renno ACM, Bossini PS, Crovace MC, Rodrigues ACM, Zanotto ED, Parizotto NA. Characterization and *in vivo* biological performance of biosilicate. Biomed Res Int 2013; 2013

[29] Merwin GE, Atkins JS, Wilson J, Hench LL. Comparison of ossicular replacement materials in a mouse ear model. Otolaryngol Head Neck Surg 1982; 90(4): 461-9.
[http://dx.doi.org/10.1177/019459988209000417] [PMID: 6817277]

[30] Stanley HR, Hall MB, Clark AE, King CJ III, Hench LL, Berte JJ. Using 45S5 bioglass cones as endosseous ridge maintenance implants to prevent alveolar ridge resorption: A 5-year evaluation. Int J Oral Maxillofac Implants 1997; 12(1): 95-105.
[PMID: 9048461]

[31] Wilson J, Pigott GH, Schoen FJ, Hench LL. Toxicology and biocompatibility of bioglasses. J Biomed Mater Res 1981; 15(6): 805-17.
[http://dx.doi.org/10.1002/jbm.820150605] [PMID: 7309763]

[32]   Hench LL, Jones JR. Bioactive glasses: Frontiers and challenges. Front Bioeng Biotechnol 2015; 3: 194.
[http://dx.doi.org/10.3389/fbioe.2015.00194] [PMID: 26649290]

[33]   Horejs C. Preventing fibrotic encapsulation. Nat Rev Mater 2021; 6(7): 554.
[http://dx.doi.org/10.1038/s41578-021-00338-4]

[34]   Oliver JN, Su Y, Lu X, Kuo PH, Du J, Zhu D. Bioactive glass coatings on metallic implants for biomedical applications. Bioact Mater 2019; 4: 261-70.
[http://dx.doi.org/10.1016/j.bioactmat.2019.09.002] [PMID: 31667443]

[35]   Atiq M, Rehman U. Bioactive glass coatings with controlled degradation of magnesium under physiological conditions. Designed for Orthopedic Implants 2020; 211-24.

[36]   Yun Y, Dong Z, Lee N, *et al.* Revolutionizing biodegradable metals. Mater Today 2009; 12(10): 22-32.
[http://dx.doi.org/10.1016/S1369-7021(09)70273-1]

[37]   Montazerian M, Hosseinzadeh F, Migneco C, Fook MVL, Baino F. Bioceramic coatings on metallic implants: An overview. Ceram Int 2022; 48(7): 8987-9005.
[http://dx.doi.org/10.1016/j.ceramint.2022.02.055]

[38]   Sergi R, Bellucci D, Cannillo V. A comprehensive review of bioactive glass coatings: State of the art, challenges and future perspectives. Coatings 2020; 10(8): 757.
[http://dx.doi.org/10.3390/coatings10080757]

[39]   Rahaman MN, Day DE, Sonny Bal B, *et al.* Bioactive glass in tissue engineering. Acta Biomater 2011; 7(6): 2355-73.
[http://dx.doi.org/10.1016/j.actbio.2011.03.016] [PMID: 21421084]

[40]   Souza EQM, Costa Klaus AE, Espósito Santos BF, *et al.* Evaluations of hydroxyapatite and bioactive glass in the repair of critical size bone defects in rat calvaria. J Oral Biol Craniofac Res 2020; 10(4): 422-9.
[http://dx.doi.org/10.1016/j.jobcr.2020.07.014] [PMID: 32775186]

[41]   Su Y, Cockerill I, Zheng Y, Tang L, Qin YX, Zhu D. Biofunctionalization of metallic implants by calcium phosphate coatings. Bioact Mater 2019; 4: 196-206.
[http://dx.doi.org/10.1016/j.bioactmat.2019.05.001] [PMID: 31193406]

[42]   Drago L, Toscano M, Bottagisio M. Recent evidence on bioactive glass antimicrobial and antibiofilm activity: A mini-review. Mater 2018; p. 11.

[43]   Posada OM, Tate RJ, Grant MH. Toxicity of cobalt–chromium nanoparticles released from a resurfacing hip implant and cobalt ions on primary human lymphocytes *in vitro*. J Appl Toxicol 2015; 35(6): 614-22.
[http://dx.doi.org/10.1002/jat.3100] [PMID: 25612073]

[44]   Madl AK, Liong M, Kovochich M, Finley BL, Paustenbach DJ, Oberdörster G. Toxicology of wear particles of cobalt-chromium alloy metal-on-metal hip implants Part I: Physicochemical properties in patient and simulator studies. Nanomedicine 2015; 11(5): 1201-15.
[http://dx.doi.org/10.1016/j.nano.2014.12.005] [PMID: 25744761]

[45]   Bekmurzayeva A, Duncanson WJ, Azevedo HS, Kanayeva D. Surface modification of stainless steel for biomedical applications: Revisiting a century-old material. Mater Sci Eng C 2018; 93: 1073-89.
[http://dx.doi.org/10.1016/j.msec.2018.08.049] [PMID: 30274039]

[46]   Elias CN, Fernandes DJ, Souza FM, Monteiro ES, Biasi RS. Mechanical and clinical properties of titanium and titanium-based alloys (Ti G2, Ti G4 cold worked nanostructured and Ti G5) for biomedical applications. J Mater Res Technol 2019; 8(1): 1060-9.
[http://dx.doi.org/10.1016/j.jmrt.2018.07.016]

[47]   Abbaspour N, Hurrell R, Kelishadi R. Review on iron and its importance for human health. J Res Med

Sci 2014; 19(2): 164-74.
[PMID: 24778671]

[48]   Zhu D, Su Y, Fu B, Xu H. Magnesium reduces blood-brain barrier permeability and regulates amyloid-β transcytosis. Mol Neurobiol 2018; 55(9): 7118-31.
[http://dx.doi.org/10.1007/s12035-018-0896-0] [PMID: 29383689]

[49]   Roohani N, Hurrell R, Kelishadi R, Schulin R. Zinc and its importance for human health: An integrative review. J Res Med Sci 2013; 18(2): 144-57.
[PMID: 23914218]

[50]   Khanmohammadi S, Ojaghi-Ilkhchi M, Farrokhi-Rad M. Evaluation of bioglass and hydroxyapatite based nanocomposite coatings obtained by electrophoretic deposition. Ceram Int 2020; 46(16): 26069-77.
[http://dx.doi.org/10.1016/j.ceramint.2020.07.100]

[51]   Farnoush H, Muhaffel F, Cimenoglu H. Fabrication and characterization of nano-HA-45S5 bioglass composite coatings on calcium-phosphate containing micro-arc oxidized CP-Ti substrates. Appl Surf Sci 2015; 324: 765-74.
[http://dx.doi.org/10.1016/j.apsusc.2014.11.032]

[52]   Li Y, Cai S, Xu G, *et al.* Synthesis and characterization of a phytic acid/mesoporous 45S5 bioglass composite coating on a magnesium alloy and degradation behavior. RSC Advances 2015; 5(33): 25708-16.
[http://dx.doi.org/10.1039/C5RA00087D]

[53]   Singh PP, Dixit K, Sinha N. A sol-gel based bioactive glass coating on laser textured 316L stainless steel substrate for enhanced biocompatability and anti-corrosion properties. Ceram Int 2022; 48(13): 18704-15.
[http://dx.doi.org/10.1016/j.ceramint.2022.03.144]

[54]   Aqib R, Kiani S, Bano S, Wadood A, Ur Rehman MA. Ag–Sr doped mesoporous bioactive glass nanoparticles loaded chitosan/gelatin coating for orthopedic implants. Int J Appl Ceram Technol 2021; 18(3): 544-62.
[http://dx.doi.org/10.1111/ijac.13702]

[55]   Bano S, Akhtar M, Yasir M, *et al.* Synthesis and characterization of silver–strontium (Ag-Sr)-doped mesoporous bioactive glass nanoparticles. Gels 2021; 7(2): 34.
[http://dx.doi.org/10.3390/gels7020034] [PMID: 33805013]

[56]   Kargozar S, Montazerian M, Hamzehlou S, Kim HW, Baino F. Mesoporous bioactive glasses: Promising platforms for antibacterial strategies. Acta Biomater 2018; 81: 1-19.
[http://dx.doi.org/10.1016/j.actbio.2018.09.052] [PMID: 30273742]

[57]   Stähli C, James-Bhasin M, Hoppe A, Boccaccini AR, Nazhat SN. Effect of ion release from Cu-doped 45S5 Bioglass® on 3D endothelial cell morphogenesis. Acta Biomater 2015; 19: 15-22.
[http://dx.doi.org/10.1016/j.actbio.2015.03.009] [PMID: 25770928]

[58]   Nawaz A, Bano SSS, Yasir M, *et al.* Ag-Mn doped mesoporous bioactive glass nanoparticles in corporated into the chitosan/gelatin coatings deposited on peek / bioactive glass layers for favorable osteogenic differentiation and antibacterial activity materials advances. Mater Adv 2020; 1: 1273-84.
[http://dx.doi.org/10.1039/D0MA00325E]

[59]   Luo J, Mamat B, Yue Z, *et al.* Multi-metal ions doped hydroxyapatite coatings *via* electrochemical methods for antibacterial and osteogenesis. Colloid Interface Sci Commun 2021; 43: 100435.
[http://dx.doi.org/10.1016/j.colcom.2021.100435]

[60]   Mahato A, De M, Bhattacharjee P, *et al.* Role of calcium phosphate and bioactive glass coating on *in vivo* bone healing of new Mg–Zn–Ca implant. J Mater Sci Mater Med 2021; 32(5): 55.
[http://dx.doi.org/10.1007/s10856-021-06510-0] [PMID: 33961158]

[61]   Qi X, Wang H, Zhang Y, *et al.* Mesoporous bioactive glass-coated 3D printed borosilicate bioactive

glass scaffolds for improving repair of bone defects. Int J Biol Sci 2018; 14(4): 471-84.
[http://dx.doi.org/10.7150/ijbs.23872] [PMID: 29725268]

[62]   Zhang M, Pu X, Chen X, Yin G. *In-vivo* performance of plasma-sprayed CaO–MgO–SiO$_2$-based bioactive glass-ceramic coating on Ti–6Al–4V alloy for bone regeneration. Heliyon 2019; 5(11): e02824.
[http://dx.doi.org/10.1016/j.heliyon.2019.e02824] [PMID: 31763479]

[63]   Souza L, Lopes JH, Encarnação D, *et al.* Comprehensive *in vitro* and *in vivo* studies of novel melt-derived Nb-substituted 45S5 bioglass reveal its enhanced bioactive properties for bone healing. Sci Rep 2018; 8(1): 12808.
[http://dx.doi.org/10.1038/s41598-018-31114-0] [PMID: 30143690]

[64]   Nathanael AJ, Oh TH. Biopolymer Coatings for Biomedical Applications. Polym 2020; p. 12.

[65]   Peter M, Binulal NS, Nair SV, Selvamurugan N, Tamura H, Jayakumar R. Novel biodegradable chitosan–gelatin/nano-bioactive glass ceramic composite scaffolds for alveolar bone tissue engineering. Chem Eng J 2010; 158(2): 353-61.
[http://dx.doi.org/10.1016/j.cej.2010.02.003]

[66]   Rehman MAU, Munawar MA, Schubert DW, Boccaccini AR. Electrophoretic deposition of chitosan/gelatin/bioactive glass composite coatings on 316L stainless steel: A design of experiment study. Surf Coat Tech 2019; 358: 976-86.
[http://dx.doi.org/10.1016/j.surfcoat.2018.12.013]

[67]   Pawlik A, Rehman MAU, Nawaz Q, Bastan FE, Sulka GD, Boccaccini AR. Fabrication and characterization of electrophoretically deposited chitosan-hydroxyapatite composite coatings on anodic titanium dioxide layers. Electrochim Acta 2019; 307: 465-73.
[http://dx.doi.org/10.1016/j.electacta.2019.03.195]

[68]   Avcu E, Yıldıran Avcu Y, Baştan FE, Rehman MAU, Üstel F, Boccaccini AR. Tailoring the surface characteristics of electrophoretically deposited chitosan-based bioactive glass composite coatings on titanium implants *via* grit blasting. Prog Org Coat 2018; 123: 362-73.
[http://dx.doi.org/10.1016/j.porgcoat.2018.07.021]

[69]   Meyer N, Rivera L, Ellis T, Qi J, Ryan M, Boccaccini A. Bioactive and antibacterial coatings based on zein/bioactive glass composites by electrophoretic deposition. Coatings 2018; 8(1): 27.
[http://dx.doi.org/10.3390/coatings8010027]

[70]   Ahmed Y, Yasir M, Atiq M, Rehman U. Fabrication and characterization of zein hydroxyapatite composite coatings for biomedical applications. 2020; 237-50.

[71]   Zhao H, Cai S, Ding Z, Zhang M, Li Y, Xu G. A simple method for the preparation of magnesium phosphate conversion coatings on a AZ31 magnesium alloy with improved corrosion resistance. RSC Advances 2015; 5(31): 24586-90.
[http://dx.doi.org/10.1039/C5RA00329F]

[72]   Höhlinger M, Heise S, Wagener V, Boccaccini AR, Virtanen S. Developing surface pre-treatments for electrophoretic deposition of biofunctional chitosan-bioactive glass coatings on a WE43 magnesium alloy. Appl Surf Sci 2017; 405: 441-8.
[http://dx.doi.org/10.1016/j.apsusc.2017.02.049]

[73]   Ramakrishna S, Mayer J, Wintermantel E, Leong KW. Biomedical applications of polymer-composite materials: A review. Compos Sci Technol 2001; 61(9): 1189-224.
[http://dx.doi.org/10.1016/S0266-3538(00)00241-4]

[74]   Atiq Ur Rehman M, Bastan FE, Haider B, *et al.* Electrophoretic deposition of PEEK/bioactive glass composite coatings for orthopedic implants: A design of experiments (DoE) study. Mater Des 2017; 130: 223-30.
[http://dx.doi.org/10.1016/j.matdes.2017.05.045]

[75]   Baştan FE, Atiq Ur Rehman M, Avcu YY, Avcu E, Üstel F, Boccaccini AR. Electrophoretic co-

deposition of PEEK-hydroxyapatite composite coatings for biomedical applications. Colloids Surf B Biointerfaces 2018; 169: 176-82.
[http://dx.doi.org/10.1016/j.colsurfb.2018.05.005] [PMID: 29772473]

[76]   Mojarad Shafiee B, Torkaman R, Mahmoudi M, *et al.* Surface modification of 316L SS implants by applying bioglass/gelatin/polycaprolactone composite coatings for biomedical applications. Coatings 2020; 10(12): 1220.
[http://dx.doi.org/10.3390/coatings10121220]

[77]   Wu J, Miao G, Zheng Z, *et al.* 3D printing mesoporous bioactive glass/sodium alginate/gelatin sustained release scaffolds for bone repair. J Biomater Appl 2019; 33(6): 755-65.
[http://dx.doi.org/10.1177/0885328218810269] [PMID: 30426864]

[78]   Mao D, Li Q, Li D, Tan Y, Che Q. 3D porous poly(ε-caprolactone)/58S bioactive glass–sodium alginate/gelatin hybrid scaffolds prepared by a modified melt molding method for bone tissue engineering. Mater Des 2018; 160: 1-8.
[http://dx.doi.org/10.1016/j.matdes.2018.08.062]

[79]   Dhinasekaran D, Vimalraj S, Rajendran AR, Saravanan S, Purushothaman B, Subramaniam B. Bio-inspired multifunctional collagen/electrospun bioactive glass membranes for bone tissue engineering applications. Mater Sci Eng C 2021; 126: 111856.
[http://dx.doi.org/10.1016/j.msec.2020.111856] [PMID: 34082925]

[80]   Hernández-González AC, Téllez-Jurado L, Rodríguez-Lorenzo LM. Alginate hydrogels for bone tissue engineering, from injectables to bioprinting: A review. Carbohydr Polym 2020; 229: 115514.
[http://dx.doi.org/10.1016/j.carbpol.2019.115514] [PMID: 31826429]

[81]   Singh BN, Pramanik K. Development of novel silk fibroin/polyvinyl alcohol/sol–gel bioactive glass composite matrix by modified layer by layer electrospinning method for bone tissue construct generation. Biofabrication 2017; 9(1): 015028.
[http://dx.doi.org/10.1088/1758-5090/aa644f] [PMID: 28332482]

[82]   Reakasame S, Trapani D, Detsch R, Boccaccini AR. Cell laden alginate-keratin based composite microcapsules containing bioactive glass for tissue engineering applications. J Mater Sci Mater Med 2018; 29(12): 185.
[http://dx.doi.org/10.1007/s10856-018-6195-5] [PMID: 30519790]

[83]   Shekhawat D, Singh A, Banerjee MK, Singh T, Patnaik A. Bioceramic composites for orthopaedic applications: A comprehensive review of mechanical, biological, and microstructural properties. Ceram Int 2021; 47(3): 3013-30.
[http://dx.doi.org/10.1016/j.ceramint.2020.09.214]

[84]   Bai X, Gao M, Syed S, Zhuang J, Xu X, Zhang XQ. Bioactive hydrogels for bone regeneration. Bioact Mater 2018; 3(4): 401-17.
[http://dx.doi.org/10.1016/j.bioactmat.2018.05.006] [PMID: 30003179]

[85]   Ensoylu M, Deliormanlı AM, Atmaca H. Tungsten disulfide nanoparticle-containing PCL and PLGA-coated bioactive glass composite scaffolds for bone tissue engineering applications. J Mater Sci 2021; 56(33): 18650-67.
[http://dx.doi.org/10.1007/s10853-021-06494-w]

[86]   Barabadi Z, Azami M, Sharifi E, *et al.* Fabrication of hydrogel based nanocomposite scaffold containing bioactive glass nanoparticles for myocardial tissue engineering. Mater Sci Eng C 2016; 69: 1137-46.
[http://dx.doi.org/10.1016/j.msec.2016.08.012] [PMID: 27612811]

[87]   Daly AC, Riley L, Segura T, Burdick JA. Hydrogel microparticles for biomedical applications. Nat Rev Mater 2019; 5(1): 20-43.
[http://dx.doi.org/10.1038/s41578-019-0148-6] [PMID: 34123409]

[88]   Buwalda SJ, Vermonden T, Hennink WE. Hydrogels for therapeutic delivery: Current developments and future directions. Biomacromolecules 2017; 18(2): 316-30.

[http://dx.doi.org/10.1021/acs.biomac.6b01604] [PMID: 28027640]

[89]   Thangavel P, Ramachandran B, Kannan R, Muthuvijayan V. Biomimetic hydrogel loaded with silk and L -proline for tissue engineering and wound healing applications. J Biomed Mater Res B Appl Biomater 2017; 105(6): 1401-8.
[http://dx.doi.org/10.1002/jbm.b.33675] [PMID: 27080564]

[90]   Dorishetty P, Balu R, Athukoralalage SS, *et al.* Tunable biomimetic hydrogels from silk fibroin and nanocellulose. ACS Sustain Chem& Eng 2020; 8(6): 2375-89.
[http://dx.doi.org/10.1021/acssuschemeng.9b05317]

[91]   Jia Y, Wang X, Huo M, Zhai X, Li F, Zhong C. Preparation and characterization of a novel bacterial cellulose/chitosan bio-hydrogel. Nanomaterials and Nanotechnology 2017; 7.
[http://dx.doi.org/10.1177/1847980417707172]

[92]   Trombino S, Servidio C, Curcio F, Cassano R. Strategies for hyaluronic acid-based hydrogel design in drug delivery. Pharmaceutics 2019; 11(8): 407.
[http://dx.doi.org/10.3390/pharmaceutics11080407] [PMID: 31408954]

[93]   Ahmadi Z, Moztarzadeh F. Synthesizing and characterizing of gelatin-chitosan-bioactive glass (58s) scaffolds for bone tissue engineering. Silicon 2018; 10(4): 1393-402.
[http://dx.doi.org/10.1007/s12633-017-9616-z]

[94]   Gao Y, Chang J. Surface modification of bioactive glasses and preparation of PDLLA/bioactive glass composite films. J Biomater Appl 2009; 24(2): 119-38.
[http://dx.doi.org/10.1177/0885328208094265] [PMID: 18801895]

[95]   De Laia AGS, Barrioni BR, Valverde TM, de Goes AM, de Sá MA, Pereira MM. Therapeutic cobalt ion incorporated in poly(vinyl alcohol)/bioactive glass scaffolds for tissue engineering. J Mater Sci 2020; 55(20): 8710-27.
[http://dx.doi.org/10.1007/s10853-020-04644-0]

[96]   Du X, Wei D, Huang L, Zhu M, Zhang Y, Zhu Y. 3D printing of mesoporous bioactive glass/silk fibroin composite scaffolds for bone tissue engineering. Mater Sci Eng C 2019; 103: 109731.
[http://dx.doi.org/10.1016/j.msec.2019.05.016] [PMID: 31349472]

[97]   Gönen SÖ, Erol Taygun M, Küçükbayrak S. Fabrication of bioactive glass containing nanocomposite fiber mats for bone tissue engineering applications. Compos Struct 2016; 138: 96-106.
[http://dx.doi.org/10.1016/j.compstruct.2015.11.033]

[98]   Alonso JM, Andrade del Olmo J, Perez Gonzalez R, Saez-Martinez V. Injectable hydrogels: From laboratory to industrialization. Polymers 2021; 13(4): 650.
[http://dx.doi.org/10.3390/polym13040650] [PMID: 33671648]

[99]   Hoffman AS. Hydrogels for biomedical applications. Adv Drug Deliv Rev 2012; 64: 18-23.
[http://dx.doi.org/10.1016/j.addr.2012.09.010] [PMID: 11755703]

[100] Singh BN, Pramanik K. Generation of bioactive nano-composite scaffold of nanobioglass/silk fibroin/carboxymethyl cellulose for bone tissue engineering. J Biomater Sci Polym Ed 2018; 29(16): 2011-34.
[http://dx.doi.org/10.1080/09205063.2018.1523525] [PMID: 30209974]

[101] Sarker B, Papageorgiou DG, Silva R, *et al.* Fabrication of alginate–gelatin crosslinked hydrogel microcapsules and evaluation of the microstructure and physico-chemical properties. J Mater Chem B Mater Biol Med 2014; 2(11): 1470-82.
[http://dx.doi.org/10.1039/c3tb21509a] [PMID: 32261366]

[102] Koivisto JT, Gering C, Karvinen J, *et al.* Mechanically biomimetic gelatin–gellan gum hydrogels for 3D culture of beating human cardiomyocytes. ACS Appl Mater Interfaces 2019; 11(23): 20589-602.
[http://dx.doi.org/10.1021/acsami.8b22343] [PMID: 31120238]

[103] Wang B, Wan Y, Zheng Y, *et al.* Alginate-based composites for environmental applications: A critical review. Crit Rev Environ Sci Technol 2019; 49(4): 318-56.

[http://dx.doi.org/10.1080/10643389.2018.1547621] [PMID: 34121831]

[104]   Shariatinia Z, Jalali AM. Chitosan-based hydrogels: Preparation, properties and applications. Int J Biol Macromol 2018; 115: 194-220.
[http://dx.doi.org/10.1016/j.ijbiomac.2018.04.034] [PMID: 29660456]

[105]   Elia R, Newhide DR, Pedevillano PD, *et al.* Silk–hyaluronan-based composite hydrogels: A novel, securable vehicle for drug delivery. J Biomater Appl 2013; 27(6): 749-62.
[http://dx.doi.org/10.1177/0885328211424516] [PMID: 22090427]

[106]   Kreller T, Distler T, Heid S, Gerth S, Detsch R, Boccaccini AR. Physico-chemical modification of gelatine for the improvement of 3D printability of oxidized alginate-gelatin hydrogels towards cartilage tissue engineering. Mater Des 2021; 208: 109877.
[http://dx.doi.org/10.1016/j.matdes.2021.109877]

[107]   Ahmed EM. Hydrogel: Preparation, characterization, and applications: A review. J Adv Res 2015; 6(2): 105-21.
[http://dx.doi.org/10.1016/j.jare.2013.07.006] [PMID: 25750745]

[108]   Sarker B, Zehnder T, Rath SN, *et al.* Oxidized alginate-gelatin hydrogel: A favorable matrix for growth and osteogenic differentiation of adipose-derived stem cells in 3D. ACS Biomater Sci Eng 2017; 3(8): 1730-7.
[http://dx.doi.org/10.1021/acsbiomaterials.7b00188] [PMID: 33429654]

[109]   Bai X, Lü S, Liu H, *et al.* Polysaccharides based injectable hydrogel compositing bio-glass for cranial bone repair. Carbohydr Polym 2017; 175: 557-64.
[http://dx.doi.org/10.1016/j.carbpol.2017.08.020] [PMID: 28917901]

[110]   Fernandes JS, Gentile P, Martins M, *et al.* Reinforcement of poly-l-lactic acid electrospun membranes with strontium borosilicate bioactive glasses for bone tissue engineering. Acta Biomater 2016; 44: 168-77.
[http://dx.doi.org/10.1016/j.actbio.2016.08.042] [PMID: 27554018]

[111]   Rescignano N, Hernandez R, Lopez LD, Kenny M, Mijangos C. Preparation of alginate hydrogels containing silver nanoparticles : A facile approach for antibacterial applications 2016.
[http://dx.doi.org/10.1002/pi.5119]

[112]   Lewandowska-Łańcucka J, Mystek K, Mignon A, Van Vlierberghe S, Łatkiewicz A, Nowakowska M. Alginate- and gelatin-based bioactive photocross-linkable hybrid materials for bone tissue engineering. Carbohydr Polym 2017; 157: 1714-22.
[http://dx.doi.org/10.1016/j.carbpol.2016.11.051] [PMID: 27987887]

[113]   Qiu M, Chen D, Shen C, Shen J, Zhao H, He Y. Preparation of *in situ* forming and injectable alginate/mesoporous Sr-containing calcium silicate composite cement for bone repair. RSC Advances 2017; 7(38): 23671-9.
[http://dx.doi.org/10.1039/C6RA28860J]

[114]   Cheng T, Qu H, Zhang G, Zhang X. Osteogenic and antibacterial properties of vancomycin-laden mesoporous bioglass/PLGA composite scaffolds for bone regeneration in infected bone defects. Artif Cells Nanomed Biotechnol 2017; 46(8): 1-13.
[http://dx.doi.org/10.1080/21691401.2017.1396997] [PMID: 29113502]

[115]   Bonetti L, Altomare L, Bono N, *et al.* Electrophoretic processing of chitosan based composite scaffolds with Nb-doped bioactive glass for bone tissue regeneration. J Mater Sci Mater Med 2020; 31(5): 43.
[http://dx.doi.org/10.1007/s10856-020-06378-6] [PMID: 32358696]

[116]   Abd El-Fattah A, Nageeb Hassan M, Rashad A, Marei M, Kandil S. Viscoelasticity, mechanical properties, and *in vivo* biocompatibility of injectable polyvinyl alcohol/bioactive glass composite hydrogels as potential bone tissue scaffolds. IJPAC Int J Polym Anal Charact 2020; 25(5): 362-73.
[http://dx.doi.org/10.1080/1023666X.2020.1790253]

[117]   Kumar A, Han SS. Enhanced mechanical, biomineralization, and cellular response of nanocomposite hydrogels by bioactive glass and halloysite nanotubes for bone tissue regeneration. Mater Sci Eng C 2021; 128: 112236.
[http://dx.doi.org/10.1016/j.msec.2021.112236] [PMID: 34474814]

[118]   Valanezhad A, Shahabi S, Hashemian A, *et al.* Preparation of a PLGA-coated porous bioactive glass scaffold with improved mechanical properties for bone tissue engineering approaches. Regen Eng Transl Med 2021; 7(2): 175-83.
[http://dx.doi.org/10.1007/s40883-021-00196-0]

CHAPTER 4

Bioactive Glasses: Structure, Properties, and Processing

David Bahati[1], Meriame Bricha[1] and **Khalil El Mabrouk[1,*]**

[1] *Euromed University of Fes, UEMF, Fes, Morocco*

Abstract: Bioactive glasses, as pioneering artificial biomaterials, uniquely establish strong bonds with hard and soft native tissues by forming a bone-like hydroxyapatite layer in contact with physiological body fluid. This hydroxyapatite layer, mimicking the inorganic phase of natural bone, adds a fascinating dimension to their biomedical significance. Comprising three primary components; network formers, network modifiers, and intermediate oxide components; bioactive glasses allow tailored properties through component variation. While extensively explored for broadening biomedical applications, especially in regenerative medicine, their use is constrained by inherent mechanical shortcomings such as brittleness, fragility, and poor elasticity. Ongoing studies focus on incorporating bioactive glasses into composite/hybrid biomaterials with biopolymers, aiming to optimize mechanical properties for diverse biomedical applications, especially in load-bearing sites of hard tissues. Despite successful applications, the mechanical limitations persist, prompting investigations into the influence of composition and processing methods on bioactive glass properties. Notably, doping bioactive glasses with metallic ions at lower concentrations emerges as a promising avenue, enhancing mechanical and biological attributes, including bioactivity, osteogenicity, osteoinductivity, and antibacterial effects. This chapter provides a comprehensive examination of three bioactive glass types, accentuating their structures, properties, and processing methods. Additionally, it delves into property modifications facilitated by metallic ion dopants, contributing valuable insights to the evolving landscape of biomaterials.

Keywords: Amorphous solids, Bioactive glass, Bioactivity, Borate, Bridging oxygen atom, Doping, Melt-quench, Network connectivity, Network formers, Network modifiers, Non-bridging oxygen atom, Phosphate, Silicate, Sol-gel.

INTRODUCTION

Bioceramics can be grouped into naturally occurring, like coral-derived apatite or synthetic ones [1]. Synthetic bioceramics can either be nearly bioinert such as

* **Corresponding author Khalil El Mabrouk:** Euromed University of Fes, UEMF, Fes, Morocco;
E-mail: k.elmabrouk@ueuromed.org

Saeid Kargozar and Francesco Baino (Eds.)

Alumina and zirconia or bioresorbable such as tricalcium phosphate, or bioactive like calcium phosphate (hydroxyapatite), bioactive glasses, and glass-ceramics [1, 2]. Bioceramics have been used in orthopedics to replace, repair, or enhance the regeneration of diseased and damaged hard tissues, including hips, knees, teeth, tendons, spinal fusion, jawbones, and other maxillofacial surgical applications such as the treatment of periodontitis disease [1]. Calcium phosphate-based compositions are preferred due to their chemical and structural similarity in composition with the main mineral phase of the bone [3, 4]. Alumina and zirconia are bioinert ceramics that have been used in orthopedic applications, mainly in total hip prostheses and teeth implants, due to their high wear resistance and chemical stability [2]. However, bioceramics are generally brittle, fragile, have low mechanical stability, poor elasticity, and low fracture toughness. In addition, their degradation rates are not very predictable [4], thus limiting their use in medical applications. Composite materials composed of polymers as bulk matrix and bioceramics as fillers have been studied recently [5 - 8]. Bioactive glasses are widely studied artificial biomaterials, especially for medical applications. This chapter highlights bioactive ceramics, focusing mainly on bioactive glasses, particularly their properties, synthesis techniques, and property enhancement through adding dopant ions.

Calcium Phosphate Ceramics

Calcium phosphate-based bioceramics (CaPs) have been primarily used in orthopedics due to their chemical and structural similarity with the inorganic phase of natural bone. The most studied CaPs are hydroxyapatite (HA) $[Ca_{10}(PO_4)_6(OH)_2]$ with a calcium to phosphate ratio of 1.67 and tricalcium phosphate (TCP). The latter can either be α-TCP or β-TCP with a similar chemical composition $[Ca_3(PO_4)_2]$ but different in crystal phases leading to different absorption characteristics [9, 10]. The solubility of α-TCP is higher than that of β-TCP resulting in faster release of Ca^{2+} and PO_4^{3-} when in contact with body fluid, facilitating fast precipitation to form HA and new bone formation compared to the similar situation employing β-TCP [9]. In this comparison, β-TCP was more soluble than synthetic HA [10]. However, rapid solubility may result in too high ionic concentration leading to ineffective cellular responses. On the other hand, too slow solubility results in low ionic concentration to trigger cellular activities for extracellular matrix deposition.

An *in vivo* comparison based on the time required for complete bone restoration between 45S5 Bioglass and HA showed that complete bone restoration could be archived in 2 weeks with 45S5 bioglass while HA required about 12 weeks to produce comparable results [11]. Yuan *et al.* [12] conducted an *in vivo* comparison between α and β-TCP on their ability to induce bone formation in the soft tissue of dogs. Bone tissue was observed after 45 and 150 days of β-TCP implantation, while no bone tissue was observed during the same period with α-TCP. It was concluded that a higher local ionic concentration of Ca^{2+} and PO_4^{3-} resulting from the rapid dissolution of α-TCP could resist bone formation. In contrast, too slow dissolution of β-TCP could be inadequate to trigger the cellular activities for new bone deposition. The rapid dissolution of α-TCP leads to supersaturating local ionic concentration of Ca^{2+} and PO_4^{3-} which may negatively impact the migration, proliferation, and differentiation of bone-forming cells (osteoblasts) and subsequently ossification process [13].

Bioactive Glasses

As an alternative to nearly bioinert substitutes, Larry Hench discovered the first bioactive glass in the 1960s. The discovery originated from a friendly discussion between Hench and a US Army colonel who had just returned from the Vietnam War in 1967. When Hench explained his previous research work with his colleagues about their research results on a glass material (vanadium phosphate, V_2O_5-P_2O_5) resistant to high radiation exposure, the colonel asked him if he could make a material resistant to human body exposure. Then Hench changed his research paradigm towards new ceramic material that could resemble or stimulate the formation of hydroxyapatite *in vivo* similar to the inorganic phase of the natural bone with the assumption that it could not be rejected by the human body [14].

About ten years later, it was found that the bioactive glass could stimulate osteogenesis when used in particulate form, which led to the concept of tissue regeneration [15, 16]. This bioactive glass had quaternary composition with the main components being silicon dioxide, calcium oxide, sodium oxide, and phosphorous pentoxide (45 wt% SiO_2, 6 wt% P_2O_5, 24.5 wt% CaO and 24.5 wt% Na_2O). It was made through the melt-quenching method and finally was termed as 45S5 Bioglass® [14, 16]. It is worth noting that the term Bioglass only stands for the original bioactive glass (45S5), and therefore it cannot be used referring to any other composition of bioactive glass. The *in vitro* test of 45S5 bioglass showed that it could develop a hydroxyapatite layer (HA) when socked in solutions that did not contain calcium or phosphate ions. The formed HA was equivalent to observed interfacial HA bonded to collagen fibrils produced by osteoblasts at the interface of the 45S5 implant and the native bones of a rat femoral in an *in vivo* study by Dr. Ted Greenlee. The first *in vivo* tests were for six weeks in which at the end, Greenlee reported, *"These ceramic implants will not come out of bone. They are bonded in place. I can push on them, I can shove them, I can hit them, and they do not move. The controls easily slide out"* [14, 16]. The *in vitro* and *in vivo* results of the 45S5 bioactive glass were published for the first time in 1971 in the Journal of Biomedical Materials Research [16]. The formation of HA on the surface of bioactive glass immersed in simulated body fluid is considered a measure of bioactivity [17, 18].

Since its discovery, bioglasses have passed through four eras, and they are now in the fourth period, termed the era of innovation, which started in 2005 and is projected to end by 2025 [19]. The four eras can be classified as follows [15, 19]:

a) The first era is the era of discovery, between 1969-1979.

b) The second era is era of clinical application between 1980-1995.

c) The third era is the era of tissue regeneration that took place between 1995-2005.

d) The fourth and current era is of innovation between 2005-2025.

Although the discovery of the bioactive glasses was on silicate-based glasses [15], borate [20–22] and phosphate-based [23–25] glasses are among the commonly studied bioactive glasses, especially for wound healing and drug delivery applications, respectively. All these glasses have been studied for the enhancement of both mechanical and biological properties. Structural properties of the mentioned bioactive glasses are discussed in the following section.

Structural Properties of Bioactive Glasses

Glasses are solids characterized by a very short-range order of atoms with no translational symmetry, thus belonging to amorphous solids [26]. Before Zachariasen's random network theory [27], glasses were considered made up of nanocrystals with the approximate size of 20 Å. Under this consideration, the crystal sizes were estimated from Sherrer's equation's breadth widening of the x-ray diffraction pattern (Equation 1).

$$Breadth = \frac{0.9\lambda}{t Cos(\theta)} \tag{1}$$

where t is the crystal particle size, λ represents the wavelength of the x-ray used, and θ is Bragg's angle. Zachariasen postulated four rules that must be satisfied in order to form a glass.

1. Oxygen can utmost be linked to 2 cations.

2. In the structure, the cation coordination number is either 3 or 4.

3. The polyhedral is formed by sharing oxygen only from vertices and not faces or edges.

4. At least three vertices must be shared for a 3D glass structure to be formed.

Generally, all bioactive glasses possess two common properties: amorphous character and intrinsic glass transition temperature (Tg), where the glass material changes between super-cooled liquid and solid glassy phases. Being amorphous solids, glasses do not possess long-range order of arrangement of atoms. Therefore, under thermal treatment, all glasses show a consistent decrease in viscosities [28 - 30].

Zachariasen's model successfully gave the foundation for understanding the structures of vitreous glasses. Depending on structural components, glasses are formed by the utmost three main components, *viz* network formers, network modifiers, and intermediate oxide component [31]:

Network Formers

The main component of the glass compositions can form a glass material without any additional component. They build a glass backbone by forming tetrahedral oxygen units connected to each other through shared vertices at the bridging oxygen atoms (BOs).

Network Modifiers

The additives that when added to the network formers, reduce network connectivity by bonding to the bridging oxygen atoms (BOs) and converting them into non-bridging oxygen atoms (NBOs). They disrupt the network connectivity by creating a terminal oxygen atom (NBO).

Intermediate Oxides

These additives can either act as a network former or modifier but can not form glass material themselves without other components. These include oxides such as magnesium oxide (MgO) and aluminum oxide (Al_2O_3).

Bioactive glasses can be classified based on their network formers as silicate, borate, and bioactive phosphate glasses with silicon dioxide (SiO_2), boron trioxide (B_2O_3), and phosphorous pentoxide (P_2O_5), respectively being glass network formers. A complex glass formed by more than one network former is usually classified by combining the network formers. These include bioactive glasses such as phosphosilicate, borosilicate, and borophosphate [32].

Silicate-based Bioactive Glass

Silica-based bioactive glasses are the oldest discovered bioactive artificial material. With his co-workers for the first time, L.L Hench introduced the silicate-based bioactive glass referred to as 45S5 Bioglass® in the 1960s [14, 16, 33]. 45S5 was the first artificial material that revealed the bonding ability with soft tissues and bones [34]. Silicon dioxide forms the main component of silica-based bioactive glasses. Other components are calcium oxide, sodium dioxide, and phosphorous pentoxide, and others with varying amounts depending on the composition of interest [35].

In silicate-based bioactive glass, silicon dioxide (SiO_2) is the network former in which a continuous 3D glass network is formed by connected SiO_4 tetrahedral networks [36]. In the glass networks, the alkali and alkaline earth oxides such as CaO, Na_2O, MgO *etc.*, in glass networks are referred to as network modifiers. The oxygen atom connecting two silicon atoms in the glass network is called a bridging oxygen atom (BO) [26]. The addition of network modifiers in the glass network causes a reduction in the degree of network connectivity by replacing the BO with non-bridging oxygen atoms (NBO), which results in an open network structure due to the formation of ionic bonds between the modifier cations and the NBO atoms [37]. Network connectivity (NC) of silicate-based bioactive glass can be estimated from Equation **2** [31], assuming that each silicon atom is bonded to the utmost four oxygen atoms. Under this consideration, phosphorous pentoxide increases the number of BO since a modifier cation is required to balance phosphate ions (PO_4^{3-}) in the composition. Some researchers have reported that glasses with NC greater than 2.4 are not bioactive [38]. The model confirms the well-noted bioactivity of some melt-quench-derived bioactive glasses such as 45S5 Bioglass [NC = 2.11] and ICIE16 bioactive glass [NC = 2.13]. However, there are commercially available bioactive glass materials such as [13-93, NC = 3.01], BonAlive, NC = 2.54 and (6P53B, NC = 3.31] that are confirmed to have good bioactivity but based on Equation **2**, the mentioned glasses could be classified as non-bioactive. This may be due to assumptions made such as assuming a 100% homogeneous glass and all additives to act either as network formers or modifiers without considering the intermediate components.

$$NC = \frac{4(SiO_2)+6(P_2O_5)-2(M^{I}O+M^{II}O)}{SiO_2} \qquad (2)$$

All values are in mol% where the oxides stand for their concentrations in the glass composition, $M'O$ and $M''O$ being the monovalent and divalent network modifiers, respectively.

The network modifiers are added to the glass network to tailor the surface reactivity of the bioactive glass in a physiological environment [39]. Generally, the structure of silicate-based bioactive glasses is described referring to "Q^n" notation in which n stands for the number of bridging oxygen atoms (BOs) per tetrahedron, with 4 being its maximum value that indicates a tetrahedron bonded to other four tetrahedral by BOs. The value of n can also be 0, indicating silicon atom bonded to four NBOs [40, 41].

The bonding ability that is also considered the measure of bioactivity depends on the composition of the bioactive glass material. For example, it was reported that by varying the percentage of the main compositions in the 45S5 Bioglass, it was noted that some resulting bioactive glass material could bond to both soft and bone tissues. In contrast, some could not bond to either soft or bone tissue [42]. Since the discovery of 45S5 Bioglass, many other studies have been performed in silicate-based bioactive glasses such as S53P4, I3-93B1, I3-93 *etc.*, with additional components and varied compositions revealed excellent bioactive properties [43 - 45].

The bonding ability of bioactive glass is firmly attributed to the formation of the hydroxycarbonate apatite layer (HA) on the surface of the bioactive material. The strong bond formed between the silica-based bioactive glass and the native bone results from a sequence of 11 reaction stages. The first five stages occur on the surface of the bioglass material [46] and are similar for both *in vitro* and *in vivo*.

First Stage

Exchange of alkali cations such as Na^+, K^+ or Ca^{2+} from the bioactive glass with protons (H^+) or hydronium (H_3O^+) from the body fluid. This stage leads to the rise in the pH of the surrounding body fluid, usually becoming higher than 7.4.

Second Stage

Dissolution of the glass network by breaking of Si-O-Si bonds due to the high alkalinity of the solution resulting from the first stage. The dissolution releases silica acid [$Si(OH)_4$] and finally production of silanols (Si-OH) at the glass-solution interface.

Third Stage

Polycondensation of silanols that lead to the formation of silica-rich layer.

Fourth Stage

Recruitment of Ca^{2+} and PO_4^{3-} to the surface of the silica-rich layer leading to the formation of an amorphous CaO-P_2O_5-rich film that includes soluble Ca^{2+} and PO_4^{3-} from the solution.

Fifth Stage

The amorphous CaO-P_2O_5 crystallizes with time to form crystalline hydroxycarbonated apatite (HA).

The formed crystalline HA layer interacts with the physiological body fluid leading to the adsorption and desorption of signals and macrophages that create the environment for tissue repair at the defect site. Finally, mesenchymal stem cells are recruited on the surface of the bioactive bone implant. They proliferate and differentiate followed by other cell functions that all together result in the deposition of native extracellular matrix and ossification of the matrix through osteoblastic activities [47].

Phosphate-based Bioactive Glass

In phosphate-based glasses, phosphate (PO_4^{3-}) tetrahedral structural unit is a network former in the glass material. A covalent bond connects it to at most other three PO_4^{3-} tetrahedral units [48]. When modifiers such as CaO and Na_2O are added to the phosphate glass network, it responds similarly to silica-based bioactive glasses. In this case, the P-O-P bonds are broken, resulting in the reduction of BO and the increase of NBO atoms [49].

As in silicate-based glasses, network modifiers can be added to the glass network. When the incorporated network modifiers in a phosphate glass are more than the network formers, the resulting phosphate glass material is called an inverted glass. In this case, the properties of the resulting glass material are dominated by the ionic bond formed between the NBO and the modifier cations instead of than the covalent bond of the network formers (P-O-P) [50].

Borate-based Bioactive Glass

Network formers in vitreous borate glasses are BO_3 groups. Unlike the silicate and phosphate-based bioactive glasses, when network modifiers are added to the vitreous borate network, the glass network connectivity first increases due to the formation of 4-coordinated BO_4^- units but further addition of the modifiers leads to a decrease in network connectivity. This first increase, followed by a decrease in glass network connectivity due to the addition of network modifiers to the vitreous borate glass, is referred to as a borate anomaly [51].

Compared to bioactive silicate glasses, bioactive borate glasses have shown a high dissolution rate and rapid hydroxycarbonated apatite (HA) formation. The high dissolution of borate glasses leading to the release of ions from the glass network makes borate-based bioactive glass fit for soft tissue and wound healing applications where the rapid release of antibacterial and other therapeutic ions is needed [52].

Processing of Bioactive Glasses

Based on the synthesis methods, bioactive glasses can be categorized into melt-quench-derived and sol-gel derived bioactive glasses [53]. The melt-quench is the older traditional method for the synthesis of bioactive glasses. Professor Larry Hench synthesized the first bioactive glass, also known as 45S5 Bioglass, through the melt-quench method in the 1970s [54]. In recent decades, sol-gel synthesis method has raised much interest to scientists due to its advantages over the traditional method, such as its low-temperature chemistry, reproducibility, and high surface to volume ratios of the final products [55].

Melt-quench-derived bioactive glasses have higher mechanical strength than their counterparts sol-gel-derived bioactive glasses. However, the second holds intrinsically exciting properties such as higher porosity and surface area to volume ratio compared to similar compositions of their counterparts. The properties of sol-gel-derived bioactive glasses are of interest in medical applications due to their enhanced bioactivity and biodegradability.

Melt Quench Method

This is the oldest and the most common technique of producing bioactive glasses [56]. In this method, the stoichiometric mixture of the raw materials (oxides) is melted at high temperatures ranging between 1300-1600 °C in platinum crucibles for a while depending on the melting temperatures of the constituents and then quenched in water or in a mold at low temperatures in order to avoid crystallization [57, 58]. The quenching of the molten glass allows it to maintain its amorphous character since atoms do not get sufficient time to rearrange into a crystalline phase. In addition, the residual internal stress in the glass material is removed through annealing at about 500 °C [59]. As a result, bioactive glasses synthesized through the melt quench method have superior mechanical properties and flexural strength but are less porous with poor pore interconnectivity [60, 61].

The melt quench method has been used for the synthesis of bioactive glasses, mostly silicate-based glasses in the form of solid and nonporous materials consisting of mainly silicon dioxide (SiO_2) and other three basic components *viz.* sodium dioxide (Na_2O), calcium oxide (CaO) and phosphate (P_2O_5) [35]. Different forms of bioactive glasses with various properties can be obtained by varying compositions of these constituents [62]. Fig. (**1**) depicts the bonding properties of silicate-based bioactive glasses as a function of the composition of the main constituents based on the melt-quench method [42]. The compositional ranges of the main components can be categorized into four classes [19, 42, 63]:

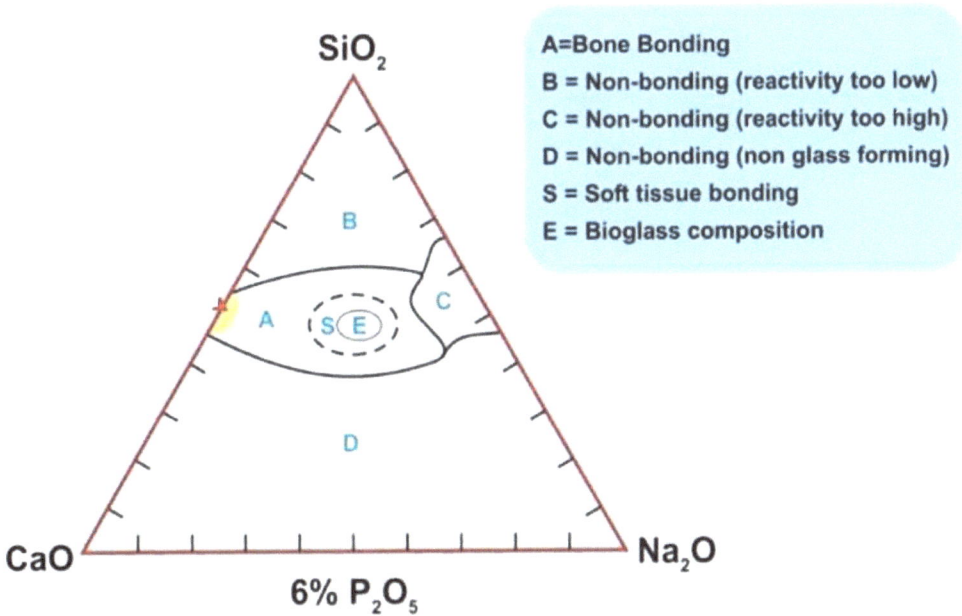

Fig. (1). Bonding properties of melt-quench derived bioactive silicate glasses as a function of composition [42].

1. 35-60 mol% SiO_2, 10-50 mol% CaO, 5-40 mol% Na_2O: this class is bioactive, form bonds with bones and some with both bones and soft tissues (**E & S**).

2. <35 mol% SiO_2: This class is nonglass forming (**D**).

3. >50 mol% SiO_2, <10 mol% CaO, <35 mol% Na_2O: This class is bioactive and is characterized by resorption ranging between 10-30 days (**A**).

4. >65 mol% SiO_2: This class is non-bioactive, and it is a nearly inert material (**B**).

Sol-Gel Method

This wet-chemical technique is used to synthesize glassy and ceramic powder or film materials from the precursor solutions, usually alkoxides. Several steps in the following chronological order are involved [64, 65]: hydrolysis and polycondensation of the alkoxide, gelation, aging, drying, and densification. In this method, compositional precursors undergo polymer-type reactions to form a wet network of formers (SiO_2, B_2O_3, and P_2O_5) covalently bonded that can be turned into the glass under thermal treatment (calcination) at a temperature ranging between 400 – 800 °C [53].

The sol-gel synthesis method is preferred for the synthesis of porous glasses because it is relatively inexpensive and can be done at low-temperature conditions, allowing control of the product's chemical composition [66, 67]. In addition, this method has many advantages over other bioactive glass synthesis methods, including its flexibility to incorporate dopants [68]. Fig. (**2**) is a schematic representation of the sol-gel synthesis method of bioactive glass nanoparticles.

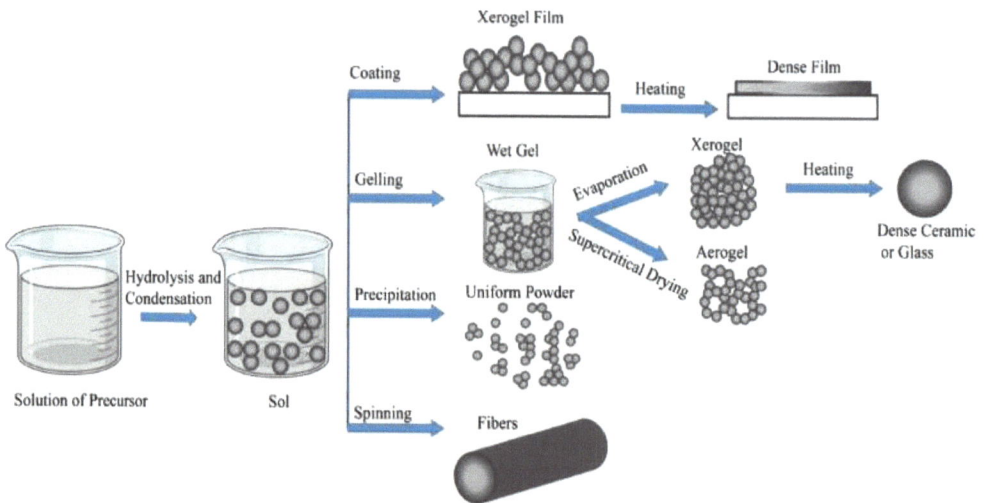

Fig. (2). Steps of the sol-gel process of materials and example of the expected microstructure of the final product [74].

The sol-gel process can be modified by adding surfactants as structural directing agents to enhance network porosity resulting in bioactive glass materials with controlled surface morphologies such as mesoporous structures [69]. Modified sol-gel methods are briefly presented in section 3.3. Mesoporous bioactive glass (MBG) has revealed preferred properties of biomedical applications, such as the high specific surface area to volume ratio and pore volume leading to enhanced bioactivity, osteogenesis, and drug delivery. Several successful studies aiming to enhance bioactive glasses' mechanical and biological properties have been done through traditional or modified sol-gel methods [70 - 73].

Sol-gel Method of Silicate Glasses

Sol-gel-derived silicate-based bioactive glasses have been widely studied for various medical applications. The sol-gel synthesis of silica-based bioactive glasses involves mainly two steps: hydrolysis [75, 76] (Equation **3**) and polycondensation (Equation **4**) of silicon alkoxides, mainly tetraethyl orthosilicate $(Si(OC_2H_5)_4$ commonly known as TEOS under basic or acidic medium. The hydrolysis of TEOS results in a whitish suspension (sol) whose polycondensation forms a rigid interconnected 3D network (gel), procedures that bear the technique name "sol-gel". The gel is left foraging followed by a drying and thermal treatment (calcination) process to burn out unrequired organic components presented in the glass composition. The resulting bioactive glasses are intrinsically characterized by porous structures and a higher surface-to-volume ratio, enhancing bioactivity and biodegradability [77]. Under the sol-gel synthesis technique, the limit of silica content in the glass composition can be increased up to about 90% by mole, contrary to a maximum of 60% by mole offered by the melt-quench method.

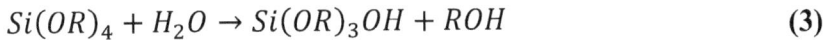

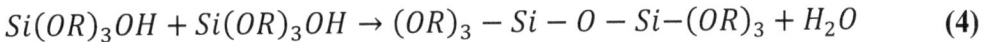

$$Si(OR)_4 + H_2O \rightarrow Si(OR)_3OH + ROH \tag{3}$$

$$Si(OR)_3OH + Si(OR)_3OH \rightarrow (OR)_3 - Si - O - Si-(OR)_3 + H_2O \tag{4}$$

Sol-gel Method of Phosphate Glasses

The sol-gel synthesis of phosphate-based glasses was studied as early as the 1990s on studies related to thin films and nanostructured glass materials [78]. The main challenge during that time was to locate reasonable precursors. Phosphoric acid (H_3PO_4) was thought an option. However, unfortunately, it could highly react with water to form metal oxide phosphate that quickly precipitated without hydrolysis and condensation reactions contrary to the sol-gel reaction stages. Since then, other precursors, including phosphate esters, have been studied, but due to the presence of a P-O-C bond whose hydrolysis is sluggish, they could not be a better option [79]. Currently, commonly phosphate precursor used in the sol-gel synthesis of phosphate-based glasses is triethyl phosphate $((C_2H_5)_3PO_4)$, commonly known as TEP. The hydrolysis reaction of TEP is similar to that of TEOS (Equations **5** and **6**) [80]. However, polycondensation of phosphoric acid is only possible at elevated temperatures, usually above 100 °C, contrary to the condensation of TEOS at room temperatures.

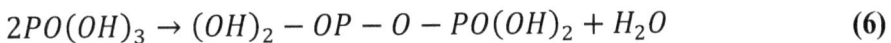

$$PO(OR)_3 + H_2O \rightarrow PO(OR)_2OH + ROH \tag{5}$$

$$2PO(OH)_3 \rightarrow (OH)_2 - OP - O - PO(OH)_2 + H_2O \tag{6}$$

Sol-gel Method of Borate Glasses

Sol-gel-derived borate glasses were studied as early as the 1980s, mainly with binary composition systems [81]. Since then, many studies have been performed on the synthesis and characterization of sol-gel-derived borate-based bioactive glasses [82, 83]. Generally, in the sol-gel process, borate alkoxides react quickly with water leading to the formation of boric acid precipitates (Equation **7**) that can hydrolyze further in the aqueous media to form borate ions (Equation **8**) [84]. The gelation process is thought to occur after the formation of polyborate that is considered the backbone of the resulting gel.

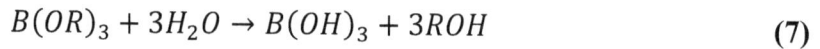

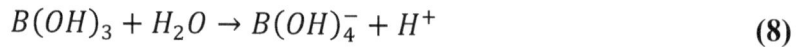

$$B(OR)_3 + 3H_2O \rightarrow B(OH)_3 + 3ROH \tag{7}$$

$$B(OH)_3 + H_2O \rightarrow B(OH)_4^- + H^+ \tag{8}$$

Modified Sol-gel Methods

Evaporation Induced Self Assembly (EISA) Method

Self-assembly can be defined as the spontaneous formation of an organized structure from pre-existing components by means of local non-covalent interactions like hydrogen bonding, van der Waals, electrostatic or hydrophobic forces [85]. It is a modified sol-gel synthesis method that employs different polymers as structural directing agents to fabricate homogeneous 2D and 3D structures [86]. EISA method can result in highly uniform, robust, and crack-free films with a controllable thickness [87].

In the EISA method, a homogenous solution of precursors and surfactant is prepared in ethanol or other volatile solvent and water with very low initial micelle concentration (C_o<<<CMC). As ethanol or volatile solvent evaporates, the concentration of the precursor and the surfactant in the solution increases, resulting in the self-assembly of micelles. Finally, an organized mesoporous structure is formed. Different mesostructures can be obtained by varying initial alcohol/water/surfactant mole ratios [88]. The structure-directing agent is finally decomposed and burned out through the calcination process [89, 90].

Stöber Method

This is an effective sol-gel modified strategy of producing uniform, monodispersed silica particles with highly controllable particle sizes and surface morphologies [91]. The Stöber method is commonly preferred for the synthesis of spherical silica particles. This process involves hydrolysis and polycondensing of TEOS catalyzed by acid or base [92]. In a traditional Stöber method, ethanol, water, and ammonium hydroxide are mixed, followed by dropwise addition of TEOS at constant stirring. The TEOS is hydrolyzed with ammonium hydroxide acting as a catalyst to form silicic acid that condenses to form spherical silica particles. By changing the ratios of water, ammonia, and TEOS in the solution, the size of the resulting silica particles can be controlled, ranging between 20-500nm [92].

The original Stöber method was introduced in the 1960s by W. Stöber. Around 30 years later, Unger's group modified the method by introducing cetyltriethylammonium bromide (CTAB) as a structure-directing agent and successfully prepared Mesoporous monodispersed silica spheres [93]. The use of a template as a structural directing agent in the Stöber method is referred to as the modified Stöber method. The morphology and size of silica nanoparticles produced through the Stöber method depend on ammonium concentrations mainly, water, and TEOS. It was observed that the average particle size increases as the concentrations of these reagents increase. In contrast, the ethanol concentration had a negative effect on the particle size [94]. Furthermore, the amount and concentration of surfactant in the Stöber method significantly affect the morphology and particle size of the silica particles. It was found that pore volume and surface area increase when the amount of surfactant (*e.g.*, CTAB) increases, whereas the pore diameter and particle size decrease [95].

PHYSICAL, CHEMICAL, AND BIOLOGICAL PROPERTIES OF BIOACTIVE GLASSES

Melt-quench Derived Bioactive Glasses

The high sintering temperature in a melt-quench synthesis method may lead to high densification and hence a non-porous glass material [96, 97]. Generally, bioactive glasses synthesized *via* melt-quench possess much higher mechanical strength than their counterparts from the sol-gel method [77, 97, 98]. However, although melt-quench derived glasses show good mechanical strength, they possess inherent brittleness and low resistance to fracture, especially in the presence of even a tiny crack.

As shown in Fig. (**1**) above, bonding ability and bioactivity in general of melt-quench derived bioactive glasses depend on the concentration of their constituents. In addition, the high densification behavior of melt-quench glasses results in glass particles with a low surface-to-volume ratio leading to low dissolution rates under contact with physiological body fluid [99 - 101].

Sol-gel Derived Bioactive Glasses

Porosity is an intrinsic property of sol-gel-derived glasses with varied pore sizes [102]. As per IUPAC systems, the porous material can be categorized into three main groups: *Microporous material*, whose pore sizes are less than 2 nm, *mesoporous material*, in which pore sizes range between 2 nm to 50 nm and *macroporous material*, where pore sizes are greater than 50 nm.

Mesoporous glasses are mainly defined by their pore sizes and structures. However, the pores may result in a homogeneous or heterogeneous pore distribution in terms of their shapes and sizes, where they can form isolated or interconnected pore structures. Mesoporous structures can also be induced and controlled in the sol-gel synthesis process by incorporating surfactant as a mesoporous structural-directing agent [103-105].

Sol-gel-derived glasses generally result in glass particles with spherical shapes [104]. For medical applications such as drug delivery systems, the morphology of glass particles such as shapes and sizes play a vital role. For example, it was found that mesoporous glass particles with smaller particle sizes showed high bioactivity compared to glasses of the same composition with larger particle sizes [106]. Furthermore, the particle sizes of the bioactive glass can be controlled by varying the sol-gel synthesis parameter, such as the type of the template and concentration of organic species included in the synthesis [103].

The bioactivity of bioactive glass is highly affected by the chemical composition of the glass material. Generally, sol-gel-derived bioactive glasses show higher bioactivity than melt-quench-derived glasses of the same compositions [107]. In order to facilitate nucleation sites for the formation of a hydroxyapatite layer, the glass should have adequate R-OH groups. In addition, the presence of calcium and phosphate ions in the glass composition plays a vital role in the formation and crystallization of the hydroxyapatite layer. However, these ions are also present in the physiological body fluid or simulated body fluid.

Generally, sol-gel-derived bioactive glasses show low mechanical strength mainly due to the inherent porosity properties of the synthesis method. However, some studies have shown that the mechanical properties of sol-gel-derived bioactive glasses could be improved by varying the processing parameters and incorporating metallic additives [108-111].

Ionic dissolutions from the bioactive glasses are responsible for stimulating the osteogenesis process and other healing activities [112]. Thus, biodegradability and dissolution rates of the glass material when in contact with body fluid are among the critical properties to be controlled for specific medical applications.

Properties Enhancement through Doping

Doping bioactive glasses with essential metallic elements can modify bioactive glasses' surface morphology, biological and mechanical properties. For example, Tabia *et al*. [70] studied the effect of magnesium (Mg) doping on the binary glass (Si-Ca), specifically on the particle texture, bio-mineralization process, and drug release. They found increasing Mg content to positively affect the enhancement of specific areas and drug release mainly contributed by the porosity of the formed biomaterial. However, the increasing Mg content hindered the crystallization ability to form a more stable apatite-like phase in the synthetic biomaterial.

The effect of various element substitutions on the properties of binary glass (Si-Ca) synthesized *via* the sol-gel method was also done by Abdul-Rahman *et al*., [71]. It was found that increasing Sr, Zr, and Zn ion content has a negative effect on mechanical properties. It was further found that the formation of the hydroxycarbonateapatite (HCA) layer on the bioactive glass was enhanced with increasing Sr and Zr content. The Sr ions in parallel increased the biodegradability of the synthesized bioactive glass material. Furthermore, the *in vitro* studies on the Sr-containing bioactive glasses revealed enhancement on osteoconductive effect with stimulated formation of HA on the surface of the bioactive glass implants [72].

Barrioni *et al*. [73] studied the effect of incorporating manganese oxide (MnO) on the structure, texture, *in vitro* bioactivity, and cytocompatibility of SiO_2-P_2O_5-CaO-MnO sol-gel derived bioactive glasses. It was found that Mn addition enhances amorphous character, high surface area, and Mesoporous structure. In addition, the formation of hydroxycarbonateapatite (HCA) was revealed after the immersion of the material in simulated body fluid (SBF).

The effect on increasing cell osteogenic activity of a melt-derived bioactive glass containing strontium (Sr) and Cobalt (Co) was studied by Kargozar *et al*. [113]. It was found that an apatite-like layer was formed on the surface of SBF-immersed samples after 3, 7, and 14 days. In addition, the *in vitro* study showed that incorporation of Sr and Co ions in the bioactive glass promotes osteogenic activity without any cytotoxicity effect. Finally, Wei *et al*. [114] studied the *in vitro* biological influence of strontium (Sr) substitution on electro-spraying derived bioactive glasses (ESBG). It was revealed that the Sr substitution on a molar basis hindered the samples' bioactivity to some extent.

Kazon *et al*. [115] studied the antibacterial effect of Ag-doped silicate-based bioactive glass. The results confirmed antibacterial effect by inhibition of growth of *E.Coli* bacterial colonies. Similarly, Marsh *et al*. [116] confirmed the strong antibacterial response of Ag-doped bioactive glass, thus suggesting Ag ions as a promising antibacterial candidate in silicate-based biomaterials.

CONCLUDING REMARKS

Bioactive glasses have drawn attention worldwide due to their peculiar properties of bonding with native tissues, thus bringing hope for life-saving in regenerative medicine. Medical specialists need to understand bioactive glasses' structures, properties, and processing to fit a specific application. Physical, chemical, mechanical, and biological properties of three prominent bioactive glasses *viz.* silicate, phosphate, and borate-based bioactive glasses, are presented. The last two are rarely studied. Thus more studies on enhancing their biological and mechanical properties are in great demand. In addition, two main processing methods of bioactive glasses, namely melt-quench and sol-gel for each of the three bioactive glasses, are presented. This book chapter is an introductory document on bioactive glasses, emphasizing their structural properties, and processing for medical applications.

REFERENCES

[1]   Pina S, Rebelo R, Correlo VM, Oliveira JM, Reis RL. Bioceramics for osteochondral tissue engineering and regeneration. Advances in Experimental Medicine and Biology. Springer New York LLC 2018; pp. 53-75.

[2]   Huang J, Best S. Ceramic biomaterials for tissue engineering. Tissue Engineering Using Ceramics and Polymers. 2nd ed. Elsevier Inc. 2014; pp. 3-34.
        [http://dx.doi.org/10.1533/9780857097163.1.3]

[3]   Rahaman MN. Bioactive ceramics and glasses for tissue engineering. Tissue Engineering Using Ceramics and Polymers. 2nd ed. Elsevier Inc. 2014; pp. 67-114.
        [http://dx.doi.org/10.1533/9780857097163.1.67]

[4]   Mohamed S, Shamaz BH. Bone tissue engineering and bony scaffolds. Int J Dent Oral Heal 2015; 1(1): 1-6.
        [http://dx.doi.org/10.25141/2471-657X-2015-1.0001]

[5]   Gentile P, Mattioli-Belmonte M, Chiono V. *et al.* Bioactive glass/polymer composite scaffolds mimicking bone tissue. J Biomed Mater Res 2012; 2654-67.

[6]   Erol-Taygun M, Unalan I, Idris MIB, Mano JF, Boccaccini AR. Bioactive glass-polymer nanocomposites for bone tissue regeneration applications: A review Advanced Engineering Materials 2019; 1900287.https://onlinelibrary.wiley.com/doi/full/10.1002/adem.201900287

[7]   Distler T, Fournier N, Grünewald A, *et al.* Polymer-bioactive glass composite filaments for 3d scaffold manufacturing by fused deposition modeling: Fabrication and characterization. Front Bioeng Biotechnol 2020; 8: 552.
        [http://dx.doi.org/10.3389/fbioe.2020.00552] [PMID: 32671025]

[8]   Liang W, Wu X, Dong Y. *et al. In vivo* behavior of bioactive glass-based composites in animal models for bone regeneration. Biomaterials Science The Royal Society of Chemistry 2021; 1924-44.https://pubs.rsc.org/en/content/articlehtml/2021/bm/d0bm01663b

[9]   Grandi G, Heitz C, Santos LA, *et al.* Comparative histomorphometric analysis between α-Tcp cement and β-Tcp/Ha granules in the bone repair of rat calvaria. Mater Res 2011; 14(1): 11-6.
        [http://dx.doi.org/10.1590/S1516-14392011005000020]

[10] Tang Z, Li X, Tan Y, Fan H, Zhang X. The material and biological characteristics of osteoinductive calcium phosphate ceramics. Regen Biomater 2018; 5(1): 43-59.
[http://dx.doi.org/10.1093/rb/rbx024] [PMID: 29423267]

[11] Oonishi H, Kushitani S, Yasukawa E, *et al.* Particulate bioglass compared with hydroxyapatite as a bone graft substitute. Clin Orthop Relat Res 1997; 334(334): 316-25.
[http://dx.doi.org/10.1097/00003086-199701000-00041] [PMID: 9005929]

[12] Yuan H, De Bruijn JD, Li Y, *et al.* Bone formation induced by calcium phosphate ceramics in soft tissue of dogs: a comparative study between porous α-TCP and β-TCP. J Mater Sci Mater Med 2001; 12(1): 7-13.
[http://dx.doi.org/10.1023/A:1026792615665] [PMID: 15348371]

[13] Chai YC, Carlier A, Bolander J, *et al.* Current views on calcium phosphate osteogenicity and the translation into effective bone regeneration strategies. Acta Biomaterialia. Elsevier Ltd 2012; pp. 3876-87.
[http://dx.doi.org/10.1016/j.actbio.2012.07.002]

[14] Hench LL. The story of bioglass. Journal of Materials Science. Materials in Medicine 2006; pp. 967-78.

[15] Hench LL, Jones JR. Bioactive glasses: Frontiers and challenges. Frontiers in bioengineering and biotechnology. Frontiers media S.A. 2015; Vol. 3.

[16] Hench LL, Hench L. The story of Bioglass®. J Mater Sci Mater Med 2006; 17(11): 967-78.
[http://dx.doi.org/10.1007/s10856-006-0432-z] [PMID: 17122907]

[17] Baino F, Yamaguchi S. The use of simulated body fluid (SBF) for assessing materials bioactivity in the context of tissue engineering: Review and challenges [Internet]. Vol. 5. Biomimetics 2020; 5(4): 57.
[http://dx.doi.org/10.3390/biomimetics5040057] [PMID: 33138246]

[18] Chajri S, Bouhazma S, Adouar I, *et al.* Synthesis, characterization and evaluation of bioactivity of glasses in the CaO-SiO2-P2O5-MgO system with different CaO/MgO ratios.Journal of Physics: Conference Series. IOP Publishing 2019; p. 12013.
[http://dx.doi.org/10.1088/1742-6596/1292/1/012013]

[19] Karasu B, Yanar AO, Koçak A, Kisacik O. Bioactive glasses fosforesans özelliğe sahip mavimsi-yeşil, sarimsi-yeşil pigmentlerin üretimi ve iii. pişirim (dekor pişirimi) duvar karosu sirlarinda ve vetroza uygulamalarinda kullanim,. Projesi View project The improvement of ceramic body properties View project 2017.www.dergipark.gov.tr

[20] Chen R, Li Q, zhang Q, *et al.* Nanosized HCA-coated borate bioactive glass with improved wound healing effects on rodent model. Chem Eng J 2021; 426: 130299.
[http://dx.doi.org/10.1016/j.cej.2021.130299]

[21] Naseri S, Nazhat SN. Bioactive and soluble glasses for wound-healing applications. Bioactive Glasses. Elsevier 2018; pp. 381-405.
[http://dx.doi.org/10.1016/B978-0-08-100936-9.00019-8]

[22] Naseri S, Lepry WC, Nazhat SN. Bioactive glasses in wound healing: hope or hype? J Mater Chem B Mater Biol Med 2017; 5(31): 6167-74.
[http://dx.doi.org/10.1039/C7TB01221G] [PMID: 32264432]

[23] Islam MT, Felfel RM, Abou Neel EA, Grant DM, Ahmed I, Hossain KMZ. Bioactive calcium phosphate–based glasses and ceramics and their biomedical applications: A review. J Tissue Eng 2017; 8.
[http://dx.doi.org/10.1177/2041731417719170] [PMID: 28794848]

[24] Christie JK, Ainsworth RI, Hernandez SER, de Leeuw NH. Structures and properties of phosphate-based bioactive glasses from computer simulation: A review. J Mater Chem B Mater Biol Med 2017; 5(27): 5297-306.

[http://dx.doi.org/10.1039/C7TB01236E] [PMID: 32264067]

[25] Pickup DM, Newport RJ, Knowles JC. Sol-gel phosphate-based glass for drug delivery applications. J Biomater Appl 2012; 26(5): 613-22.
[http://dx.doi.org/10.1177/0885328210380761] [PMID: 20819917]

[26] Henderson GS, Calas G, Stebbins JF. The structure of silicate glasses and melts the tetrahedra have a well-defined geometry and are linked 2006; 1: 269-73.

[27] Zachariasen WH. The atomic arrangement in glass. J Am Chem Soc 1932; 54(10): 3841-51.
[http://dx.doi.org/10.1021/ja01349a006]

[28] Ojovan MI. Viscosity and glass transition in amorphous oxides. Adv Condens Matter Phys 2008; 2008: 1-23.
[http://dx.doi.org/10.1155/2008/817829]

[29] Coon E, Whittier AM, Abel BM, Stapleton EL, Miller R, Fu Q. Viscosity and crystallization of bioactive glasses from 45S5 to 13-93. Int J Appl Glass Sci 2021; 12(1): 65-77.
[http://dx.doi.org/10.1111/ijag.15837]

[30] Garcia-Valles M, Hafez HS, Cruz-Matías I, et al. Calculation of viscosity–temperature curves for glass obtained from four wastewater treatment plants in Egypt. J Therm Anal Calorim 2013; 111(1): 107-14.
[http://dx.doi.org/10.1007/s10973-012-2232-7]

[31] Angewandte Chemie. International Edition. Brauer DS. Bioactive glasses - Structure and properties.Wiley-VCH Verlag 2015; 54: pp. 4160-81.

[32] Varshneya AK, Mauro JC. Fundamentals of inorganic glasses [Internet]. Fundamentals of inorganic glasses. 2019; 1-735.https://books.google.com/books?hl=en&lr=&id=MSgXBQAAQBAJ&oi=fnd&pg=PP1&dq=+Fundamentals+of+Inorganic+Glasses&ots=Bv_mpLi4-O&sig=aBVrTe6FA m6x5-0pBxRw3_VCLgY

[33] Hench LL, Splinter RJ, Allen WC, Greenlee TK. Bonding mechanisms at the interface of ceramic prosthetic materials. J Biomed Mater Res 1971; 5(6): 117-41.
[http://dx.doi.org/10.1002/jbm.820050611]

[34] Al-Harbi N, Mohammed H, Al-Hadeethi Y. et al. Silica-based bioactive glasses and their applications in hard tissue regeneration: A review. Pharmaceuticals 2021; 14: 1-20.

[35] Van Vugt TA, Geurts JAP, Arts JJ, Lindfors NC. Biomaterials in treatment of orthopedic infections. Management of Periprosthetic Joint Infections (PJIs). Elsevier Inc. 2017; pp. 41-68.
[http://dx.doi.org/10.1016/B978-0-08-100205-6.00003-3]

[36] Islam MT, Felfel RM, Abou Neel EA, Grant DM, Ahmed I, Hossain KMZ. Bioactive calcium phosphate–based glasses and ceramics and their biomedical applications: A review. J Tissue Eng 2017.

[37] Elgayar I, Aliev AE, Boccaccini AR, Hill RG. Structural analysis of bioactive glasses. J Non-Cryst Solids 2005; 351(2): 173-83.
[http://dx.doi.org/10.1016/j.jnoncrysol.2004.07.067]

[38] Hill RG, Brauer DS. Predicting the bioactivity of glasses using the network connectivity or split network models. J Non-Cryst Solids 2011; 357(24): 3884-7.
[http://dx.doi.org/10.1016/j.jnoncrysol.2011.07.025]

[39] Fábián M, Kovács Z, Lábár JL, et al. Network structure and thermal properties of bioactive (SiO2–CaO–Na2O–P2O5) glasses. J Mater Sci 2020; 55(6): 2303-20.
[http://dx.doi.org/10.1007/s10853-019-04206-z]

[40] Rao KJ. Structural Chemistry of Glasses Structural Chemistry of Glasses. Elsevier 2002; p. 568.

[41] Shelby J. Introduction to glass science and technology. 2020. Available from: https://books.google.com/books?hl=en&lr=&id=s-kCEAAAQBAJ&oi=fnd&pg=PP1&dq= Introduction+to+Glass+Science+and+Technology,+2nd+ed&ots=wrJ5FYOKAW&sig=nG1yZ4ESaS

Cy_4X_gLWJ54htVNk

[42] Khalid MD, Khurshid Z, Zafar MS, Farooq I, Khan RS, Najmi A. Bioactive glasses and their applications in dentistry. J Pak Dent Assoc 2017; 26(1): 32-8.
[http://dx.doi.org/10.25301/JPDA.261.32]

[43] Luo G, Ma Y, Cui X, *et al.* 13-93 bioactive glass/alginate composite scaffolds 3D printed under mild conditions for bone regeneration. RSC Advances 2017; 7(20): 11880-9.
[http://dx.doi.org/10.1039/C6RA27669E]

[44] Mehatlaf AA, Farid SBH, Atiyah AA. Synthesis and characterisation of bioactive glass 13-93 scaffolds for bone tissue regeneration. IOP Conf Ser Mater Sci Eng 2021.
[http://dx.doi.org/10.1088/1757-899X/1067/1/012136]

[45] Kiran P, Ramakrishna V, Trebbin M, Udayashankar NK, Shashikala HD. Effective role of CaO/P$_2$O$_5$ ratio on SiO$_2$-CaO-P$_2$O$_5$ glass system. J Adv Res 2017; 8(3): 279-88.
[http://dx.doi.org/10.1016/j.jare.2017.02.001] [PMID: 28337345]

[46] Hench LL, Polak JM. Third-generation biomedical materials. Science American Association for the Advancement of Science. 2002.
[http://dx.doi.org/10.1126/science.1067404]

[47] Filgueiras MR, La Torre G, Hench LL. Solution effects on the surface reactions of a bioactive glass. J Biomed Mater Res 1993; 27(4): 445-53.
[http://dx.doi.org/10.1002/jbm.820270405] [PMID: 8385143]

[48] Moustafa YM, El-Egili K. Infrared spectra of sodium phosphate glasses. J Non-Cryst Solids 1998; 240(1-3): 144-53.
[http://dx.doi.org/10.1016/S0022-3093(98)00711-X]

[49] Fernandes HR, Gaddam A, Rebelo A, Brazete D, Stan GE, Ferreira JMF. Bioactive glasses and glass-ceramics for healthcare applications in bone regeneration and tissue engineering. Materials 2018; 11https://www.mdpi.com/380192

[50] Walter G, Vogel J, Hoppe U, Hartmann P. The structure of CaO–Na2O–MgO–P2O5 invert glass. J Non-Cryst Solids 2001; 296(3): 212-23.
[http://dx.doi.org/10.1016/S0022-3093(01)00912-7]

[51] Lepry WC, Nazhat SN. The anomaly in bioactive sol–gel borate glasses 2020. Available from: https://pubs.rsc.org/en/content/articlehtml/2020/ma/d0ma00360c
[http://dx.doi.org/10.1039/D0MA00360C]

[52] Wang H, Zhao S, Zhou J, *et al.* Evaluation of borate bioactive glass scaffolds as a controlled delivery system for copper ions in stimulating osteogenesis and angiogenesis in bone healing. J Mater Chem B Mater Biol Med 2014; 2(48): 8547-57.
[http://dx.doi.org/10.1039/C4TB01355G] [PMID: 32262213]

[53] Jones JR. Review of bioactive glass: From Hench to hybrids. Acta Biomaterialia. Elsevier 2013; pp. 4457-86.

[54] Nandi SK, Mahato A, Kundu B, Mukherjee P. Doped bioactive glass materials in bone regeneration. Advanced Techniques in Bone Regeneration. InTech 2016. Internet
[http://dx.doi.org/10.5772/63266]

[55] Esposito S. Traditional" sol-gel chemistry as a powerful tool for the preparation of supported metal and metal oxide catalysts. Materials 2019; 12www.mdpi.com/journal/materials

[56] Kaur G, Pandey OP, Singh K, Homa D, Scott B, Pickrell G. A review of bioactive glasses: Their structure, properties, fabrication and apatite formation. J Biomed Mater Res A 2014; 102(1): 254-74.
[http://dx.doi.org/10.1002/jbm.a.34690] [PMID: 23468256]

[57] Khurshid Z, Husain S, Alotaibi H. *et al.* Novel techniques of scaffold fabrication for bioactive glasses. Biomedical, Therapeutic and Clinical Applications of Bioactive Glasses. Elsevier 2018; pp. 497-519.

[58]   Karmakar B. Fundamentals of glass and glass nanocomposites. Glass Nanocomposites: Synthesis, Properties and Applications. Elsevier Inc. 2016; pp. 3-53.
[http://dx.doi.org/10.1016/B978-0-323-39309-6.00001-8]

[60]   Kaur G, Pickrell G, Sriranganathan N, Kumar V, Homa D. Review and the state of the art: Sol–gel and melt quenched bioactive glasses for tissue engineering. Journal of Biomedical Materials Research - Part B Applied Biomaterials. John Wiley and Sons Inc 2016; pp. 1248-75.

[61]   Carta D, Pickup DM, Knowles JC, Ahmed I, Smith ME, Newport RJ. A structural study of sol–gel and melt-quenched phosphate-based glasses. J Non-Cryst Solids 2007; 353(18-21): 1759-65.
[http://dx.doi.org/10.1016/j.jnoncrysol.2007.02.008]

[62]   Giannoudis P V, Dinopoulos H, Tsiridis E. Bone substitutes: an update 2005; 20-7.
[http://dx.doi.org/10.1016/j.injury.2005.07.029]

[63]   Hench LL, Jones JR. Bioactive glasses: Frontiers and Challenges. Frontiers in Bioengineering and Biotechnology. Frontiers Media S.A. 2015; Vol. 3: p. 194.www.frontiersin.org

[64]   Neacşu IA, Nicoară AI, Vasile OR, Vasile BŞ. Inorganic micro- and nanostructured implants for tissue engineering. Nanobiomaterials in Hard Tissue Engineering: Applications of Nanobiomaterials. Elsevier Inc. 2016; pp. 271-95.
[http://dx.doi.org/10.1016/B978-0-323-42862-0.00009-2]

[65]   Rahim S, Jan Iftikhar F, Malik MI. Biomedical applications of magnetic nanoparticles. Metal Nanoparticles for Drug Delivery and Diagnostic Applications. Elsevier Inc. 2019; pp. 301-28.

[66]   Makhlouf ASH, Rodriguez R. Bioinspired smart coatings and engineering materials for industrial and biomedical applications. Advances in Smart Coatings and Thin Films for Future Industrial and Biomedical Engineering Applications. Elsevier 2019; pp. 407-27.

[68]   Palanisamy P, Chavali M, Kumar EM, Etika KC. Hybrid nanocomposites and their potential applications in the field of nanosensors/gas and biosensors. Nanofabrication for Smart Nanosensor Applications. Elsevier 2020; pp. 253-80.
[http://dx.doi.org/10.1016/B978-0-12-820702-4.00011-8]

[69]   Zemke F, Schölch V, Bekheet MF, Schmidt F. Surfactant-assisted sol–gel synthesis of mesoporous bioactive glass microspheres. Nanomater Energy 2019; 8(2): 126-34.
[http://dx.doi.org/10.1680/jnaen.18.00020]

[70]   Tabia Z, El Mabrouk K, Bricha M, Nouneh K. Mesoporous bioactive glass nanoparticles doped with magnesium: drug delivery and acellular *in vitro* bioactivity. RSC Advances 2019; 9(22): 12232-46.
[http://dx.doi.org/10.1039/C9RA01133A] [PMID: 35515868]

[71]   Abdul-Rahman F, Motawea I, Shoreibah E. The effect of various elements substitution on properties of bioactive glass scaffolds for bone tissue engineering. Al-Azhar Dent J Girls 2017; 4(3): 255-70.
[http://dx.doi.org/10.21608/adjg.2017.5266]

[72]   Bellucci D, Cannillo V, Anesi A, *et al.* Bone regeneration by novel bioactive glasses containing strontium and/or magnesium: A preliminary *in-vivo* study. Materials 2018; 11(11): 2223.
[http://dx.doi.org/10.3390/ma11112223] [PMID: 30413108]

[73]   Barrioni BR, Oliveira AC, de Fátima Leite M, de Magalhães Pereira M. Sol–gel-derived manganese-releasing bioactive glass as a therapeutic approach for bone tissue engineering. J Mater Sci 2017; 52(15): 8904-27.
[http://dx.doi.org/10.1007/s10853-017-0944-6]

[74]   Aboualigaledari N, Rahmani M. A review on the synthesis of the TiO2-based photocatalyst for the environmental purification. Journal of Composites and Compounds 2021; 2(5): 25-42.
[http://dx.doi.org/10.52547/jcc.3.1.4]

[75]   Deshmukh K, Kovářík T, Křenek T, Docheva D, Stich T, Pola J. Recent advances and future perspectives of sol-gel derived porous bioactive glasses: A review. RSC Advances Royal Society of

Chemistry. 2020; pp. 33782-835.
[http://dx.doi.org/10.1039/D0RA04287K]

[76]   Vafa E, Bazargan-Lari R, Bahrololoom ME. Synthesis of 45S5 bioactive glass-ceramic using the sol-gel method, catalyzed by low concentration acetic acid extracted from homemade vinegar. J Mater Res Technol 2021; 10: 1427-36.
[http://dx.doi.org/10.1016/j.jmrt.2020.12.093]

[77]   Baino F, Fiume E, Miola M, Verné E. Bioactive sol-gel glasses: Processing, properties, and applications. Int J Appl Ceram Technol 2018; 15(4): 841-60.
[http://dx.doi.org/10.1111/ijac.12873]

[78]   Benhamza H, Barboux P, Bouhaouss A, Josien F-A, Livage J. Sol–gel synthesis of $Zr(HPO_4)_2 \cdot H_2O$. J Mater Chem 1991; 1(4): 681-4.
[http://dx.doi.org/10.1039/JM9910100681]

[79]   Willinger MG, Clavel G, Di W, Pinna N. A general soft-chemistry route to metal phosphate nanocrystals. J Ind Eng Chem 2009; 15(6): 883-7.
[http://dx.doi.org/10.1016/j.jiec.2009.09.017]

[80]   Lepry WC, Nazhat SN. A review of phosphate and borate sol–gel glasses for biomedical applications. Adv NanoBiomed Res 2021; 1(3): 2000055.
[http://dx.doi.org/10.1002/anbr.202000055]

[81]   Tohge N, Mackenzie JD. Preparation of $20Na2O \cdot 80B2O3$ glasses by sol-gel method. J Non-Cryst Solids 1984; 68(2-3): 411-8.
[http://dx.doi.org/10.1016/0022-3093(84)90021-8]

[82]   Lepry WC, Nazhat SN. Highly bioactive sol-gel-derived borate glasses. Chem Mater 2015; 27(13): 4821-31.
[http://dx.doi.org/10.1021/acs.chemmater.5b01697]

[83]   Lepry WC, Rezabeigi E, Smith S, Nazhat SN. Dissolution and bioactivity of a sol-gel derived borate glass in six different solution media 2019. Available from: https://www.degruyter.com/document/doi/10.1515/bglass-2019-0009/html
[http://dx.doi.org/10.1515/bglass-2019-0009]

[85]   Brinker CJ. Evaporation-induced self-assembly: Functional nanostructures made easy. MRS Bulletin. Materials Research Society; 2004; pp. 631-40.

[86]   Shah AT, Ain Q, Chaudhry AA, *et al*. Acid catalysed synthesis of bioactive glass by evaporation induced self assembly method. J Non-Cryst Solids 2018; 479: 1-8.
[http://dx.doi.org/10.1016/j.jnoncrysol.2017.09.041]

[87]   Mahoney L, Koodali R. Versatility of evaporation-induced self-assembly (EISA) method for preparation of mesoporous TiO2 for energy and environmental applications. Materials 2014; 7(4): 2697-746.
[http://dx.doi.org/10.3390/ma7042697] [PMID: 28788590]

[88]   Carreon MA, Guliants VV. Mesostructuring of metal oxides through EISA. Fundamentals and applications. Ordered Porous Solids. Elsevier 2009; pp. 413-39.
[http://dx.doi.org/10.1016/B978-0-444-53189-6.00016-0]

[89]   Shih CC, Chien CS, Kung JC, *et al*. Effect of surfactant concentration on characteristics of mesoporous bioactive glass prepared by evaporation induced self-assembly process. Appl Surf Sci 2013; 264: 105-10.
[http://dx.doi.org/10.1016/j.apsusc.2012.09.134]

[90]   Zhao S, Li Y, Li D. Synthesis of CaO–SiO2–P2O5 mesoporous bioactive glasses with high P2O5 content by evaporation induced self assembly process. J Mater Sci Mater Med 2011; 22(2): 201-8.
[http://dx.doi.org/10.1007/s10856-010-4200-8] [PMID: 21170573]

[91]   Ghimire PP, Jaroniec M. Renaissance of Stöber method for synthesis of colloidal particles: New

developments and opportunities. Journal of Colloid and Interface Science 2021; 838-65.

[92]   Greasley SL, Page SJ, Sirovica S, *et al.* Controlling particle size in the Stöber process and incorporation of calcium. J Colloid Interface Sci 2016; 469: 213-23.
[http://dx.doi.org/10.1016/j.jcis.2016.01.065] [PMID: 26890387]

[93]   Liu S, Lu L, Yang Z, Cool P, Vansant EF. Further investigations on the modified Stöber method for spherical MCM-41. Mater Chem Phys 2006; 97(2-3): 203-6.
[http://dx.doi.org/10.1016/j.matchemphys.2005.09.003]

[94]   Fernandes RS, Raimundo IM Jr, Pimentel MF. Revising the synthesis of Stöber silica nanoparticles: A multivariate assessment study on the effects of reaction parameters on the particle size. Colloids Surf A Physicochem Eng Asp 2019; 577: 1-7.
[http://dx.doi.org/10.1016/j.colsurfa.2019.05.053]

[95]   Kachbouri S, Mnasri N, Elaloui E, Moussaoui Y. Tuning particle morphology of mesoporous silica nanoparticles for adsorption of dyes from aqueous solution. J Saudi Chem Soc 2018; 22(4): 405-15.
[http://dx.doi.org/10.1016/j.jscs.2017.08.005]

[96]   Woignier T, Prassas M, Duffours L. Sintering of aerogels for glass synthesis. J Sol-Gel Sci Technol 2019; 90(1): 76-86.
[http://dx.doi.org/10.1007/s10971-018-4826-4]

[97]   Ben-Arfa BAE, Pullar RC. A comparison of bioactive glass scaffolds fabricated by robocasting from powders made by sol-gel and melt-quenching methods. Processes Multidisciplinary Digital Publishing Institute 2020; 615.https://www.mdpi.com/2227-9717/8/5/615/htm

[98]   Fiume E, Migneco C, Verné E, Baino F. Comparison between bioactive sol-gel and melt-derived glasses/glass-ceramics based on the multicomponent SiO_2-P_2O_5-CaO-MgO-Na_2O-K_2O System. Materials 2020; 13: 3.

[99]   Nandi SK, Mahato A, Kundu B, Mukherjee P. Doped bioactive glass materials in bone regeneration. IntechOpen 2016.https://www.intechopen.com/chapters/50915
[http://dx.doi.org/10.5772/63266]

[100] Martin RA, Yue S, Hanna JV. *et al.* Characterizing the hierarchical structures of bioactive sol-gel silicate glass and hybrid scaffolds for bone regeneration [Internet]. Vol. 370, philosophical transactions of the royal society a: Mathematical, physical and engineering sciences. The Royal Society Publishing 2012; 1422-43.https://royalsocietypublishing.org/doi/abs/10.1098/rsta.2011.0308

[101] Sepulveda P, Jones JR, Hench LL. *In vitro* dissolution of melt-derived 45S5 and sol-gel derived 58S bioactive glasses. J Biomed Mater Res 2002; 61(2): 301-11.
[http://dx.doi.org/10.1002/jbm.10207] [PMID: 12007211]

[102] Available from: https://pubs.rsc.org/en/content/articlehtml/2020/ra/d0ra04287k

[103] Letaief N, Lucas-Girot A, Oudadesse H, Dorbez-Sridi R. Influence of synthesis parameters on the structure, pore morphology and bioactivity of a new mesoporous glass. J Biosci Med 2014; 02(02): 57-63.

[104] Zemke F, Schölch V, Bekheet MF, Schmidt F. Surfactant-assisted sol–gel synthesis of mesoporous bioactive glass microspheres. Nanomaterials and Energy 2019; 8(2): 126-34.
[http://dx.doi.org/10.1680/jnaen.18.00020]

[105] Jones JR, Hench LL. Effect of surfactant concentration and composition on the structure and properties of sol-gel-derived bioactive glass foam scaffolds for tissue engineering. J Mater Sci 2003; 38(18): 3783-90.
[http://dx.doi.org/10.1023/A:1025988301542]

[106] Dang TH, Bui TH, Guseva EV, *et al.* Characterization of bioactive glass synthesized by sol-gel process in hot water. Crystals 2020; 10(6): 529.
[http://dx.doi.org/10.3390/cryst10060529]

[107] Haque J, Habib Munna A, Sidrat Rahman Ayon A, *et al.* A comprehensive overview of the pertinence and possibilities of bioactive glass in the modern biological world. Journal of Biomaterials 2020; 4(2): 23-38.
[http://dx.doi.org/10.11648/j.jb.20200402.11]

[108] Fang W, Huang BR, Liu TH, Chen JA, Chang CF, Huang CL. Fabrication of bioactive glass with titanium ion doping *via* various reactive environments. Key Engineering Materials. Trans Tech Publications Ltd 2019; pp. 21-6. Internet

[109] Vyas VK, Kumar AS, Singh SP, Pyare R. Effect of nickel oxide substitution on bioactivity and mechanical properties of bioactive glass. Bull Mater Sci 2016; 39(5): 1355-61.
[http://dx.doi.org/10.1007/s12034-016-1242-7]

[110] Ali A, Ershad M, Hira S, Pyare R, Singh SP. Mechanochemical and *in vitro* cytocompatibilityevaluation of zirconia modified silver substituted1393 bioactive glasses. Bol la Soc Esp Ceram y Vidr 2020.https://www.elsevier.es/en-revista-boletin-sociedad-espanola-ceramica--idrio-26-avance-resumen-mechanochemical-in-vitro-cytocompatibility-evaluation-S0366317520-300704

[111] Amudha S, Ramya JR, Arul KT, *et al.* Enhanced mechanical and biocompatible properties of strontium ions doped mesoporous bioactive glass. Compos, Part B Eng 2020; 196: 108099.
[http://dx.doi.org/10.1016/j.compositesb.2020.108099]

[112] Fiume E, Barberi J, Verné E, Baino F. Bioactive glasses: From parent 45s5 composition to scaffold-assisted tissue-healing therapies. Journal of Functional Biomaterials Multidisciplinary Digital Publishing Institute 2018; 9.

[113] Kargozar S, Lotfibakhshaiesh N, Ai J, *et al.* Synthesis, physico-chemical and biological characterization of strontium and cobalt substituted bioactive glasses for bone tissue engineering. J Non-Cryst Solids 2016; 449: 133-40.
[http://dx.doi.org/10.1016/j.jnoncrysol.2016.07.025]

[114] Hong W, Zhang Q, Jin H. Roles of strontium and hierarchy structure on the *in vitro* biological response and drug release mechanism of the strontium-substituted bioactive glass microspheres. Mater Sci Eng C 2020; 107: 110336.
[http://dx.doi.org/10.1016/j.msec.2019.110336] [PMID: 31761170]

[115] Available from: https://www.mdpi.com/1996-1944/9/4/225/htm

[116] Marsh AC, Mellott NP, Crimp M, Wren A, Hammer N, Chatzistavrou X. Ag-doped bioactive glass-ceramic 3D scaffolds: Microstructural, antibacterial, and biological properties. J Eur Ceram Soc 2021; 41(6): 3717-30.
[http://dx.doi.org/10.1016/j.jeurceramsoc.2021.01.011]

CHAPTER 5

On the Biocompatibility of Bioactive Glasses (BGs)

Saeid Kargozar[1,*], **Francesco Baino**[2] and **Fabian Westhauser**[3]

[1] Department of Radiation Oncology, Simmons Comprehensive Cancer Center, UT Southwestern Medical Center, 5323 Harry Hines Blvd, Dallas, TX75390, USA

[2] Department of Applied Science and Technology (DISAT), Institute of Materials Physics and Engineering, Politecnico di Torino, 10129 Torino, Italy

[3] Department of Orthopaedics, Heidelberg University Hospital, Schlierbacher Landstraße 200a, 69118 Heidelberg, Germany

Abstract: Bioactive glasses (BGs) form a versatile class of biocompatible materials that can be utilized for various therapeutic strategies, including bone tissue engineering, soft tissue healing, and cancer therapy. Commonly, BGs are classified into three distinct categories, namely silicate, phosphate, and borate glasses. Several commercial BG-based products are now available on the market, and new generations with unique therapeutic features are also expected to introduce them in the near future. Due to their clinical significance, the biological behaviors of BGs have been one of the most interesting topics in tissue engineering and regenerative medicine. Although BGs are generally recognized as biocompatible materials in medicine, any new composition and formulation should be carefully tested through a series of standard *in vitro* and *in vivo* tests provided by international agencies (*e.g.*, Food and Drug Administration (FDA)) and regulatory bodies (*e.g.*, the International Organization for Standardization (ISO)). As a rule of thumb, the release of ionic dissolution products from BGs into the surrounding biological environment is regarded as the main parameter that modulates cellular and molecular phenomena. This process is even more crucial when specific elements (strontium, copper, *etc.*) are added to the basic composition of BGs to improve their physico-chemical properties, mechanical strength, and biological performance. Moreover, it is now well-established that some physical (*e.g.*, the topography) aspects of BGs can directly affect their compatibility with the living systems (cells and tissues). Therefore, a multifaceted design and testing approach should be applied while synthesizing BGs in the laboratory, and the collaboration of materials and chemical engineers with biologists and medical experts can be really helpful for producing optimized formulations.

* **Corresponding author Saeid Kargozar:** Department of Radiation Oncology, Simmons Comprehensive Cancer Center, UT Southwestern Medical Center, 5323 Harry Hines Blvd, Dallas, TX75390, USA; E-mail: kargozarsaeid@gmail.com

Keywords: Angiogenesis, Antibacterial activity, Bioactive glasses (BGs), Bioactivity, Biocompatibility, Borate bioactive glasses, Bone tissue engineering, Cytotoxicity, Tissue compatibility, Genotoxicity, Hemocompatibility, Glass-ceramics, Inflammatory response, International Organization for Standardization (ISO), *In vitro* study, *In vivo* animal study, Phosphate bioactive glasses, Polymers, Silicate bioactive glasses, Soft tissue engineering, Three-dimensional (3D) printing, Wound healing.

BIOLOGICAL EVALUATION OF MEDICAL DEVICES

One of the most critical aspects of medical devices is related to their compatibility with living systems (cells, tissues, and organs). Accordingly, several attempts have been made to determine a comprehensive definition of "biocompatibility" over the years. The first definition of biocompatibility was issued as "the ability of a material to perform with an appropriate host response in a specific application" [1]. However, the emergence of new technologies (*e.g.*, tissue engineering) in the concept of modern medicine led to a redefinition of this term. For instance, two important features of materials, *i.e.* bioactivity and biodegradation, were later specified and explained; then, they indeed needed to be considered in the new definition of biocompatibility. Nowadays, biocompatibility is identified as "the ability of a material to perform its desired function concerning a medical therapy, without eliciting any undesirable local or systemic effects in the recipient or beneficiary of that therapy, but generating the most appropriate beneficial cellular or tissue response in that specific situation, and optimizing the clinically relevant performance of that therapy" [2]. Over the years, the list of criteria for biocompatibility of materials is constantly being updated. Governmental agencies and regulatory bodies are indeed the developers of these rigid criteria. In this regard, the Food and Drug Administration (FDA) and the European Medicines Agency (EMA) are known as the leading international agencies. In addition, the International Organization for Standardization (ISO), American Society for Testing and Materials (ASTM), and the United States Pharmacopeia (USP) are recognized as regulatory bodies that provide valid procedures, protocols, guidelines, and standards to appraise all medical devices before implantation into the human body.

The ISO-10993 is among the most well-known standards that set a series of procedures for assessing the biological risk of any medical device to the human body. In fact, the ISO-10993 comprises different parts in which assays and tests are classified according to the nature of body contact (surface device, external communicating device, and implant device) (Table **1**).

Table 1. The ISO-10993 standard provides a framework for the biocompatibility evaluation of biomaterials and medical devices.

Medical Device Categorization by			Biological Effects							
Nature of Body Contact		Contact duration A= Limited (≤24h) B= Prolonged (>24 to 30 d) C= Permanent (>30 d)	Cytotoxicity	Sensitization	Irritation or intracutaneous reactivity	Systemic toxicity (acute)	Subchronic toxicity (subacute)	Genotoxicity	Implantation	Hemocompatibility
Category	Contact									
Surface Device	Skin	A	X	X	X	-	-	-	-	-
		B	X	X	X	-	-	-	-	-
		C	X	X	X	-	-	-	-	-
	Mucosal membrane	A	X	X	X	-	-	-	-	-
		B	X	X	X	-	-	-	-	-
		C	X	X	X	-	X	X	-	-
	Breached or compromised surface	A	X	X	X	-	-	-	-	-
		B	X	X	X	-	-	-	-	-
		C	X	X	X	-	X	X	-	-
External Communicating Device	Blood path, indirect	A	X	X	X	X	-	-	-	X
		B	X	X	X	X	-	-	-	X
		C	X	X	-	X	X	X	-	X
	Tissue/ Bone/ Dentine	A	X	X	X	-	-	-	-	-
		B	X	X	X	X	X	X	X	-
		C	X	X	X	X	X	X	X	-
	Circulating blood	A	X	X	X	X	-	-	-	X
		B	X	X	X	X	X	X	X	-
		C	X	X	X	X	X	X	X	-
Implant Device	Tissue/ Bone	A	X	X	X	-	-	-	-	-
		B	X	X	X	X	X	X	X	-
		C	X	X	X	X	X	X	X	-
	Blood	A	X	X	X	X	X	-	X	X
		B	X	X	X	X	X	X	X	X
		C	X	X	X	X	X	X	X	X

Cell culture systems represent the most common types of biocompatibility assays utilized for identifying cytotoxicity, cell adhesion, cell activation, or cell death. In fact, the compatibility of new biomaterials with cells is among the most extensively used tests before further biological evaluations [3]. The investigator is strongly suggested to use a cell type for which the device and biomaterial under examination are planned for clinical use. For instance, the cytocompatibility of materials designed for bone tissue repair and regeneration should be assessed by using osteoblast cells rather than "osteoblast-like" cell lines that are derived from the tissue of bone tumors, mostly osteosarcoma [4]. Most of these cells show a significant reduction in cell viability upon exposure to the well-known 45S5-bioactive glass composition when compared to the effects on primary bone marrow derived stromal cells (bone precursor cells) or primary human osteoblasts that tolerate exposure to 45S5-BG well [4]. Thus, the selection of cell types might have significant implications on the observed biocompatibility of the investigated biomaterial. *In vitro* cytotoxicity tests can be used in three different formats, including direct contact test, indirect contact test, and test on extracts [5]. In direct contact tests, material samples under investigation are carefully placed on cell monolayers (*e.g.*, L-929 mammalian fibroblast cells) and incubated for 24 h at 37 °C. At the end of the test, the culture medium and samples are removed from the incubator; then, the cells are fixed and stained with suitable histological stains and eventually observed by light microscopy. The changes in morphology and the number of cells are evaluated after their direct contact with solid medical devices and materials. In indirect assays (*e.g.*, agar diffusion test), sub-confluent cells (*e.g.*, fibroblasts) are overlaid by a thin layer of agar, and the given material is centrally placed on top of the agar to cover around 10% of the surface of the cell culture. In indirect cytotoxicity assays, the degree of reactivity of test material is only assessed by grading the size of the zone of dead cells around the sample using the cell morphology and/or *via* live/dead cell staining [6]. In the case of tests on extracts, a series of extract dilutions are prepared for evidencing toxicity and cellular interaction with intended materials. The most common use of this kind of cell culture assays is in providing information for regulatory compliance.

Generally, there are two problems in the use of *in vitro* biocompatibility tests, including (I) the short duration of the assays (hours to a few days) and (II) the use of immortalized cell lines instead of primary cells. Accordingly, the assessment of materials biocompatibility is followed up by performing a series of *in vivo* assays. In fact, any non-cytotoxic material should be checked for tissue toxicity *in vivo* to determine its biocompatibility under conditions simulating clinical use. Similar to *in vitro* tests, the selection of proper *in vivo* assay is of utmost importance for biocompatibility evaluation of materials. In this issue, medical devices and biomaterials are classified according to the nature of body contact and the duration of contact. Given the body contact, materials can be categorized into surface devices (*e.g.*, skin and mucosal membranes), external communicating devices (*e.g.*, circulating blood and tissue/bone/dentin communication), and implant devices (*e.g.*, bone implants). On the other hand, materials contact duration is sorted as limited (24 h), prolonged (>24 h and <30 days), and permanent (>30 days). The items appraised during *in vivo* biocompatibility tests include sensitization, irritation, intracutaneous (intradermal) reactivity, systemic toxicity (acute toxicity) and subacute and subchronic toxicity), genotoxicity, implantation, chronic toxicity, carcinogenicity, reproductive and developmental toxicity, biodegradation, and inflammation, wound healing, and the foreign body response. It is mentioned that *in vivo* biocompatibility can be conducted on a prototype (materials in the early stage of development) or the final product.

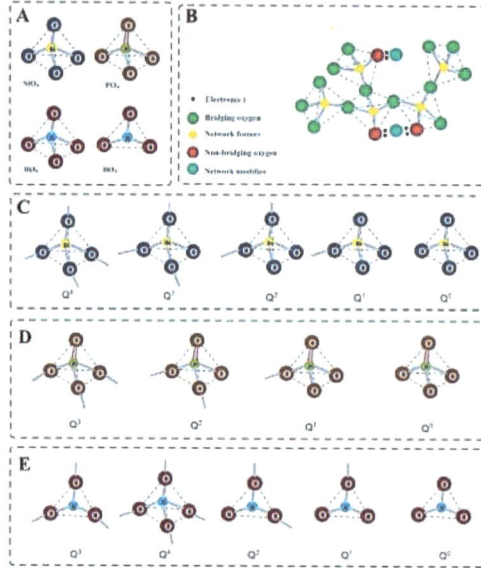

Fig. (1). Schematic representation of **(A)** structural units of silicate, phosphate, and borate-based BGs; **(B)** bridging oxygens and non-bridging oxygens in the structure of silicate BGs. Representative schematic of **(C, D,** and **E)** of the possible Q^n species in silicate, phosphate, and borate BGs, respectively. The value of n is the number of bridging oxygens. Reproduced with permission from [11].

BIOACTIVE GLASSES (BGs): FROM BENCH TO CLINIC

Bioactive glasses (BGs) represent versatile synthetic materials in medicine and hold great promise in tissue engineering and regenerative medicine. Historically, the invention of BGs dates back to the original work of Prof. Larry Hench in 1969 at the University of Florida [7]. 45S5 Bioglass® with a composition of $45SiO_2–24.5Na_2O–24.5CaO–6P_2O_5$ (wt%) is regarded as a parent of a wide family of biocompatible glasses and the first example of the third-generation biomaterials, owing to the biological role of its ionic dissolution by-products released into the physiological environment [8]. The most distinctive feature of BGs is related to their ability to bind to living tissues, which is known as "bioactivity". Indeed, bioactivity refers to the formation of a hydroxycarbonated apatite (HCA) layer on the BG surface upon contact with physiological fluids [9]. BGs are generally produced through mixing different inorganic oxides that act as network-forming (*e.g.*, SiO_2, B_2O_3, and P_2O_5), network-modifying (*e.g.*, Na_2O, CaO, MgO, and K_2O), and intermediate (*e.g.*, Al_2O_3, ZnO, ZrO_2, and TiO_2) oxides. In silicate glasses, $(SiO_4)^{4-}$ tetrahedron is regarded as the structural unit see Fig. (**1**), which can be linked to a maximum of four other silica tetrahedra *via* its corner oxygens. Although silicate BGs were originally developed for managing hard tissue injuries (*e.g.*, bone), numerous innovative compositions have been later prepared to tune BG characteristics for specific clinical applications in contact with non-osseous tissues [10]. In this regard, phosphate and borate BGs show great promise thanks to their unique physico-chemical and biological properties. Borate glasses are known for their high dissolution rates and apatite-forming ability in physiological fluids, making them useful substances in both hard and soft tissue engineering. In this sort of glasses, boron trioxide (B_2O_3) is the network forming oxide, and the planar $(BO_3)^{3-}$ trigonal group forms the structural unit. Additionally, phosphate BGs show less bioactivity but higher solubility in physiological conditions as compared with silicate BGs. Similar to silicate glasses, a tetrahedral unit (PO_{43}-) is the building block of phosphate glasses; however, this orthophosphate tetrahedron can attach to a maximum of three neighboring units and form a 3D network (Fig. **1**), leading to generating a terminal double bond due to the presence of the oxygen atoms that are not shared between phosphate tetrahedra. This limits the connectivity of phosphate glasses as compared to their silicate counterparts.

Traditionally, two main synthesis routes for producing BGs are applied, *i.e.* the melt-quenching and the sol-gel approaches (Fig. **2**). These synthesis methods lead to the production of BGs with different physico-chemical (*e.g.* porosity and bioactivity) and mechanical (*e.g.* hardness and flexural strength) properties. The melt-quenched glasses are produced by mixing the stoichiometric amounts of different reagents (oxides, nitrates, sulfates, or carbonates) in the form of powder and heating them at high temperatures in an electrical furnace. The resultant homogenous melt can be poured into suitable molds and cooled for producing desired bulk pieces (*e.g.*, rods and monoliths) or quenched in cold water for producing a glass frit. In addition, the final product can be fabricated as glass fibers by drawing the viscous melt on a rotating drum. The melt-quenching route is fairly simple and suitable for the large-scale production of BGs. Nevertheless, this method suffers from some limitations, including the need for high temperatures (typically 1200-1600 °C), compositional limitations, and the lack of precise control on properties of the final product (*e.g.*, particles shape and size). On the other side, the sol-gel represents a wet chemical method for producing homogenous BGs at lower processing temperatures (typically <900 °C). In this technique, organometallic alkoxides are used as the precursors of glass network formers in the solution. For example, tetraethyl orthosilicate (TEOS) and triethyl phosphate (TEP) serve as the precursors of network formers in silicate and silicophosphate glasses. It is stated that nitrate or carbonate salts can be used as precursors of network modifiers (*e.g.*, Na_2O and CaO). The next step is the gelation process of the prepared sol, which is followed by thermal treatment (500–900°C) for the removal of sub-products (*e.g.*, nitrate ions). The BGs produced by the sol-gel process possess an inherent mesoporous structure with a pore diameter in the range of 2–50 nm [12]. In 2004, Yan *et al.* introduced a new generation of sol-gel silicate BGs, called mesoporous BGs (MBGs), based on the sol-gel chemistry and supramolecular chemistry principles [13]. These nanostructured BGs have well-ordered pores in their structure, providing the possibility of loading and delivering therapeutic molecules (*e.g.*, drugs and growth factors). Actually, MBGs are synthesized in the presence of a structure-directing agent such as cetyltrimethylammonium bromide (CTAB), F127 (EO_{106}- PO_{70}-EO_{106}), or P123 (EO_{20}-PO_{70}-EO_{20}). The mixture of BGs and structure-directing agents is subjected to an evaporation-induced self-assembly (EISA) process in order to form self-organized spherical or cylindrical micelle structures. MBGs have shown an enhanced *in vitro/in vivo* bioactivity and cytocompatibility in comparison to other types of glasses [14, 15].

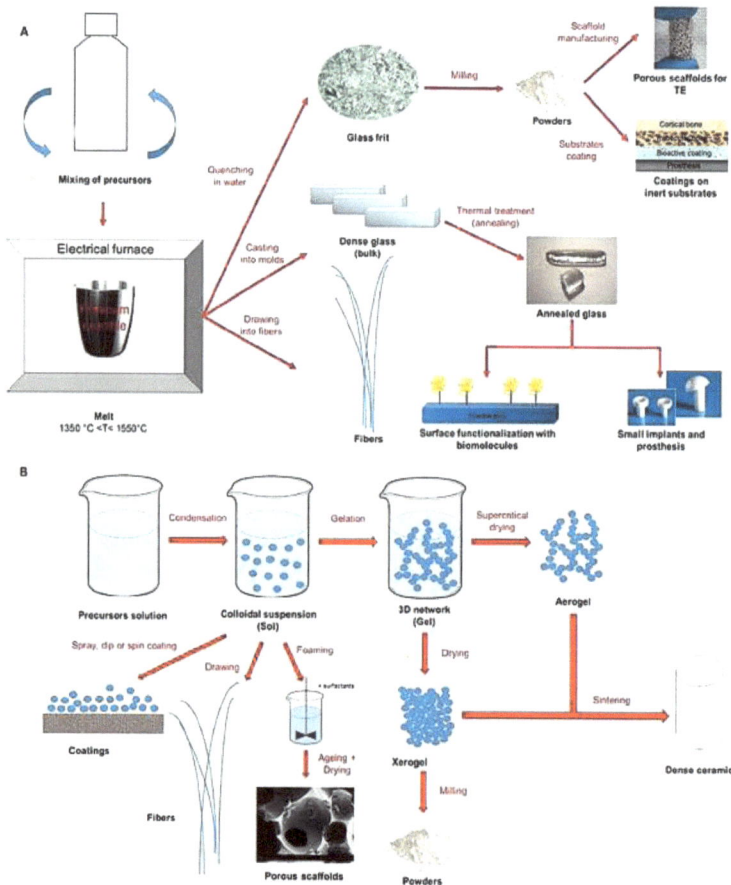

Fig. (2). Schematic illustration of (**A**) melt-quenching and (**B**) sol-gel methods utilized for synthesizing BGs. Reproduced with permission from [16].

Up to now, numerous metallic and non-metallic elements have been added as dopants to the basic composition of BGs to improve their inherent properties.

Previously, monovalent (Ag^+, Li^+, and F^-), divalent (Cu^{2+}, Co^{2+} Zn^{2+}, Mg^{2+}, and Sr^{2+}), trivalent (B^{3+}, Eu^{3+}, Ce^{3+}, and Ga^{3+}), tetravalent (Si^{4+}), and pentavalent (P^{5+} and Nb^{5+}) therapeutic ions were successfully incorporated into BGs composition. One of the most critical aspects of the doping process is to evaluate the release kinetics of ions into the surrounding biological environment. The released ions can trigger various signaling pathways in mammalian cells in favor of cell proliferation, migration, and differentiation. Furthermore, the dopants can impart specific biological performance to BGs, including antibacterial, anti-inflammatory, immunomodulatory, and pro-angiogenic activities [17]. For example, adding silver to the basic composition of BGs was reported to significantly increase their antibacterial activity against both Gram-positive and Gram-negative species. On the other hand, BGs doped with copper or cobalt showed stimulatory effects on human umbilical vein endothelial cells (HUVECs), leading to enhanced neovascularization [18, 19]. It is clear that the ions added to the BG network directly affect the biological outcome, providing extra-functionalities for clinical use ranging from bone regeneration and wound healing to cancer theranostics [20]. It should be mentioned that the dopant concentration can hinder or encourage the bioactivity of BGs. Apart from therapeutic ions, natural and synthetic pharmaceutical biomolecules (*e.g.*, phytochemicals, antibiotics, and anticancer agents) can be loaded into MBGs for delivery to desirable locations in the body [21]. The success rate of this procedure depends on some physico-chemical parameters of MBGs, including pore diameter, volume and surface area, as well as surface functionalization and charge. However, the formation of the HCA layer on the surface may interfere with drug release from MBGs into the surrounding biological environment. Also, the solvents used throughout the time of the MBG preparation may degrade the loaded biomolecules (*e.g.*, proteins) [15].

From a tissue engineering perspective, BGs in different shapes and sizes can be utilized for treating various tissue damages and injuries. For example, BG granules can be used as bone fillers for healing bone cavities caused by surgery [22, 23]. In the damaged site, BGs can integrate into the host bone and provide a suitable substrate for cell attachment (known as osteoconductivity) and induce cell differentiation towards osteogenic lineages (known as osteoinductivity). Regarding these beneficial properties, three-dimensional (3D) constructs (scaffolds) are being routinely fabricated from BGs through various fabrication techniques, including conventional methods (*e.g.*, sponge replication) and additive manufacturing approaches (*e.g.*, 3D printing). Each fabrication technique has its pros and cons in biomedical engineering strategies. It is also worth mentioning that BGs are being commonly added to polymeric matrices to generate bio-composites with improved physico-chemical, mechanical, and biological properties. For hard tissue engineering, BGs are commonly incorporated into biopolymers to enhance their mechanical strength and generate valuable substitutes for timely repair and regeneration. In addition, BGs could activate specific signaling pathways (the mitogen-activated protein kinase (MAPK or MAP kinase)) in osteoblast cells, resulting in enhanced osteogenic activity and new bone formation [24]. Recent studies have shown that mixing BGs to polymers may lead to novel remedies for soft tissue injuries (*e.g.*, skin wounds) [25, 26]. In this case, BGs are expected to add up specific biological capacities to the polymeric substrate, ranging from antibacterial activity to pro-angiogenesis potential. Given the nature of skin wound healing, biocompatible glasses with higher dissolution rates seem to be more desirable additives for rendering biological performance in a short time (nearly 28 days). Therefore, the selection of an appropriate BG formulation is a critical step before utilizing these materials in soft tissue healing strategies. More importantly, it should be taken into account that the ion release from BGs into the surrounding environment is significantly decreased due to incorporating BGs into the polymeric substrate.

BIOCOMPATIBILITY OF BGs

Silicate BGs

There is a long successful history of silicate materials (silicate nanoparticles, silicate glasses, *etc.*). Up to now, numerous types of silicate glasses have been developed that exhibit bioactive features in the physiological environment. The Hench's 45S5 Bioglass® is the first example of silicate BGs and belongs to class A bioactive material (having osteoconductive and osteo-productive effects). Over time, several other types of silicate glasses have been developed and utilized for hard tissue engineering, including 42S5, S534P, 55S4, 58S, 70S30C, 45S5F, 13-93, *etc.* [27]. Currently, several commercial products of BGs are available on the market for managing different defects of hard tissues (bone and dental), such as BonAlive® (Turku, Finland) [28]. Concerning the biocompatibility issue, numerous experimental studies have confirmed the biosafety of silicate glasses for living systems. Indeed, the basic components of silicate glasses, *i.e.*, silicon, sodium, calcium, and phosphorous, are trace elements that are naturally found in certain concentrations in the human body. However, there are some specific parameters, also related to material processing, that can affect the compatibility of BGs with living systems. For instance, heat treatment parameters were reported as an effective factor in the blood compatibility of BGs. In this sense, different shapes of BGs were produced by applying various temperatures; BGs subjected to 600°C resulted in cubical glass particles that have superior compatibility with erythrocytes as compared with their flake-like counterparts produced at 800°C [29]. The catalysts used in producing sol-gel glasses can also alter their nano-structure, crystallization, and *in vitro* behavior. As an illustration, 45S5 BG synthesized with hydrochloride acid (HCl) as a catalyst showed a lesser lysis percentage in comparison with its counterpart prepared with nitric acid (HNO_3) [30].

On the other hand, the high reactivity of BGs in water-based solutions and the burst release of alkaline ions into the surrounding biological environment can cause an undesired increase in the pH under static *in vitro* conditions [31]. However, this is not a problematic issue *in vivo* due to the presence of a large volume of fluids (*e.g.*, blood) favoring dilution. Another important factor that affects the biocompatibility of BGs is the nature of used dopants; for example, cobalt that is helpful for enhancing angiogenesis is a toxic heavy metal [22]. In this kind of situations, caution should be taken into account while calculating the dopant final concentration in the glass composition. As a rule of thumb, dopants with higher atomic radius are released into physiological fluids faster than elements with smaller atomic radii. Therefore, this item should be also considered when designing a biocompatible glass.

Up to date, a broad range of metallic and non-metallic elements were added to silicate BGs to potentiate their therapeutic impacts. Previously published studies have emphasized that the cytotoxicity of silicate glasses is directly determined by the dopant nature and concentration. For example, it has been reported that the toxicity of ytterbium-doped silicate glasses is less than gadolinium-doped counterparts at the same concentrations [32]. Emphasizing the importance of dopant concentration on toxicity, it has been reported that Cu-doped 45S5-based glasses have no toxicity against osteoblast cells at low concentrations (200 µg mL^{-1}) while they can cause a significant reduction in cell vitality at higher concentrations (1500 µg mL^{-1}) [33]. As a matter of fact, the rate of BG dissolution may greatly affect the implant's biological behavior, such as toxicity. In a research work under the supervision of Prof. Boccaccini, the dissolution behavior of boron-containing ICIE16 glasses (49.46 SiO_2, 36.27 CaO, 6.6 K_2O, 6.6 Na_2O, 1.07 P_2O_5 mol%) was investigated under static and dynamic conditions in different media [27]. The authors of this study reported that the dissolution of the ICIE16-based glasses becomes faster with increasing boron concentrations (from 0, 4, 8 mol%). The compatibility of ion-doped silicate glasses with blood compartments was also evaluated in a series of experimental studies. For example, SiO_2–CaO–P_2O_5 glasses doped with 0-2.5 mol% of SrO or/and Al_2O_3 were evaluated for their cytocompatibility and hemocompatibility [34]. This composition did not negatively affect the viability of U2-OS osteoblast-like cells during 72 h post-incubation and was well-tolerated to white blood cells (WBC) and red blood cells (RBC) without causing any significant loss of viability or hemolysis. The determinant effects of the used precursors on the biocompatibility and hemostatic properties of BGs were also investigated elsewhere [35]. It was reported that the use of different sodium precursors (sodium nitrate and sodium hydroxide) can greatly influence the mechanical and biological performance of 45S5-based glasses; better blood- and cytocompatibility with elevated protein adsorption rate were recorded when sodium hydroxide is used. In contrast, higher mechanical stability and controlled degradation rate were achieved when using sodium nitrate as the precursor of sodium.

Silicate MBGs have also been evaluated for their compatibility with the living systems. In this regard, ternary SiO_2-CaO-P_2O_5 MBGs were prepared in the presence of surfactants of CTAB, PEG, and Pluronic P123 and implanted into New Zealand White (NZW) rabbits (Oryctolagus cuniculus) [36]. The histological evaluations revealed no major necrotic or degenerative changes in the kidney along with normal hepatic parenchyma containing a proper number of hepatocytes and other supporting cells in the liver at 45 and 90 days post-implantation. The same observations, *i.e.*, lack of adverse effects, were noted in the hearts of the animals receiving the MBGs after 45 and 90 days of implantation. Incorporation of specific dopants can impart anti-inflammatory activity to BGs, which is regarded as one of the important criteria in identifying the biocompatibility of materials. In this regard, boron incorporation into binary silicate MBG nanoparticles led to the downregulation of the expression of pro-inflammatory genes in macrophages as a critical indicator of reduced inflammation [37]. It is worth mentioning that silicate glasses can show toxicity once directly incubated with cancer cells (*e.g.*, semi-malignant giant cell tumors of bone (GCTB)) [38]. However, more studies should be performed to identify the exact mechanisms behind such toxicity, especially in terms of signaling pathways involved.

Phosphate BGs

Phosphorous, in the form of phosphorous pentoxide (P_2O_5), is the network former of phosphate BGs. Phosphorous represents the second most abundant element in the human body, which is mostly located in hard tissues (bones and teeth). In addition, this mineral element is found in the extracellular fluid and soft tissues. Previously, phosphate glasses were presented as useful materials for a wide range of biomedical applications, ranging from hard tissue engineering to soft tissue healing applications (*e.g.*, muscle, ligaments, and tendon tissue engineering) [39, 40]. This type of BGs can dissolve upon contact with physiological fluids; for example, the P_2O_5–CaO–Na_2O system is degraded in aqueous-based fluids during hours to several weeks based on the glass composition. In addition to the composition, new findings have revealed the critical role of solution pH in the biodegradation of phosphate glasses; in general, lower pH can accelerate the degradation [41].

Generally, phosphate glasses are recognized as safe substances for living systems (cells, tissues, and organs). Parameters affecting the biocompatibility of phosphate glasses can be summarized into physical characteristics and chemical compositions. As an illustration, the topography and ionic dissolution products of phosphate glass fiber (PGF) were reported to influence osteoblast behaviors, including adhesion, spreading, proliferation, and osteogenic differentiation [42]. Regarding the chemical structure of phosphate glasses, it is feasible to add various elements to their basic composition for improving their properties. On this matter, elements such as Ag, Zn, Al, Cu, and Co have been incorporated into the parent glass for generating potent substitutes for hard tissue engineering approaches [43 - 46]. Moreover, it has been interesting for researchers to study the impact of a series of rare earth elements, including Y, La, Nd, Sm, and Gd, on structural and physico-chemical properties of phosphate-based glasses [47]. From a biomedical point of view, the nature of the doped elements can directly dictate the biocompatibility of phosphate glasses. On the other hand, the concentration of dopants may greatly affect the compatibility of phosphate glasses with mammalian cells as already seen in silicate-based glasses. Phosphate glasses with the composition of $48P_2O_5-(32-x)CaO-20Na_2O-xCaF_2$ ($x = 0$, 1, 2, 3 and 4 mol%) were developed for examining the influence of fluoride (F) in enhancing the material bioactivity [48]. The 3 mol% F-doped phosphate glass showed appropriate bioactivity without eliciting any toxic effect against osteoblast cells; however, 4 mol% F-doped samples caused high toxicity against the cells after 24 h of incubation. In another study, $8ZnO-22Na_2O-[24-x)CaO-46P_2O_5-xAl_2O_3$ (where $x = 0$, 2, 4, 6, 8, and 10 mol%) glass system was reported to be compatible with rat mesenchymal stem cells (rMSCs) in formulations containing up 6 mol% Al_2O_3 [49]. However, a higher amount of Al_2O_3 in this compositional glass system led to a slight decrease in cell viability because of the network-forming action of Al^{3+} ions. The biocompatibility and osteogenic potential of phosphate glasses have also been investigated in large animals. In this regard, melt-derived porous calcium phosphate glass microspheres ($40P_2O_5 -16CaO- 24MgO- (20-x)NaO- xTiO_2$ ($x=0$, 2.5 mol%)) were prepared and implanted into skeletally mature English male ewes for 13 weeks [50]. The histological evaluations confirmed the presence of some lymphocyte-like cells and foreign body multinucleated giant cells in the animals treated with glasses containing 2.5 mol% Ti, either alone or in combination with autologous bone marrow concentrate (BMC), indicating that this more durable composition might be connected to an inflammatory response. Focusing on soft tissue healing applications, phosphate glasses with a formulation of $(CuO)x-(KPO_3)(79.5-x)-(ZnO)20-(Ag_2O)0.5$ ($x = 0$, 0.5, 1, 3, 5, 10 mol %) were produced using the melt-quench method [51]. The *in vitro* and *in vivo* results have shown that 5 mol% Cu-containing phosphate glasses have no significant toxic effects on HaCaT keratinocytes and can improve angiogenesis, collagen synthesis, and re-epithelialization in full-thickness wounds as compared to non-treated wounds in rats. It is important to mention that most cytocompatibility assays on phosphate glasses have been performed on cell lines (cancer cells) instead of normal cells; therefore, there are serious concerns about the reproducibility of outcomes and interpretation for real clinical cases.

Borate and Borosilicate BGs

Boron (B) is a necessary trace element for the performance of various metabolic enzymes and its deficiency can lead to impaired growth and abnormal bone development [52]. As mentioned earlier in this chapter, B can serve as the network former of BGs and generate a specific class of biocompatible glasses. Up to date, different shapes and forms of borate and borosilicate BGs were utilized for managing bone defects and damages [53 - 56]. Diverse therapeutic elements have been added to the basic composition of borate glasses to boost their therapeutic capacity in the frame of bone tissue engineering concepts [57, 58]. For instance, Sr was successfully added to 13-93B2 glass (18SiO$_2$, 36B$_2$O$_3$, 22CaO, 6Na$_2$O, 8K$_2$O, 8MgO, 2P$_2$O$_5$; mol.%) and led to improving the proliferation of MG-63 cells after 24 and 72 h of incubation [59]. On the other hand, there is sufficient scientific evidence that supports the beneficial role of borate BGs in soft tissue healing, too (*e.g.*, skin and nerve, *etc.*) [60, 61]. In fact, this sort of synthetic biomaterials can accelerate the wound healing process through different cellular and molecular mechanisms such as stimulating neovascularization [62]. Importantly, borate glasses (*e.g.*, 1393-B3 glass) were demonstrated to effectively eradicate pre-formed biofilms originating from a variety of clinically relevant Gram-positive, Gram-negative, and fungi organisms [63]. This issue is of utmost importance in wound healing because bacterial infections can hinder the repair and regeneration of desired tissues.

Biocompatibility of borate and borosilicate BGs has been the main goal of several *in vitro* and *in vivo* experimental studies. It is generally accepted that borate BGs are compatible with living organisms and can improve the proliferation rate of mammalian cells [64]. On this matter, a series of bioactive borate glasses with the composition of xM$_2$O–20-x Na$_2$O–20CaO–60B$_2$O$_3$, (wt %) (M$_2$O= Li$_2$O or K$_2$O) were investigated for determining their biocompatibility in Sprague Dawley adult male rats [65]. Histological evaluations have revealed no sign of inflammatory cell infiltration at the implantation site (femur bone) after 6 weeks of surgery. In addition, biochemical analyses showed no significant changes in the activities of serum (alanine transaminase (ALT) and aspartate transaminase (AST)) as well as serum levels of urea and creatinine, demonstrating the lack of toxic side effects of BGs on liver and kidney functions. Still, it is strongly suggested to assess the biocompatibility of any novel formulations of bioactive borate glasses, especially in the case of doped glasses. In this regard, it has been reported that the addition of Cu at dosages higher than 0.5 mol% to 1393B3-based scaffolds can reduce the proliferation of L929 cells *in vitro* [66]. The dose-dependent toxicity of ion-doped borate BGs was also reported in other experimental studies [67, 68]. It is worth stating that the degradation of borate glasses can greatly influence the release profile of ions from their structure into physiological environments and thereby alter biological behaviors such as biocompatibility. It has been reported that the dissolution rate of borate glasses is higher than borosilicate and silicate glasses after incubation in tris(hydroxymethyl)aminomethane (TRIS) buffer and simulated body fluid (SBF), which can affect the release of dopants (*e.g.*, Cu and Zn) [69]. Different formulations and forms of borate BGs were utilized for wound healing applications with promising *in vitro* and *in vivo* outcomes [70]. The 13-93B3 borate glass is regarded as one of the most promising biocompatible glasses in managing chronic, non-healing wounds in the clinic without scarring. Regarding multiple roles of adipose-derived stem cells (ASCs) in wound repair, the interaction of 1393-B3 glass with this type of cells has been interesting for researchers [71]. The 1393-B3 glass did not reduce the ASC viability at concentrations ≤ 10 mg/mL under static cell culture conditions and also enhanced their migratory capacity [71]. Some innovative approaches have been proposed for controlling the degradation of borate BGs in physiological solutions and subsequently improving their biocompatibility. For example, borate glasses ($53B_2O3$–$20CaO$–$6Na_2O$–$5MgO$–$12K_2O$–$4SrO$; wt%) coated with nanosized HCA were demonstrated to be better substances for wound healing in rodent model as compared with non-coated counterparts [72]. In a case series study, bioactive borate glass fibers (BBGFs) were applied for managing non-healing wounds (diabetic foot ulcer (DFU) and pressure ulcer) [73]. The clinical outcomes have revealed outstanding therapeutic potential of BBGFs (MIRRAGEN Advanced Wound Matrix, ETS Wound Care, Rolla, Missouri), which led to a sharp reduction in the healing process as well as costs.

CONCLUSION AND FUTURE PROSPECTIVE

Although BGs were invented half a century ago and demonstrated a long story of clinical success, there will still be some important challenges ahead for researchers and industry in terms of both materials design and materials testing.

Recently, the incorporation of metallic dopants has been emphasized as a valuable, relatively simple and affordable strategy to potentiate the therapeutic effect of BGs in terms of both quantity of achievable properties (*i.e.* multifunctionality) and quality of therapeutic action (*e.g.* some antibacterial ions could replace the use – and abuse – of antibiotics, thus contributing to overcome the problem of bacterial resistance). However, besides the intended therapeutic benefits, the introduction of dopants in the glass composition can also elicit detrimental effects on glass biocompatibility, the prediction of which is difficult especially when multiple ions are simultaneously released. Furthermore, ion-related toxic effects may also be unknown or become evident only in the long term. Hence, we can see an important limitation of *in vitro* biological tests applied to BGs and biomaterials in general, *i.e.* the incubation times currently recommended are usually too short (commonly 7 days, in some cases up to 1 month) to reveal mid- to long-term cytotoxicity. Testing duration is often dictated by an unavoidable balance between the completeness of results and the cost of the experiment (in terms of money and facilities allocated). However, it seems more than necessary to assess the possible effects of released therapeutic ions for longer times in order to have a clearer picture of BG biocompatibility and effectiveness.

Other limitations of current *in vitro* assessment of BGs include the absence of an immune or inflammatory response and the lack of the same cascade of events resulting from *in vivo* implantation. Furthermore, *in vitro* tests are often performed in a 2D system (*e.g.* using BG flat samples like discs or slices) that does not properly mimic the 3D environment of tissues and organs.

Another important challenge is associated with comprehensively determining the biomolecular mechanisms behind the interactions between BGs and cells, which can be a highly complex task if, for example, multiple ions are released from the material which can activate mutually-interlocking signaling pathways. When BGs are intended to deliver both therapeutic ions and drugs, such as MBGs, this task becomes even more complex due to the synergistic effects between the factors involved.

We also have to consider that the range of possible usages of BGs has expanded dramatically over the last few years from orthopedic and dental fields, where BGs are used to fill and repair bone/dental defects, to a wider range of applications in contact with soft tissues (*e.g.* peripheral nerve regeneration, cardiac tissue engineering, skin repair, opthlamology); therefore, *in vitro* and *in vivo* models should be refined and "specialized" accordingly to make them more consistent with the target clinical use. This often involves rethinking the experimental procedure, starting from the form of material to use (*e.g.* pure BG fine powder, granules, bulk blocks, 3D porous scaffolds and fibers, or BG/polymer composites like pliable sheets, tubules, injectable pastes, *etc.*), as well as the involvement of specialized biologists and surgeons to perform implantation in the different anatomic districts.

Lastly, increasing attention is addressed to applying the 3Rs principle (Reduce, Refine and Replace the use of animals for *in vivo* studies), which means that a well-designed experiment helps to reduce the number of animals used in the research. Hence, scientists are recommended to collect data *via* the minimum number of animals needed, provided that reliable and statistically significant results are achieved in order to prevent the repetition of additional experiments with the use of more animals in a later stage.

REFERENCES

[1] Anderson JM. Biological responses to materials. Annu Rev Mater Res 2001; 31(1): 81-110.
[http://dx.doi.org/10.1146/annurev.matsci.31.1.81]

[2] Williams DF. On the mechanisms of biocompatibility. Biomaterials 2008; 29(20): 2941-53.
[http://dx.doi.org/10.1016/j.biomaterials.2008.04.023] [PMID: 18440630]

[3] Mozafari M. Principles of biocompatibility. In: Mozafari M, Ed. Handbook of Biomaterials Biocompatibility. Woodhead Publishing 2020; pp. 3-9.
[http://dx.doi.org/10.1016/B978-0-08-102967-1.00001-3]

[4] Wilkesmann S, Fellenberg J, Nawaz Q, *et al.* Primary osteoblasts, osteoblast precursor cells or osteoblast-like cell lines: Which human cell types are (most) suitable for characterizing 45S5-bioactive glass? J Biomed Mater Res A 2020; 108(3): 663-74.
[http://dx.doi.org/10.1002/jbm.a.36846] [PMID: 31747118]

[5] Anderson JM. 9.19 - Biocompatibility. In: Matyjaszewski K, Möller M, Eds. Polymer Science: A Comprehensive Reference. Amsterdam: Elsevier 2012; pp. 363-83.
[http://dx.doi.org/10.1016/B978-0-444-53349-4.00229-6]

[6] Pusnik M, Imeri M, Deppierraz G, Bruinink A, Zinn M. The agar diffusion scratch assay - A novel method to assess the bioactive and cytotoxic potential of new materials and compounds. Sci Rep 2016; 6(1): 20854.
[http://dx.doi.org/10.1038/srep20854] [PMID: 26861591]

[7] Hench LL, Splinter RJ, Allen WC, Greenlee TK. Bonding mechanisms at the interface of ceramic prosthetic materials. J Biomed Mater Res 1971; 5(6): 117-41.
[http://dx.doi.org/10.1002/jbm.820050611]

[8] Xynos ID, Hukkanen MVJ, Batten JJ, Buttery LD, Hench LL, Polak JM. Bioglass 45S5 stimulates osteoblast turnover and enhances bone formation *in vitro*: Implications and applications for bone tissue engineering. Calcif Tissue Int 2000; 67(4): 321-9.
[http://dx.doi.org/10.1007/s002230001134] [PMID: 11000347]

[9] Hench LL. The story of bioglass®. J Mater Sci Mater Med 2006; 17(11): 967-78.
[http://dx.doi.org/10.1007/s10856-006-0432-z] [PMID: 17122907]

[10] Baino F, Hamzehlou S, Kargozar S. Bioactive glasses: Where are we and where are we going? J Funct Biomater 2018; 9(1): 25.
[http://dx.doi.org/10.3390/jfb9010025] [PMID: 29562680]

[11] Mehrabi T, Mesgar AS, Mohammadi Z. Bioactive glasses: A promising therapeutic ion release strategy for enhancing wound healing. ACS Biomater Sci Eng 2020; 6(10): 5399-430.
[http://dx.doi.org/10.1021/acsbiomaterials.0c00528] [PMID: 33320556]

[12] Kaur G, Pickrell G, Sriranganathan N, Kumar V, Homa D. Review and the state of the art: Sol-gel and melt quenched bioactive glasses for tissue engineering. J Biomed Mater Res B Appl Biomater 2016; 104(6): 1248-75.
[http://dx.doi.org/10.1002/jbm.b.33443] [PMID: 26060931]

[13] Yan X, Yu C, Zhou X, Tang J, Zhao D. Highly ordered mesoporous bioactive glasses with superior *in vitro* bone-forming bioactivities. Angew Chem Int Ed 2004; 43(44): 5980-4.
[http://dx.doi.org/10.1002/anie.200460598] [PMID: 15547911]

[14]   Vallet-Regi M, Salinas AJ. Mesoporous bioactive glasses for regenerative medicine. Mater Today Bio 2021; 11: 100121.
[http://dx.doi.org/10.1016/j.mtbio.2021.100121] [PMID: 34377972]

[15]   Kargozar S, Baino F, Hamzehlou S, Hill RG, Mozafari M. Bioactive glasses entering the mainstream. Drug Discov Today 2018; 23(10): 1700-4.
[http://dx.doi.org/10.1016/j.drudis.2018.05.027] [PMID: 29803626]

[16]   Kargozar S, Montazerian M, Fiume E, Baino F. Multiple and promising applications of strontium (Sr)-containing bioactive glasses in bone tissue engineering. Front Bioeng Biotechnol 2019; 7: 161.
[http://dx.doi.org/10.3389/fbioe.2019.00161] [PMID: 31334228]

[17]   Zheng K, Niu W, Lei B, Boccaccini AR. Immunomodulatory bioactive glasses for tissue regeneration. Acta Biomater 2021; 133: 168-86.
[http://dx.doi.org/10.1016/j.actbio.2021.08.023] [PMID: 34418539]

[18]   Dai Q, Li Q, Gao H, *et al.* 3D printing of Cu-doped bioactive glass composite scaffolds promotes bone regeneration through activating the HIF-1α and TNF-α pathway of hUVECs. Biomater Sci 2021; 9(16): 5519-32.
[http://dx.doi.org/10.1039/D1BM00870F] [PMID: 34236062]

[19]   Kermani F, Mollazadeh Beidokhti S, Baino F, Gholamzadeh-Virany Z, Mozafari M, Kargozar S. Strontium-and cobalt-doped multicomponent mesoporous bioactive glasses (MBGs) for potential use in bone tissue engineering applications. Materials 2020; 13(6): 1348.
[http://dx.doi.org/10.3390/ma13061348] [PMID: 32188165]

[20]   Delpino GP, Borges R, Zambanini T, *et al.* Sol-gel-derived 58S bioactive glass containing holmium aiming brachytherapy applications: A dissolution, bioactivity, and cytotoxicity study. Mater Sci Eng C 2021; 119: 111595.
[http://dx.doi.org/10.1016/j.msec.2020.111595] [PMID: 33321639]

[21]   Schuhladen K, Roether JA, Boccaccini AR. Bioactive glasses meet phytotherapeutics: The potential of natural herbal medicines to extend the functionality of bioactive glasses. Biomaterials 2019; 217: 119288.
[http://dx.doi.org/10.1016/j.biomaterials.2019.119288] [PMID: 31252243]

[22]   Kargozar S, Lotfibakhshaiesh N, Ai J, *et al.* Synthesis, physico-chemical and biological characterization of strontium and cobalt substituted bioactive glasses for bone tissue engineering. J Non-Cryst Solids 2016; 449: 133-40.
[http://dx.doi.org/10.1016/j.jnoncrysol.2016.07.025]

[23]   Kargozar S, Lotfibakhshaiesh N, Ai J, *et al.* Strontium- and cobalt-substituted bioactive glasses seeded with human umbilical cord perivascular cells to promote bone regeneration *via* enhanced osteogenic and angiogenic activities. Acta Biomater 2017; 58: 502-14.
[http://dx.doi.org/10.1016/j.actbio.2017.06.021] [PMID: 28624656]

[24]   Gong W, Dong Y, Wang S, Gao X, Chen X. A novel nano-sized bioactive glass stimulates osteogenesis *via* the MAPK pathway. RSC Advances 2017; 7(23): 13760-7.
[http://dx.doi.org/10.1039/C6RA26713K]

[25]   Kargozar S, Mozafari M, Hamzehlou S, Baino F. Using bioactive glasses in the management of burns. Front Bioeng Biotechnol 2019; 7: 62.
[http://dx.doi.org/10.3389/fbioe.2019.00062] [PMID: 30984751]

[26]   Solanki AK, Lali FV, Autefage H, *et al.* Bioactive glasses and electrospun composites that release cobalt to stimulate the HIF pathway for wound healing applications. Biomater Res 2021; 25(1): 1-16.
[http://dx.doi.org/10.1186/s40824-020-00202-6] [PMID: 33451366]

[27]   Arango-Ospina M, Nawaz Q, Boccaccini AR. Silicate-based nanoceramics in regenerative medicine. Nanostructured Biomaterials for Regenerative Medicine. Elsevier 2020; pp. 255-73.
[http://dx.doi.org/10.1016/B978-0-08-102594-9.00009-7]

[28] Maçon ALB, Kim TB, Valliant EM, *et al.* A unified *in vitro* evaluation for apatite-forming ability of bioactive glasses and their variants. J Mater Sci Mater Med 2015; 26(2): 115.
[http://dx.doi.org/10.1007/s10856-015-5403-9] [PMID: 25665841]

[29] Shivalingam C, Purushothaman B, R RC, Subramanium B. Thermal treatment stimulus on erythrocyte compatibility and hemostatic behavior of one-dimensional bioactive nanostructures. J Biomed Mater Res A 2020; 108(11): 2277-90.
[http://dx.doi.org/10.1002/jbm.a.36985] [PMID: 32363715]

[30] Durgalakshmi D, Rakkesh RA, Aruna P, Ganesan S, Balakumar S. Bioactivity and hemocompatibility of sol-gel bioactive glass synthesized under different catalytic conditions. New J Chem 2020; 44(48): 21026-37.
[http://dx.doi.org/10.1039/D0NJ02445G]

[31] Hohenbild F, Arango-Ospina M, Moghaddam A, Boccaccini AR, Westhauser F. Preconditioning of bioactive glasses before introduction to static cell culture: what is really necessary? Methods Protoc 2020; 3(2): 38.
[http://dx.doi.org/10.3390/mps3020038] [PMID: 32397550]

[32] Zambanini T, Borges R, Faria PC, *et al.* Dissolution, bioactivity behavior, and cytotoxicity of rare earth-containing bioactive glasses (RE = Gd, Yb). Int J Appl Ceram Technol 2019; 16(5): 2028-39.
[http://dx.doi.org/10.1111/ijac.13317]

[33] Koohkan R, Hooshmand T, Tahriri M, Mohebbi-Kalhori D. Synthesis, characterization and *in vitro* bioactivity of mesoporous copper silicate bioactive glasses. Ceram Int 2018; 44(2): 2390-9.
[http://dx.doi.org/10.1016/j.ceramint.2017.10.208]

[34] Tripathi H, Rath C, Kumar AS, Manna PP, Singh SP. Structural, physico-mechanical and *in-vitro* bioactivity studies on SiO_2–CaO–P_2O_5–SrO–Al_2O_3 bioactive glasses. Mater Sci Eng C 2019; 94: 279-90.
[http://dx.doi.org/10.1016/j.msec.2018.09.041] [PMID: 30423710]

[35] S C, S B. Insight into the impingement of different sodium precursors on structural, biocompatible, and hemostatic properties of bioactive materials. Mater Sci Eng C 2021; 123: 111959.
[http://dx.doi.org/10.1016/j.msec.2021.111959]

[36] Anand A, Lalzawmliana V, Kumar V, *et al.* Preparation and *in vivo* biocompatibility studies of different mesoporous bioactive glasses. J Mech Behav Biomed Mater 2019; 89: 89-98.
[http://dx.doi.org/10.1016/j.jmbbm.2018.09.024] [PMID: 30267993]

[37] Zheng K, Fan Y, Torre E, *et al.* Incorporation of boron in mesoporous bioactive glass nanoparticles reduces inflammatory response and delays osteogenic differentiation. Part Part Syst Charact 2020; 37(7): 2000054.
[http://dx.doi.org/10.1002/ppsc.202000054]

[38] Westhauser F, Arango-Ospina M, Losch S, *et al.* Selective and caspase-independent cytotoxicity of bioactive glasses towards giant cell tumor of bone derived neoplastic stromal cells but not to bone marrow derived stromal cells. Biomaterials 2021; 275: 120977.
[http://dx.doi.org/10.1016/j.biomaterials.2021.120977] [PMID: 34175562]

[39] Abou Neel EA, Pickup DM, Valappil SP, Newport RJ, Knowles JC. Bioactive functional materials: A perspective on phosphate-based glasses. J Mater Chem 2009; 19(6): 690-701.
[http://dx.doi.org/10.1039/B810675D]

[40] Alshomer F, Chaves C, Serra T, Ahmed I, Kalaskar DM. Micropatterning of nanocomposite polymer scaffolds using sacrificial phosphate glass fibers for tendon tissue engineering applications. Nanomedicine 2017; 13(3): 1267-77.
[http://dx.doi.org/10.1016/j.nano.2017.01.006] [PMID: 28115252]

[41] Oosterbeek RN, Margaronis KI, Zhang XC, Best SM, Cameron RE. Non-linear dissolution mechanisms of sodium calcium phosphate glasses as a function of pH in various aqueous media. J Eur

Ceram Soc 2021; 41(1): 901-11.
[http://dx.doi.org/10.1016/j.jeurceramsoc.2020.08.076]

[42] Lin X, Chen Q, Xiao Y, *et al.* Phosphate glass fibers facilitate proliferation and osteogenesis through Runx2 transcription in murine osteoblastic cells. J Biomed Mater Res A 2020; 108(2): 316-26.
[http://dx.doi.org/10.1002/jbm.a.36818] [PMID: 31628823]

[43] Mohan Babu M, Syam Prasad P, Hima Bindu S, *et al.* Investigations on physico-mechanical and spectral studies of Zn2+ doped p2o5-based bioglass system. J Composi Sci 2020; 4(3): 129.
[http://dx.doi.org/10.3390/jcs4030129]

[44] Kyffin BA, Foroutan F, Raja FNS, *et al.* Antibacterial silver-doped phosphate-based glasses prepared by coacervation. J Mater Chem B Mater Biol Med 2019; 7(48): 7744-55.
[http://dx.doi.org/10.1039/C9TB02195G] [PMID: 31750507]

[45] Foroutan F, McGuire J, Gupta P, *et al.* Antibacterial copper-doped calcium phosphate glasses for bone tissue regeneration. ACS Biomater Sci Eng 2019; 5(11): 6054-62.
[http://dx.doi.org/10.1021/acsbiomaterials.9b01291] [PMID: 33405659]

[46] Raja FNS, Worthington T, Isaacs MA, *et al.* The antimicrobial efficacy of hypoxia mimicking cobalt oxide doped phosphate-based glasses against clinically relevant gram positive, gram negative bacteria and a fungal strain. ACS Biomater Sci Eng 2019; 5(1): 283-93.
[http://dx.doi.org/10.1021/acsbiomaterials.8b01045] [PMID: 33405859]

[47] Li H, Yi J, Qin Z, *et al.* Structures, thermal expansion, chemical stability and crystallization behavior of phosphate-based glasses by influence of rare earth. J Non-Cryst Solids 2019; 522: 119602.
[http://dx.doi.org/10.1016/j.jnoncrysol.2019.119602]

[48] Rajkumar G, Dhivya V, Mahalaxmi S, Rajkumar K, Sathishkumar GK, Karpagam R. Influence of fluoride for enhancing bioactivity onto phosphate based glasses. J Non-Cryst Solids 2018; 493: 108-18.
[http://dx.doi.org/10.1016/j.jnoncrysol.2018.04.046]

[49] Babu MM, Venkateswara Rao P, Veeraiah N, Prasad PS. Effect of Al^{3+} ions substitution in novel zinc phosphate glasses on formation of HAp layer for bone graft applications. Colloids Surf B Biointerfaces 2020; 185: 110591.
[http://dx.doi.org/10.1016/j.colsurfb.2019.110591] [PMID: 31704606]

[50] MLaren JS, Macri-Pellizzeri L, Hossain KMZ, *et al.* Porous phosphate-based glass microspheres show biocompatibility, tissue infiltration, and osteogenic onset in an ovine bone defect model. ACS Appl Mater Interfaces 2019; 11(17): 15436-46.
[http://dx.doi.org/10.1021/acsami.9b04603] [PMID: 30990301]

[51] Taulescu CA, Taulescu M, Suciu M, *et al.* A novel therapeutic phosphate-based glass improves full-thickness wound healing in a rat model. Biotechnol J 2021; 16(9): 2100031.
[http://dx.doi.org/10.1002/biot.202100031] [PMID: 34242476]

[52] Khaliq H, Juming Z, Ke-Mei P. The physiological role of boron on health. Biol Trace Elem Res 2018; 186(1): 31-51.
[http://dx.doi.org/10.1007/s12011-018-1284-3] [PMID: 29546541]

[53] Sengupta S, Michalek M, Liverani L, Švančárek P, Boccaccini AR, Galusek D. Preparation and characterization of sintered bioactive borate glass tape. Mater Lett 2021; 282: 128843.
[http://dx.doi.org/10.1016/j.matlet.2020.128843]

[54] Lepry WC, Nazhat SN. Highly bioactive sol-gel-derived borate glasses. Chem Mater 2015; 27(13): 4821-31.
[http://dx.doi.org/10.1021/acs.chemmater.5b01697]

[55] Lepry WC, Smith S, Nazhat SN. Effect of sodium on bioactive sol-gel-derived borate glasses. J Non-Cryst Solids 2018; 500: 141-8.
[http://dx.doi.org/10.1016/j.jnoncrysol.2018.07.042]

[56] Tainio JM, Salazar DAA, Nommeots-Nomm A, *et al.* Structure and *in vitro* dissolution of Mg and Sr containing borosilicate bioactive glasses for bone tissue engineering. J Non-Cryst Solids 2020; 533: 119893.
[http://dx.doi.org/10.1016/j.jnoncrysol.2020.119893]

[57] Pan HB, Zhao XL, Zhang X, *et al.* Strontium borate glass: Potential biomaterial for bone regeneration. J R Soc Interface 2010; 7(48): 1025-31.
[http://dx.doi.org/10.1098/rsif.2009.0504] [PMID: 20031984]

[58] Da Silva LCA, Neto FG, Pimentel SSC, *et al.* The role of Ag2O on antibacterial and bioactive properties of borate glasses. J Non-Cryst Solids 2021; 554: 120611.
[http://dx.doi.org/10.1016/j.jnoncrysol.2020.120611]

[59] Yin H, Yang C, Gao Y, *et al.* Fabrication and characterization of strontium-doped borate-based bioactive glass scaffolds for bone tissue engineering. J Alloys Compd 2018; 743: 564-9.
[http://dx.doi.org/10.1016/j.jallcom.2018.01.099]

[60] Balasubramanian P, Büttner T, Miguez Pacheco V, Boccaccini AR. Boron-containing bioactive glasses in bone and soft tissue engineering. J Eur Ceram Soc 2018; 38(3): 855-69.
[http://dx.doi.org/10.1016/j.jeurceramsoc.2017.11.001]

[61] Marquardt LM, Day D, Sakiyama-Elbert SE, Harkins AB. Effects of borate-based bioactive glass on neuron viability and neurite extension. J Biomed Mater Res A 2014; 102(8): 2767-75.
[http://dx.doi.org/10.1002/jbm.a.34944] [PMID: 24027222]

[62] Bi L, Rahaman MN, Day DE, *et al.* Effect of bioactive borate glass microstructure on bone regeneration, angiogenesis, and hydroxyapatite conversion in a rat calvarial defect model. Acta Biomater 2013; 9(8): 8015-26.
[http://dx.doi.org/10.1016/j.actbio.2013.04.043] [PMID: 23643606]

[63] Jung S, Day T, Boone T, Buziak B, Omar A. Anti-biofilm activity of two novel, borate based, bioactive glass wound dressings. Biomed Glass 2019; 5(1): 67-75.
[http://dx.doi.org/10.1515/bglass-2019-0006]

[64] Wei X, Xi T, Zheng Y, Zhang C, Huang W. *In vitro* comparative effect of three novel borate bioglasses on the behaviors of osteoblastic MC3T3-E1 cells. J Mater Sci Technol 2014; 30(10): 979-83.
[http://dx.doi.org/10.1016/j.jmst.2014.07.007]

[65] Omar AE, Ibrahim AM, Abd El-Aziz TH, Al-Rashidy ZM, Farag MM. Role of alkali metal oxide type on the degradation and *in vivo* biocompatibility of soda-lime-borate bioactive glass. J Biomed Mater Res B Appl Biomater 2021; 109(7): 1059-73.
[http://dx.doi.org/10.1002/jbm.b.34769] [PMID: 33274827]

[66] Ali A, Singh BN, Yadav S, *et al.* CuO assisted borate 1393B3 glass scaffold with enhanced mechanical performance and cytocompatibility: An *in vitro* study. J Mech Behav Biomed Mater 2021; 114: 104231.
[http://dx.doi.org/10.1016/j.jmbbm.2020.104231] [PMID: 33276214]

[67] Schuhladen K, Stich L, Schmidt J, Steinkasserer A, Boccaccini AR, Zinser E. Cu, Zn doped borate bioactive glasses: Antibacterial efficacy and dose-dependent *in vitro* modulation of murine dendritic cells. Biomater Sci 2020; 8(8): 2143-55.
[http://dx.doi.org/10.1039/C9BM01691K] [PMID: 32248211]

[68] Lepry WC, Griffanti G, Nazhat SN. Bioactive sol-gel borate glasses with magnesium. J Non-Cryst Solids 2022; 581: 121415.
[http://dx.doi.org/10.1016/j.jnoncrysol.2022.121415]

[69] Schuhladen K, Wang X, Hupa L, Boccaccini AR. Dissolution of borate and borosilicate bioactive glasses and the influence of ion (Zn, Cu) doping in different solutions. J Non-Cryst Solids 2018; 502: 22-34.

[http://dx.doi.org/10.1016/j.jnoncrysol.2018.08.037]

[70]   Hu H, Tang Y, Pang L, *et al.* Angiogenesis and full-thickness wound healing efficiency of a copper-doped borate bioactive glass/poly (lactic-co-glycolic acid) dressing loaded with vitamin E *in vivo* and *in vitro*. ACS Appl Mater Interfaces 2018; 10(27): 22939-50.
[http://dx.doi.org/10.1021/acsami.8b04903] [PMID: 29924595]

[71]   Thyparambil NJ, Gutgesell LC, Bromet BA, *et al.* Bioactive borate glass triggers phenotypic changes in adipose stem cells. J Mater Sci Mater Med 2020; 31(4): 35.
[http://dx.doi.org/10.1007/s10856-020-06366-w] [PMID: 32206916]

[72]   Chen R, Li Q, zhang Q, *et al.* Nanosized HCA-coated borate bioactive glass with improved wound healing effects on rodent model. Chem Eng J 2021; 426: 130299.
[http://dx.doi.org/10.1016/j.cej.2021.130299]

[73]   Buck DW II. Innovative bioactive glass fiber technology accelerates wound healing and minimizes costs: A case series. Adv Skin Wound Care 2020; 33(8): 1-6.
[http://dx.doi.org/10.1097/01.ASW.0000672504.15532.21] [PMID: 32697477]

CHAPTER 6

Bioinert Ceramics for Biomedical Applications

Amirhossein Moghanian[1,*] and **Saba Nasiripour**[2]

[1] *Department of Materials Engineering, Imam Khomeini International University, Qazvin, 34149-16818, Iran*

[2] *School of Metallurgy and Materials Engineering, Faculty of Engineering, University of Tehran, Tehran, Iran*

Abstract: Bioinert ceramics are a form of bioceramics that is characterized based on how they react biologically in the human body. Bioinert ceramics are often classified as biologically inert nature or bioinert ceramics that do not elicit a suitable reaction or interact with nearby living tissues when implanted into a biological system. In other words, exposing bioinert ceramics to the human environment will not cause any chemical interactions between the implant and the bone tissue. Bioinert ceramic materials have been used in the form of medical devices and implants to replace or re-establish the function of degenerated or traumatized organs or tissue of the human body due to their excellent chemical stability, biocompatibility, mechanical strength, corrosion restriction behavior, and wear resistance. Materials based on titanium, alumina, and zirconia are used in bioinert nanoceramics., In a biological environment, they are bioinert, fracture-tough, and have high mechanical strength. Because of their corrosion resistance, titanium and titanium-based alloys are widely used in bone tissue repair.

Keywords: Alumina, Biocompatibility, Biomaterials, Bioceramics, Bio-inert, Carbon, Coating, First-generation, Non-oxide bio-inert ceramics, Oxide bioinert ceramics, Repair, Titanium, Tissue, Implant, Zirconia.

INTRODUCTION

Biomaterials are materials that are used to examine, treat, improve, restore, or replace biological tissues or organs. Biomaterials were initially developed in the 1960s. Its goal was to have the biomaterial work as well as the replacement tissue while causing the lowest amount of toxicity to the host [1].

Any inorganic, nonmetallic solids are classified as ceramics. Ceramics are a group of materials composed of inorganic, non-metallic materials. This material cate-

---

* **Corrosponding author Amirhossein Moghanian:** Department of Materials Engineering, Imam Khomeini International University, Qazvin, 34149-16818, Iran; Tel: +989123816103; E-mail: moghanian@eng.ikiu.ac.ir

Saeid Kargozar and Francesco Baino (Eds.)

gory can be subdivided in a variety of ways [2]. Ceramic materials can be divided into two categories, traditional and advanced. These ceramics can have a crystalline or non-crystalline structure [3]. Ceramics are synthesized in different ways, the most widely used method is the synthesis method at high temperatures [3].

One of the types of ceramics is bioceramic, which is used as a biomaterial because of its good properties. It is used as one of the most biocompatible material in the body of living organisms. Its applications can be mentioned as implants in the tooth, bone tissue, and so on [4]. Fig. (**1**) shows the bioinert ceramic material category [3].

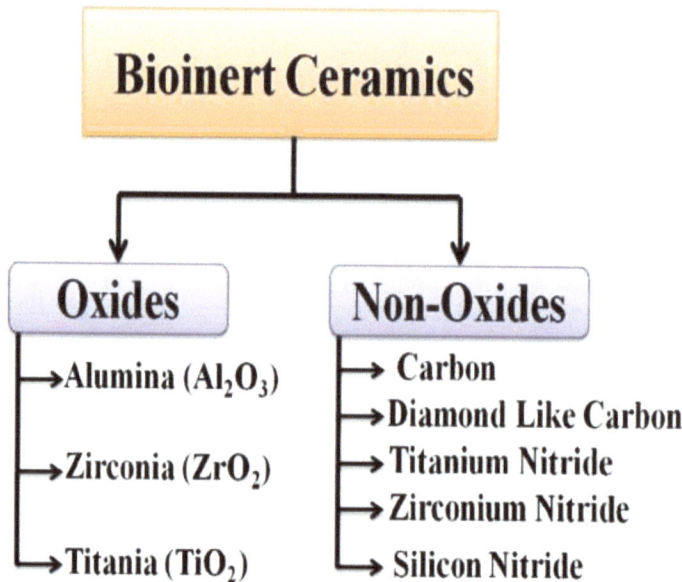

Fig. (1). Bioinert ceramic material category [3].

Brittleness in ceramics is caused by the absence of deformation tolerance in their lattice structures' covalent or ionic connections. Because of the stress concentration effect, structural defects are the preferred locations of deformation. As a result, flaws in ceramics have a significant impact on their mechanical performance. As a matter of fact, the fracture of ceramics is always initiated by the unavoidable microscopic flaws (microcracks and micropores) that result during cooling after the melt, with particular sensitivity to surface defects. Chemical bonds are broken as a result of the concentrated stresses (*i.e.* deformation) surrounding these faults, which propagate as linear fractures, which commonly run along crystal planes. However, minuscule imperfections cannot be eradicated during manufacture, and their position, whether within the material or on its surface, is random, resulting in a wide range of fracture strength in ceramic materials (scatter). The compressive strength of a material is usually ten times that of its tensile strength. Because of this, ceramics are good structural materials under compressive stresses (*e.g.*, bricks in homes, stone blocks in pyramids), but not under tensile stress (*e.g.*, flexure). To summarize, the stress concentration effect, and hence the existence of material defects, has a significant impact on the mechanical performance of ceramics. Ceramics should not be employed in situations where tensile, bending or concentrating stresses exist. They are primarily employed at compressive load-bearing places where stresses are dispersed uniformly throughout the bulk material [2].

Bioceramics have different sub-categories that can be referred to as oxide and nitride-based bioinert ceramics, bioresorbable calcium phosphate-based materials, and bioactive glasses/glass-ceramics. Ceramics (crystalline inorganic, nonmetal materials), glasses (amorphous inorganic, nonmetal materials), and glass-ceramics are the three subgroups based on crystallinity (partially crystalline inorganic, nonmetal materials) [2]. Each of which has different applications based on its interaction with the biological environment [5].

Bioinert Ceramics

Biomaterials are substances that interact with the biological environment around them. These materials exist naturally and synthetically and also have various applications in medicine, especially tissue engineering. The reactivity of bioceramics in the living body and the initial responses to them are considered a criteria for classifying bioceramics [1]. Biomaterials are divided into 4 categories bioactive, biodegradable, bioinert, and/or biotolerant based on the degree of biocompatibility they have in a living organism [4]. Bioinert ceramics are known as first-generation bioceramics and bioactive and absorbable ceramics are known as second-generation bioceramics [1].

Bioinerts are a group of biomaterials. A good feature of bioinert materials is that they are chemically stable, compatible with hard tissues, and have good mechanical strength [4]. These materials have stable physicochemical properties. When implanted into the body, the materials will not trigger a physiological reaction or produce an immune response. Hosts are capable of retaining their physiochemical and biomechanical properties. Furthermore, there is no interaction between the implant and the tissue. This leads to no adhesion between the tissues and the implant [6]. Corrosion and wear are repelled by them. There are no fractures due to their strong strength. Bone screws and bone plates, for instance, are structurally-supporting implants made out of bio-inert materials [18, 19].

Although these materials are stable in living beings, they can have a direct structural and functional relationship with the environment without any change in them. Bio-inert materials are used as femoral heads of hip implants due to their lower coefficient of friction and wear [4].

Bioinert ceramics have high chemical stability and high mechanical strength *in vivo* arthroplasty [7]. The fact remains that no material is 100% bio-inert since all materials cause some kind of reaction after implantation. In dentistry and orthopedics, zirconia and alumina are considered traditional bio-inert materials [8]. Reactivity, rather than chemical composition or crystallinity, is a superior criterion for identifying bioceramics. For example, in the field of amorphous ceramics, it is possible to generate glasses with the same chemical system that behave as bioinert, bioactive, or resorbable due to composition differences [9]. Furthermore, glasses with similar compositions may be found that behave as bioinert when melted and bioactive when manufactured *via* the sol-gel process [10]. Biomaterials are mostly bioinert, meaning they have no interaction much with the tissues around them. Metals (like titanium or titanium alloys) are synthetic polymers (like PMMA and PEEK) [1]. Moreover, most bioceramics used in tissue engineering are bio-inert and contain metal oxides such as alumina and zirconia [7]. Titanium and its alloys can also be cited as examples of bioinert materials [4].

Bioinert Ceramics Category

Ceramics that are bioinert do not react biologically with the tissues around them. The bioinert ceramics that are currently in use are the result of more than 60 years of research and improvement [11]. In a physiological setting, it has the capacity to withstand corrosion. Oxide bioinert ceramics include alumina and partly stabilized zirconia, whereas non-oxide bioinert ceramics include carbon and nitride-based ceramics [5].

Carbon-based ceramics and oxide ceramics, such as alumina and zirconia, have great hardness and wear and corrosion resistance. These bioinert and hard ceramics offer a viable alternative to the metallic materials commonly utilized in hard-bearing joint replacement components. As a result, one strategy to address the debris problem would be to integrate metal fracture toughness with the wear performance of these bioinert ceramics.

Alumina, Al_2O_3, zirconia, ZrO_2, and various types of carbon, such as low-temperature isotropic (LTI) pyrolytic carbon (PyC), glassy (vitreous) carbon, ultralow-temperature isotropic (ULTI) pyrolytic carbon (PyC), and carbon fibers, are the most extensively used virtually bioinert ceramics [9].

Main Metal-based Bioinert Ceramic

Alumina and zirconia have comparable medical uses and are discussed together. The manufacture of femoral heads is the primary use for these oxides as bioceramics [12].

Alumina

Aluminum oxide (Al_2O_3), generally called alumina, is a hard material with low friction and great wear and corrosion resistance. Since 1975, alumina (Al_2O_3) has proved to be bio-inert [13]. High hardness and resistance to abrasion characterize the alumina. Because of their excellent mechanical and biological qualities, aluminum derivatives as oxides have been a possible option for use in dentistry and orthopedic purposes [14].

The current generation of bioinert ceramics is the outcome of more than 60 years of research and development. The alpha-alumina invention went unnoticed until 1965 when Sandhaus launched the CBS dental implant system into practical usage [15]. Bioinert ceramics were used for the first time in a load-bearing medical device (Al_2O_3, alpha-alumina, corundum). Alumina was formerly employed as a component of Cerosium, a composite of alumina, aluminates, and epoxy resin developed by Lyman Smith to replace bone deformities [16]. Complex aluminates (*e.g.* SiAlON) were studied as bone replacements during these years, which are known as the "Dawn of Bioceramics [17]. The use of alumina in orthopedics and dentistry is widespread since it is highly mechanically strong, does not perform poorly under friction, and is biocompatible with the human body [18]. Because monolithic alumina is bioinert, it inhibits osseointegration, leading to implant rejection over time. Therefore, Anodized porous ceramic coatings are used to solve this problem [14]. Aluminosilicate-based bioinert glasses have been utilized or studied. Flame spheroidization was utilized to create bioinert aluminosilicates microspheres of 1030 m containing $90Y_2O_3$, which were effectively employed to treat hepatocellular carcinoma (liver cancer), with patients living for more than five years following therapy [19].

As an insoluble crystalline solid at room temperature, alumina cannot be dissolved in regular chemical reagents. Artificial femur heads have been manufactured using alumina since the 70s [20]. Since aluminum and oxygen form covalent and ionic bonds, alumina is a very stable compound. Alumina is protected from galvanic reactions and corrosion because of these strong connections. Alumina's smooth surface and high surface energy make it an excellent material for corrosion resistance, minimal wear, and low friction. In the hexagonal structure of Al_2O_3, aluminum ions reside in the interstitial regions [21].

Hard tissue engineering benefits from alumina's abrasion resistance, strength, and chemical inertness. In the case of bone marrow implanted with alumina, the circumferential tissues will not be harmed [22]. Increased density and reduced grain size can improve tensile strength. Pure alumina can be manufactured as a superior replacement for surgical metal alloys. Alumina implants have a high chance of long-term survival due to their superior mechanical properties [8].

Biocompatibility and Corrosion Resistance

The production of a thick and chemically stable thin coating of aluminum oxide or alumina (Al_2O_3) is generally recognized to be responsible for aluminum's superior corrosion resistance. In reality, Al_2O_3 has outstanding corrosion resistance as a material. Al_2O_3's chemical inertness makes it exceedingly biocompatible, with essentially little chemical influence on the body. The human body has an effective defensive mechanism that can detect and eliminate both macro and tiny foreign substances. Microscopic particles will be engulfed and digested by specialized cells. When bigger objects, such as a splinter, are present, the surrounding tissue will build a fibrous tissue barrier around the item. This separates the substance from the surface, avoiding surface release or tissue reactivity. It also inhibits subsequent biological reactions including immunological reactions, allergic reactions, or sepsis. Eventually, the splinter gets eliminated, aided by the formation of new tissue. The most typical response of tissue to an inert implant material like Al_2O_3 is the creation of a nonadherent fibrous capsule. Mechanical properties, high wear resistance and strength are other advantages of extremely dense, highly pure Al_2O_3 bulk materials. The International Organization for Standardization (ISO) specifies the physical qualities necessary for alumina. Alumina implants that meet or surpass ISO requirements have exceptional fatigue resistance and can withstand subcritical crack development and implant failure. Al_2O_3 implants must be manufactured to the highest quality assurance requirements [2].

"Bioinert" ceramic composites have mechanical and wear properties that have enabled the development of some novel medical devices in recent years, as well as more in the future. Alumina and zirconia have long been known to be biologically safe materials. Because of their strong oxidative condition, both ceramics have significant chemical resistance, particularly to alumina, which is widely used as a negative control in biocompatibility studies of potential biomaterials [23].

Medical Applications of Al_2O_3

Acetabular component loosening occurs most frequently with total hip replacement systems due to wear between the load-bearing surfaces. Heads are usually composed of Alumina, stems are made from stainless steel (Orthinox), Ti alloys, or CoCr alloys, the line is made of ultrahigh-molecular-weight polyethylene (UHMWPE), and the cup is made of Ti alloys. Several hip replacement devices now use Al_2O_3 or Co-based alloys for their heads due to advancements in ceramics and strong wrought alloys. Charnley's acetabular cup was made of polytetrafluoroethylene (PTFE) between 1958 and 1962. However, the beneficial effects of the PTFE cup were offset by the risks of revision surgery and PTFE-induced bone loss in the short term. The retrieved PTFE cup is extensively worn. Since 1962, Charnley's double-cup design has been made with UHMWPE. Hard-on-soft joints are made up of Al_2O_3 on UHMWPE and metal on UHMWPE. The key issues with a hard-on-UHMWPE system continue to be the wear of the polymer component and the related aseptic loosening. This disadvantage prompted the invention of hard-on-hard bearings. Together with newer surface polishing technologies, Al_2O_3 and CoCrMo ceramics have demonstrated improved mechanical properties, enabling hard-on-hard joint systems, which include Al_2O_3-on-Al_2O_3 and CoCrMo-on-CoCrMo.

Hard-on-hard systems, on the other hand, have their issues. Al_2O_3-on-Al_2O_3 is brittle, which results in a high fracture rate in the ceramics. Patients may encounter systemic toxicities when they use a CoCrMo-on-CoCrMo joint system. Despite the high prevalence of soft-on-hard systems, the CrCoMo-on-UHMWPE system still represents 60% of total hip replacements and the Al_2O_3-on-UHMWPE system accounts for 20%. There are mainly CoCrMo-on-CoCrMo applications in the hard-on-hard system, which account for over 20% of the remaining 20%, with only a few Al_2O_3-on-Al_2O_3 applications [2].

In addition, elbow prostheses, shoulder prostheses, wrist prostheses, finger prostheses, alveolar ridge and maxillofacial reconstruction, ossicular bone substitutes, keratoprostheses (corneal replacements), segmental bone replacements and dental implants are other applications of Al_2O_3 [2].

Zirconia

Zirconium dioxide (ZrO_2), commonly called as zirconia, is a ceramic biomaterial that is bioinert. Zirconia, which is composed of 97% zirconia oxide and 3% yttria oxide is an amorphous white powder made from zirconium [24].

Martin Heinrich Klaproth initially discovered zirconia in 1789, and it has long been utilized as a ceramic pigment [25]. The zirconia crystal structure can be classified as monoclinic (naturally occurring), cubic, or tetragonal for higher temperatures. During low temperatures, monoclinic is a stable phase, an increase in toughness is observed when tetragonal to monoclinic phases shift. By performing phase transformations, the mechanical characteristics can be lowered, which can lead to cracking. This means that implant fabrication can be based on partially stable zirconia with normal mechanical strength. Yttria stabilized zirconia is the most well-known example of partly stabilized zirconia. Stabilizing yttria helps the zirconia phase (tetragonal and cubic) remain stable. In addition to the enhancement of static and fatigue strength of zirconia implants, yttria is utilized as a support material for zirconia-based implants [26]. Surface roughness and microcracking can be caused by a single water molecule causing a phase transition from tetragonal to monoclinic. Following many years of implantation, femoral heads begin to degrade in the body [25]. The zirconia is stabilized with nonmetallic chemicals such as MgO, CaO, and Y_2O_3 It has several benefits over other ceramic materials, including phase change processes that improve toughness, as shown in components manufactured of it. Zirconia is a superior ceramic over alumina because of its excellent mechanical and wears properties [27]. It's a brittle transition metal oxide that promotes bone formation and development in the early stages. It has excellent mechanical qualities and is biocompatible. For a quarter-century, experts have been studying Zirconia as a biomaterial, and Zirconia is now being used in clinical trials for full hip replacement (THR) However, enhancements are in the works for use in other medical devices [8].

Zirconia is one of the most difficult materials to work with when creating components that can resist higher loads. When a zirconium-based alloy is oxidized at temperatures exceeding 500 °C, an oxidized layer of ZrO_2 is formed, which is known as oxidized zirconium (OxZr). This OxZr is a surface-modified layer, not a ceramic surface coating, with higher hardness and longevity characteristics. The purpose of this method was to improve the resistance, roughening, and frictional properties of brittle monolithic ceramics, as well as biocompatibility. OxiniumTM describes the transformation of metal surfaces into ceramic zirconia, generally referred to as certification. (Smith & Nephew Orthopaedics, Memphis, TN). As a result, the surface continues to exhibit the desired tribological behavior while the metallic substrate still maintains its mechanical performance. In contrast to yttria-stabilized zirconium components, the zirconium-based prosthesis is covered by a dense ceramic that is primarily monoclinic ZrO_2, which does not undergo phase transformation in an aqueous environment.

A well-known impact of the transformation of stabilized tetragonal zirconia into monoclinic zirconia is the drop in strength and surface degradation, resulting in lower wear performance. Due to grains that are columnar, staggered, nanostructural in size, and typically perpendicular to the surface, a tailored oxidation technique in terms of time and temperature generates a fine-grained layer with uniform thickness and no internal faults, demonstrating outstanding integrity. By using the diffusion method, heterogeneous layers can be coated over complex shapes, which gradually transition into the underlying metal without abrupt interphases, improving the oxide's strength and adhesion under mechanical or shear strains. OxZr-bearing surfaces for TKA have been demonstrated to significantly reduce wear against PE *in vitro* simulator experiments when compared to CoCrMo-bearing surfaces. This is due to the creation of a lubricating coating, which reduces friction in ceramic oxides. Because of its long-term superior wear behavior, this type of bearing (OxZr) is used in youthful and energetic patients. Despite no significant differences found between OxZ-femoral implants and CoCr implants in TKA clinical studies, OxZ-femoral implants do not cause any negative side effects

It is widely used for fixing partial dentures and other prosthetic devices due to its robustness and great wear resistance. Dental implants have also been coated with zirconia and are used as an implant material in the field of dental ceramics because of their strong inert features such as little contact with neighboring tissues and relatively good aesthetic properties [24]. Moreover, zirconia implants have been proven to acquire fewer bacteria *in vivo* than commercially pure Ti implants [4].

An additional ceramic known as zirconium, when used with polymeric scaffolds, enhances apatite growth. Rough topography at the nanoscale responds to bioinert ceramics to increase ALP and COL content within the early osteoblast differentiation marker within a shorter period, thereby promoting osteoblast differentiation and maturation [14].

Bioceramics constructed of zirconium oxide, also known as zirconia (ZrO_2), offer benefits over Al_2O_3 ceramics. They have increased fracture toughness, increased flexural strength, and decreased Young's modulus. Furthermore, ZrO_2 applications have three primary drawbacks: 1. In physiological fluids, strength degrades with time. 2. Wear and friction qualities that are unpleasant, 3 ZrO_2's potential radioactivity is frequently accompanied by radioactive elements with extremely long half-lives, such as Th and U. Separating these components from ZrO_2 is challenging and costly. Gamma radiation levels in generally accessible ZrO_2 bioceramics are not a serious problem. Moreover, because of short-range absorption and mutagenesis in the tissue, alpha emission from zirconium femoral head prosthesis is potentially harmful [2].

SUMMARY AND REMARKS ON AL$_2$O$_3$ AND ZRO$_2$

It is difficult to use these two ceramics in high-load bearing areas because of their poor fracture toughness, their loss of toughness over time, and their susceptibility to tensile stresses. Bioceramics based on Al_2O_3 and ZrO_2 should be used only under compressive or extremely low tensile stresses.

These vertically inert bioceramics are also limited in their use inside the body by their high elastic moduli. The modulus mismatch between bone and Al_2O_3 implants is quite significant. This problem can be addressed by composting materials. A material with a high modulus of elasticity like Al_2O_3 is capable of being an articulating surface [2].

The fabrication of alumina/zirconia composites might also increase the bioinertness of these virtually bioinert ceramics. Investigation of non-oxide bioinert ceramics like nitrides and carbides including Si_3N_4 or SiC, might also lead to breakthroughs [9].

Zirconia-Alumina

Many researchers are interested in the fracture toughness and mechanical characteristics of zirconia and alumina. The existence of oxygen vacancy in zirconia was filled by water molecules, which caused the breaking and instability of surface grains. Surface instability and cracking are no longer a problem thanks to the creation of zirconia-alumina composites. Using zirconia-alumina composites resulted in better mechanical and morphological qualities. Alumina doping aids in the stabilization of zirconia in both monoclinic and tetragonal phases [8].

Titanium

The majority of metals are biotolerant. Titanium and its alloys, on the other hand, can be bioinert in certain situations. Most biocompatible metals are employed in orthopedic applications due to their strength and fatigue resistance. They can, however, be employed in vascular and dentistry treatments as well [4].

The development of thin-film and coating technologies allowed this substance to be used in biomedicine as a bioinert, high-degree resistance ceramic that met the requirements of protective biomedical coatings. It was considered, for instance, as a ceramic coating material for titanium alloy total hip replacement implants to reduce wear debris accumulation from the soft titanium surface [19].

The mechanical features of bioinert ceramic coatings make them appropriate for load-bearing and wear-resistant clinical applications. Due to the significant differences in thermal physical characteristics between ceramic and metal, it is challenging to produce bioinert ceramic coatings on metal substrates *via* laser cladding. As a result, few studies show bioinert ceramic coatings on Ti alloys generated through laser cladding having a high wear resistance [24].

In the upcoming years, the bioinert ceramic coating generated by laser cladding with TiC nanoceramic powder might be used for load-bearing and wear-resistant surfaces of implants such as artificial joints.

Through laser cladding with preplaced TiC nanoceramic powders on the substrate surfaces, a bioinert TiC ceramic covering was produced on Ti6Al4V substrate surfaces. By employing Ti and B_4C mixed powders as precursor materials, TiC/TiB composite bioinert ceramic coatings were produced *in situ* using laser cladding, Even though it has been stated that the TiC/TiB composite bioinert ceramic exhibits better characteristics than its component phases individually (TiC or TiB), a study on single-phase ceramic coatings is still important and can be utilized as a foundation for composite ceramic coatings development.

Laser cladding of TiC ceramic nanoparticles resulted in a bioinert TiC ceramic covering. The coating was made up of TiC (rock-salt structure), 2-Ti3Al (DO19 structure), and -Ti (A3 structure) phases, with the TiC phase accounting for 63.48 wt% of the total [28].

In a study, TiC/TiB composite bioinert ceramic coatings were produced *in-situ* through laser cladding utilizing Ti and B_4C mixed powders as precursor materials to increase the wear resistance of titanium alloy. Another study employed laser cladding to create a bioinert ceramic coating on the surfaces of a biomedical titanium alloy with good wear resistance. In comparison to its constituent phases (TiC or TiB), bioinert ceramic exhibits superior characteristics. (TiB + TiC) was produced by Kim *et al.* Additionally, the laser cladding procedure was used to create TiC/TiB composite bioinert ceramic coatings *in situ* on the Ti6Al4V substrate using a mixed powder made up of Ti and B_4C.

Laser cladding of Ti and B_4C mixed powders was used to create the TiC/TiB composite bioinert ceramic coatings *in situ*. TiC, TiB, TiVC2, Ti_3Al, and -Ti phases were primarily used in the coatings. Because only a limited quantity of acicular TiB was precipitated, the phase composition and microstructure of the post-heat-treated coatings were not appreciably affected [29].

Titanium Nitride Coatings

In the medical field, titanium nitride (TiN) coatings are commonly used as an alternative to improve the wear properties of various metals used in arthroplasty due to their hardness, scratch resistance, and low friction coefficient. In addition, they are wettable with synovial fluids. In reality, the strong ceramic covering has the longest clinical applicability for complete hip and knee replacements Commercially available TiN coatings for joint replacement articulating surfaces are manufactured by Corin Medical and Endotec, Inc., and are designed to fit over UHMWPE or other surfaces. PVD techniques, such as cathodic arc deposition or magnetron sputtering, and CVD are the most common types of thin film deposition for TiN. In low-temperature PVD, there are no chemical reactions and diffusion phenomena to address the poor adhesion between the substrate and the hard layer, hence the poor adhesion between the substrate and the hard layer in modern PVD techniques. The UltraCoat®TiN coating is deposited on the articulating surfaces of Endotec, Inc.'s hip, shoulder, and knee systems with a patented PVD process. *In vitro* and *in vivo*, these TiN coatings exhibit good wear behavior, as assessed by the performance against PE and by the greater resistance to abrasive third body wear compared to CoCr, resulting in a reduction in metal ions released from the substrate. An adhesive layer of TiN is more effective on a hard CoCr substrate than on a softer Ti6Al4V substrate. The number of joint replacements recovered following the operative revision or *in vivo* data on the wear of TiN coatings is rather limited, despite the hardness of the TiN layer compared to the substrate, especially with titanium-based alloys that are softer. Delamination and coating breakage are the primary problems observed. When a load is applied to the TiN coating, the abrupt differences in hardness and elastic modulus at the coating/substrate contact produce plastic deformation in the substrate material. The deformation and fractures are too large for the TiN coating to handle. In addition to the formation of deposition flakes and polishing flaws, these materials also have coating flaws such as pinholes and embedded microparticles which reduce wear performance.

Titanium Nitride and Zirconia Coatings

This research will look at titanium nitride deposited onto Ti6Al4V and CoCr alloys, two of the most common components of THAs and TKAs.

Ti, Zn, TiO_2, MgO, ZnO, ZrO_2, TiB_2, and Al_2O_3 are bioactive and bioinert metallic and ceramic reinforcements. They have appropriate mechanical characteristics and are largely reliant on processing techniques [30].

Metallic components are often coated with nitrides such as titania nitride (TiN), titania niobium nitride (TiNbN), chromium nitride (CrN), and zirconia nitride (ZrN) to improve the hardness of the surface and increase wettability. The development of silicon nitride (Si_3N_4) ceramics for spinal surgery and joint replacements is ongoing despite coating [5].

Carbon-based Bioinert Ceramics

Carbons are a class of chemicals that fall under the category of virtually bioinert ceramics and may be synthesized in a variety of allotropic forms. Graphite, diamond, nanocrystalline glassy carbon, and PyC are examples of these materials [9].

Carbons, on the other hand, are mostly employed as coatings in applications requiring blood contact or as fibers in reinforced composites. First-generation ceramics, like metallic and polymeric biomaterials, evoke a foreign body reaction due to their almost bioinert nature. Consequently, despite their biocompatibility, organisms will respond to them through a variety of biochemical reactions involving macrophages, giant cells, cytokines, and collagen fibers.

Bioinert coatings made of carbon-based compounds are common. The ability of these coatings to improve the surface of medium- and heavy-load bearing components and worn parts is what makes them interesting by introducing numerous substantial material enhancements, such as increased wear resistance, increased protection against corrosion, and a significantly reduced frictional impact. In addition, because of the hydrophobicity of the surface, carbon-based coatings have shown little protein adherence and good blood compatibility [19]. The substrates were initially treated using silane to allow graphene oxide immobilization. Silanization is one of the most frequent bioinert substrate functionalization processes [6].

Carbon and nitride ceramics, for example, can also be generated from biomaterials. The effectiveness of carbon-based orthopedic devices was not very high. Although carbon and carbon nanotubes (CNTs) are very strong, they are also highly fractured resistant and durable. To prevent blood clotting, carbon is capable of inhibiting it at material-tissue interfaces. In addition to heart valve replacements, these are also used as a coating on the substrates. In addition to soft tissue engineering and implants, carbon nanostructures are also of interest in the field of nerve growth. For example, graphene and CNTs are utilized to fill in large nerve gaps.

The principle of reactivity, rather than the kind of bioceramic, is a good way to categorize bioceramics. For example, in the field of amorphous ceramics, glasses with slightly varying compositions might behave as bioinert, bioactive, or resorbable in the same chemical environment. It is also conceivable to discover glasses with the same composition that serves as bioinert when melted or as bioactive when made using the sol-gel process [19].

Yrolytic Carbon (PyC), Nanocrystalline Diamond, and Diamond-like Carbon (DLC) Coatings in Biomedical Applications

PyC is a turbostratic carbon that has a graphite-like structure: Graphene sheets with random layers are formed by sp2-hybridized atoms. PyC's biocompatibility and thromboresistant properties make it ideal for use in prosthetic heart valves. PyC coatings are often coated at high temperatures by the thermal breakdown ("pyrolysis") of hydrocarbon gas, such as propane, acetylene, or methane, in an inert N2 or He environment. In a fluidized bed reactor, the procedure is carried out.

In biomedical applications, PyC has the greatest potential for being deposited in a variety of structures (laminar, isotropic, columnar, or granular), with the most intriguing structure being isotropic PyC. To generate isotropic PyC coatings using this approach, precise control of the deposition conditions is required. Because PyC works effectively in heart valves, we are confident in its biocompatibility. We also generalized PyC to orthopedics. PyroCarbon has been marketed for this application (These orthopedic implants are employed in the replacement of tiny joints, including metacarpophalangeal (knuckles) or trapezium bones (wrist joints), as well as arthroplasty of the proximal interphalangeal joints. Medical implant coatings based on diamond-related materials have been studied, such as nanocrystalline diamond and DLC coatings.

CVD is a technique for developing NCD films by using a carbon-containing gas emitted from a microwave plasma and a carrier gas such as hydrogen, argon, or nitrogen at temperatures ranging from 450 to 900 °C. When extremely high nucleation densities and proper process parameters are applied, it is feasible to create very smooth, high-quality NCD films of varying thicknesses. NCD films have a variety of characteristics that make them suitable for a wide range of useful devices. Unlike conventional coatings, NCD coatings combine very low surface roughness, the remarkable qualities of a diamond, and their nature of being biologically compatible with blood, making them ideal for use in cardiovascular medical equipment and wear-resisting implants. To reduce the danger of infections, NCD can be used as a hard antibacterial coating. In an experiment with gram-negative Pseudomonas aeruginosa, NCD coatings proved to be bactericides and anti-adhesive due to their semiconducting properties: NCD surfaces react and form chemical interactions with the cell wall or membrane, which inhibits bacteria from adhering and colonizing them. Researchers demonstrated that NCD coatings can kill or repel bacteria in an experiment involving Gram-negative Pseudomonas aeruginosa, thanks to their semiconducting properties: NCD surfaces react and create chemical contacts with the cell wall or membrane, preventing bacteria from attaching or colonizing. Due to the existence of varying ratios of sp2/sp3-bonded carbon and variable quantities of hydrogen, DLC is a category of materials based on metastable amorphous carbon coatings that have a range of characteristics.

The qualities of DLC, including inertness, hydrophobicity, and smoothness, are critical for their high blood compatibility, decreasing platelet activation in contact with the blood, which might lead to thrombosis. DLC can also operate as a protective layer in human blood environs, limiting the discharge of metal ions from metallic implants, such as nickel, which is a frequent contact allergy. The methods of deposition are biocompatible and do not cause any inflammatory reactions in the bodies of animals. After seven days, DLC coatings produced by PVD on coronary stent devices promote quick and complete endothelialization with extremely little platelet activation. Two proteins, Fibrinogen and Human Serum Albumin play an important role in thrombus formation and formation prevention, respectively, and their adsorption onto DLC films revealed a link between their composition and bonded carbons with blood compatibility. Researchers have also examined the effectiveness of DLC coatings on polymer-based medical devices for providing additional stiffness and an antibacterial effect. The properties of DLC enable it to be used as a covering for a variety of implant devices, such as heart valves, cardiovascular stents, optical surgical needles, contact lenses, and medical wires.

Because of its high hardness, minimal wear and friction, excellent corrosion resistance, and smoothness, DLC can be employed as a protective coating material for articulating joint replacements.

The results of clinical and scientific investigations of the suitability of DLC coating materials for joint replacement surfaces are still inconclusive. Some studies show that DLC-coated materials are promising, while others suggest that they are inappropriate. More research into the DLC coating composition, deposition processes, and the presence of interlayers is required [19].

Because the chemical bonds of ceramic materials lack tolerance to structural deformation, their mechanical performance can be sensitively compromised by stress concentration around defects. As a general rule, ceramics cannot be used when tensile, bending or concentrated stressing occurs. They are typically used at compressive load-bearing sites where the stresses are homogeneously distributed. There are two types of motile joints: congruent and incongruent. In incongruent joints such as hip and shoulder, a ball-shaped head is on a cuplike socket, and the stress is distributed evenly. In incongruent joints such as the knee and ankle joints, contact between two incongruent hard surfaces is like a ball on a flat plate. At the contact point, there are highly concentrated (heterogeneous) stresses. The choice of materials is made for joint replacements: For congruent joints: a. CoCrMo-on-UHMWPE, b. Al_2O_3-on-UHMWPE, c. CoCrMo-on-CoCrMo, d. Al_2O_3-on-Al_2O_3. For incongruent joints: a. CoCrMo-on-UHMWPE Ceramics are completely excluded from incongruent joints. Dental porcelains must fulfill three requirements: a. Have the appearance of natural teeth b. Withstand the oral environment and c. Not excessively abrade the opposing host teeth [2].

Application of Bioinert Ceramics Materials

The qualities of inert bioceramics, such as their tribological properties, make them appropriate as protective coatings when coated onto metallic substrates, limiting metal ion release and minimizing wear debris discharge during artificial joint replacement. When employed in cardiovascular devices like prosthetic heart valves or vascular stents, the blood compatibility of carbon-based materials can reduce thrombus development [19].

Bioceramic materials for dentistry are bioinert, bioactive, and biodegradable ceramic materials. Bioinert ceramics are the first generation of biomaterials and are commonly utilized in hip and knee replacements [10].

Bioinert materials are great choices for orthopedics and dentistry due to their superior mechanical qualities, attractive appearance, strong biocompatibility, and chemical inertia [31]. It will also function as a fixed substitute for broken or diseased bone in orthopedic and dental procedures [3].

Moreover, it is used in dental implants and crowns because of remarkable qualities like high mechanical properties such as tensile, compressive, hardness, low wear, toughness, and effective anticorrosion in biological fluids [3]. In Table **1**, the applications of bioinert ceramic materials are shown [9].

Table 1. Application of bioinert ceramic materials [9].

Materials		Applications
Oxides	Alumina	Ball heads and inserts for THR bearings.
	Zirconia	Dental implants, Dental blanks for CAD/CAM.
	ZTA/ATZ	Ball heads and inserts for THR bearings, knee replacements, components for disc replacements. Hip resurfacing (experimental).

(Table 1) cont.....

Non Oxides	Carbon	Heart valves, hemocompatible coatings.
	Diamond Like Carbon	Coatings of bearing components in joint replacements (experimental).
	Titanium Nitride	Coatings in joint replacements.
	Zirconium Nitride	Coatings in knee replacements.
	Silicon Nitride	Spinal cages. Ball heads and inserts for THR bearings, components for knee replacements, dental implants (experimental).

Dental Ceramics

Dental ceramic materials are classified into two types: implant ceramics and dental porcelains, both of which are made of vitrified feldspar with metallic oxide pigments added to approximate the color of real tooth enamel [2].

Dental Implant Ceramics

Since titanium is the most often used material for dental implants, there could be some attraction in using alumina ceramics. Bioceramic dental implants can be formed like tapered cylinders, with the taper provided by a sequence of circumferential stops. The purpose of this program is to get the best possible load transmission to the implant. Zirconia and bioinert glasses are two more ceramic materials that have been utilized in dental implants. Alumina, on the other hand, is utilized far more frequently than these other ceramics [2]. Musculoskeletal applications have mainly used three types of bioinert ceramics: alumina, zirconia, and titania [23]. The nitride and carbon forms of ceramics are used as biomaterials in a wide variety of applications [17].

A bioinert ceramic can be defined as one made from oxides and used as a biomaterial. There is an agreement among surgeons that alumina's biological safety, corrosion resistance, and the low wear volume of its bearings, as well as the absence of local and systemic reactions to the wear debris, all make it an attractive material. Bioinert ceramics, including ball heads and inserts for THR bearings, have gained significant market shares, making them the "go-to" product for orthopedic applications.

Even though metal implant surfaces are often passivated to reduce undesirable host reactions. To inhibit the release of cobalt, chromium, nickel, aluminum, or vanadium from metal alloy implants, bioinert coatings have been frequently used as barriers which interact with human tissues in allergic, carcinogenic, and general harmful ways, such as metallosis. Furthermore, these coatings must have appropriate fatigue, wear, and adhesion qualities to withstand the high cycle loads experienced. Carbon-based coatings are the most widely utilized bioinert coatings because they may function as barrier diffusion layers and/or auto-lubricant coatings.

Methods for creating wear-resistant bearings include high-temperature nitrogen diffusion or ion implantation. The surface modifies a layer that develops from the surface. This ensures good conformance and adhesion, as well as a more progressive transition that avoids an abrupt mismatch. The production of carbide, nitride (TiN and Ti_2N), and oxide compounds reduces friction and increases wear resistance by orders of magnitude. Nitrogen-implantation also increases titanium's bone conductivity and corrosion resistance, particularly at low doses. By combining these two methods of layer preparation, a duplex process, combined with graded substrates that exhibit a gradual shift in hardness, has been proposed as a way to improve the coating's ability to withstand highly localized stresses. When plasma nitriding a titanium-based substrate is combined with plasma-enhanced CVD deposition of TiN, wear performance and adhesion to the substrate can be improved. In addition, TiN/Ti(C, N) multilayers and Ti(C, N) graded coatings have been deposited to increase the adhesion and functional properties of these hard coatings. Other modifications to TiN coatings have been considered, including the use of chromium nitride (CrN) and chromium carbonitride (CrCN) coatings, which also demonstrated a lower ion release from the CoCr alloy, as well as improved cohesive strength and toughness, and a lower volume of wear debris when articulating against TiN or conventional PE.

Dental Porcelains

The following three conditions must be met by dental porcelains:1 Excessive abrasion of opposing host teeth is not permitted; 2. attempt to mimic the look of real teeth, and 3. tolerate the oral environment. To support the glass matrix, all ceramic materials can include 90% crystalline components. A denture tooth is shaped by melting porcelain in metal molds using an ingredient base made up of feldspar and 15% quartz, 4% kaolin, and color oxides. The addition of kaolin $(Al_2Si_2O_5(OH)_4)$ to the mixture increases its molding properties. The teeth of dentures are used to replace all of the teeth in a full denture or to replace just one tooth in a partial denture. The inherent translucency of porcelain denture teeth mimics that of real teeth. The geometric appearance of denture teeth without the flaws and asymmetries of real teeth, which makes them appear artificial, is an aesthetic concern. Denture teeth are fragile as well, generating a distinctive clicking sound while biting. In addition to the three parameters described above, the glass-ceramics utilized in porcelain fused metal (PFM) restorations must meet two further requirements, 1. mimic the look of real teeth. 2. Tolerate the oral environment. Excessive abrasion of opposing host teeth is not permitted. The coefficients of thermal expansion are comparable to those of metals used to bind ceramics with metals. The coefficients of thermal expansion of most porcelains range between 13.0 and 14.0 10–6/°C, whereas those of metals, they range between 13.5 and 14.5 10–6/°C. Material compositions are typically based on SiO_2, Al_2O_3, Na_2O, and K_2O in glass matrix, mainly with crystalline phases, with opacifiers like TiO_2, ZrO_2, and SnO_2. In addition to fluorescent pigments (*e.g.*, CeO_2), small amounts of fluorescent dyes are also added to match the tooth structure. Natural teeth's color and translucency are created by pigments and opacifiers in combination [2].

TYPES OF JOINTS

Aside from the hip, the skeletal system comprises numerous additional types of movable joints connecting long bones (*e.g.*, knee, ankle, shoulder, elbow, and inter-digital joints) as well as static but somewhat flexible joints (*e.g.*, skull, wrist, and tooth). Joints with low mobility are rib-cage joints (congruent, partially curled articulation, primarily fibrous) and intervertebral joints (congruent but not articulated, mostly fibrous, compressive load absorption). Wear is a significant problem in every artificial joint, regardless of material, and this, together with the kind of movable joint, influences the prosthetic material selection. Depending on how closely the opposing bone surfaces fit together, mobile joints fall into two categories: congruent and incongruent. At congruent joints such as the hip and the shoulder, a ball-shaped head fits into a cup-like socket, distributing stress equally. Any powerful material, including fragile ceramics, can withstand such mechanical stresses. When two opposing hard surfaces of incongruent joints, such as the knee and ankle, come into contact, it is similar to a ball rolling on a flat surface. Highly concentrated (heterogeneous) strains exist at the contact site. Consequently, brittle ceramics are not capable of withstanding such stresses, causing metallic and polymeric materials to be favored for such junctions [2].

Methods for Reducing Bearing Surface Wear

The measures listed below can help reduce joint wear:

• To choose a decent joint system while selecting a joint system.
• To obtain exceedingly low surfaceroughness in manufacturing.
• It is critical that the alumina ball and socket have avery smooth, polished surface with a roundness deviation of 0.1 to 1 m.
• Insurgery, lubricants should be used promptly.

Because of alumina's high surfaceenergy, biological molecules adsorb quickly and strongly. This layer of adsorbedmolecules acts as a lubricant by providing a liquid-like coating that inhibits thedirect contact between the articulating solid surfaces [2].

CONCLUSION

There are four main categories of biomaterials: bioactive, biodegradable, bioinert, and/or biotolerant, according to their abilities to be biocompatible with living organisms. The first generation of materials are bioinert, a characteristic aimed to ensure they are not harmful to human tissue [2].

Bioinert Ceramic composites exhibit mechanical and wear behavior that has enabled the development of several modern medical devices in recent years, as well as more in the future [23].

Bioinert ceramics mainly include alumina (Al_2O_3) and zirconia (ZrO_2) ceramics, which are also entitled as biotolerant materials that do not encourage any interfacial bonding amongst bone and implants. Bioceramic composites are developed for orthopedic applications [2].

As an alternative, titanium and its alloys can be bioinert when used in certain conditions. A biocompatible metal is commonly used in orthopedics because of its strength and fatigue resistance.

There are a variety of shapes and sizes available in nearly bioinert carbon ceramics. Materials such as graphite, diamond, nanocrystalline glassy carbon, and PyC are examples of these.

The tribological features of inert bioceramics make them suitable as protective coatings, making them excellent alternatives for orthopedics and dentistry.

REFERENCES

[1]   Qu H, Fu H, Han Z, Sun Y. Biomaterials for bone tissue engineering scaffolds: a review. RSC Advances 2019; 9(45): 26252-62.
[http://dx.doi.org/10.1039/C9RA05214C] [PMID: 35531040]

[2]   Qizhi Chen GT. Biomaterials: A Basic Introduction 2014.

[3]   Udduttula A, Zhang JV, Ren P-G. Bioinert ceramics for biomedical applications. Biomedical Sci. and Tech. Series 2019.

[4]   Ødegaard KS, Torgersen J, Elverum CWJM. Structural and biomedical properties of common additively manufactured biomaterials: A concise review 2020; 10(12): 1677.
[http://dx.doi.org/10.3390/met10121677]

[5]   Punj S, Singh J, Singh KJCI. Ceramic biomaterials: Properties, state of the art and future prospectives. 2021; 47(20): 28059-74.

[6]   Desante G, Labude N, Rütten S. *et al.* Graphene oxide nanofilm to functionalize bioinert high strength ceramics. 2021; 566: 150670.
[http://dx.doi.org/10.1016/j.apsusc.2021.150670]

[7]   Marin E, Boschetto F, Zanocco M, *et al.* Biological responses to silicon and nitrogen-rich PVD silicon nitride coatings 2021; 19: 100404.
        [http://dx.doi.org/10.1016/j.mtchem.2020.100404]

[8]   Kumar P, Dehiya BS, Sindhu AJIJAER. Bioceramics for hard tissue engineering applications. RE:view 2018; 13(5): 2744-52.

[9]   Vallet-Regí M, Salinas AJ. Ceramics as bone repair materials Bone repair biomaterials. Elsevier 2019; pp. 141-78.
        [http://dx.doi.org/10.1016/B978-0-08-102451-5.00006-8]

[10]   Sanz J, Rodríguez-Lozano F, Llena C, Sauro S, Forner L. Bioactivity of bioceramic materials used in the dentin-pulp complex therapy. Materials 2019; 12(7): 1015.
        [http://dx.doi.org/10.3390/ma12071015] [PMID: 30934746]

[11]   Kaur M, Singh KJMS. Review on titanium and titanium based alloys as biomaterials for orthopaedic applications. 2019; 102: 844-62.

[12]   Pattnaik S, Nethala S, Tripathi A, Saravanan S, Moorthi A. Selvamurugan NJIjobm. Chitosan scaffolds containing silicon dioxide and zirconia nano particles for bone tissue engineering 2011; 49(5): 1167-72.

[13]   Ni S, Li C, Ni S, Chen T. Webster TJJIJoN. Understanding improved osteoblast behavior on select nanoporous anodic alumina 2014; 9: 3325.

[14]   Sethu SN, Namashivayam S, Devendran S. *et al.* Nanoceramics on osteoblast proliferation and differentiation in bone tissue engineering. 2017; 98: 67-74.
        [http://dx.doi.org/10.1016/j.ijbiomac.2017.01.089]

[15]   Sandhaus S. Nouveaux aspects de l'implantologie: Sandhaus. 1969.

[16]   Smith L. Peltier lfjco, research® R. Ceramic–Plastic Material as a Bone Substitute 1992; 282: 4-9.

[17]   Piconi C, Ed. Bioinert ceramics: State-of-the-art. Publ T, Ed. Key Engineering Materials 2017.

[18]   Walpole AR, Xia Z, Wilson CW, Triffitt JT. Wilshaw PRJJoBMRPAAOJoTSfB, The Japanese Society for Biomaterials. Biomaterials TASf, et al A novel nano-porous alumina biomaterial with potential for loading with bioactive materials 2009; 90(1): 46-54.

[19]   Vallet-Regi M. Bio-ceramics with clinical applications. John Wiley & Sons 2014.
        [http://dx.doi.org/10.1002/9781118406748]

[20]   Miller J, Talton J. Total hip replacement: metal-on-metal systems. 1996; 41-56.

[21]   Hong Z, Reis RL. Mano JFJJoBMRPAAOJoTSfB, The Japanese Society for Biomaterials,, Biomaterials TASf. Biomaterials tKSf. Preparation and *in vitro* characterization of novel bioactive glass ceramic nanoparticles. 2009; 88(2): 304-13.

[22]   Al-Khateeb KA, Mustafa AA. Faris bin Ismail A, Sutjipto AGE, editors. Use of porous alumina bioceramic to increase implant osseointegration to surrounding bone. Advanced Materials Research. Trans Tech Publ. 2012.

[23]   Porporati CPaAA. Bioinert Ceramics: Zirconia and Alumina Springer International Publishing Switzerland, Handbook of Bioceramics and Biocomposites. 2016.

[24]   Sharanraj V, Ramesha C, Kavya K. *et al.* Zirconia: as a biocompatible biomaterial used in dental implants. 2021; 120(2): 63-8.
        [http://dx.doi.org/10.1080/17436753.2020.1865094]

[25]   Piconi C, Maccauro GJB. Zirconia as a ceramic biomaterial 1999; (261): 171-85.
        [http://dx.doi.org/10.1016/S0142-9612(98)00010-6]

[26]   Nasser S, Campbell PA, Kilgus D, Kossovsky N. Amstutz HCJCo, research r. Cementless total joint arthroplasty prostheses with titanium-alloy articular surfaces A human retrieval analysis 1990; (261):

171-85.

[27] Ingham E, Green TR, Stone MH, Kowalski R, Watkins N, Fisher JJB. Production of TNF-α and bone resorbing activity by macrophages in response to different types of bone cement particles 2000; 21(10): 1005-13.
[http://dx.doi.org/10.1016/S0142-9612(99)00261-6]

[28] Chen T, Deng Z, Liu D, Zhu X, Xiong YJS, Technology C. Bioinert TiC ceramic coating prepared by laser cladding: Microstructures, wear resistance. and cytocompatibility of the coating 2021; 423: 127635.

[29] Chen T, Li W, Liu D, Xiong Y, Zhu XJCI. Effects of heat treatment on microstructure and mechanical properties of TiC/TiB composite bioinert ceramic coatings *in-situ* synthesized by laser cladding on Ti6Al4V. 2021; 47(1): 755-68.

[30] Ali M, Hussein M. Compounds. magnesium-based composites and alloys for medical applications: A review of mechanical and corrosion properties. 2019; 792: 1162-90.

[31] Engelmann T, Desante G, Labude N. *et al.* Coatings based on organic/non-organic composites on bioinert ceramics by using biomimetic co-precipitation. 2019; 2(2): 260-70.

CHAPTER 7

Bioresorbable Ceramics: Processing and Properties

Amirhossein Moghanian[1,*], **Saba Nasiripour**[2] and **Niloofar Kolivand**[1]

[1] *Department of Materials Engineering, Imam Khomeini International University, Qazvin, 34149-16818, Iran*

[2] *School of Metallurgy and Materials Engineering, Faculty of Engineering, University of Tehran, Tehran, Iran*

Abstract: In synthetic ceramic materials, the types of interactions that occur in the physiological environment during body implants and tissues are defined as bioinert, bioactive, and bioresorbable. Bioresorbable materials, whether polymers, ceramics, or composite-based systems, are widely used in a variety of biomedical applications. Designing a bioresorbable device requires careful consideration of an accurate way of forecasting the biosorption of this class of materials. Bioresorbable ceramics possess the ability to undergo *in vivo* absorption and consequent replacement by the newly formed bone. They have a bonding pattern that is similar to bioactive ceramics. However, the fact that bioresorbable ceramics frequently fail to make solid contact with bone limits their potential medical uses. Bioactive and bioresorbable ceramics have a narrower application range than bioinert ceramics.

Keywords: Absorb, Bioresorbable, Bone, Bioresorbable implant, Ceramics, CaP, Degradation, DCPD, Host response, Hydroxyapatite, Inflammatory response, OCP, Resorption process, TCP, Tissue engineering.

INTRODUCTION

The ability of a bioceramic to form a bond with living tissue following transplantation is used to categorize the material, According to the statement, bioinert ceramics such as alumina and zirconia exhibit no interaction with the adjacent tissue post-implantation; (a) The material in question exhibits favorable fracture toughness, as well as resistance to corrosion and wear. (b) Bioactive ceramics (*e.g.*, bioglassesand glass-ceramics) form direct bonds with living tissues, following the pattern of bonding osteogenesis. (c) Bioresorbable ceramics

---

[*] **Corresponding author Amirhossein Moghanian:** Department of Materials Engineering, Imam Khomeini International University, Qazvin, 34149-16818, Iran; E-mail: moghanian@eng.ikiu.ac.ir

Saeid Kargozar and Francesco Baino (Eds.)

(*e.g.*, calcium phosphates (CaPs), calcium phosphate cements (CPCs), calcium carbonates, and calcium silicates) undergo gradual absorption within the living organism and are ultimately substituted by osseous tissue [1].

Many materials, including ceramics, polymers, and metals, have recently been studied for bone repair and substitution. The high proportion of inorganic apatite (70%) and organic collagen (30%) in bone makes ceramics a popular choice for bone repair. In terms of how they react in the body, ceramics used to repair or replace bones are classified into three categories. Biologically inert ceramics generate a thin, non-adherent fibrous layer where they come into contact with the bone. Implanted artificial materials often develop immunoreactions, leading to fibrous tissue enclosing them and isolating them from the surrounding bone. The second category pertains to bioactive ceramics that possess the ability to adhere immediately to the bone. Bioresorbable ceramic is the third kind. The bioresorbable ceramic progressively dissolves over time, eventually being replaced by natural bone [2]. Fig. (**1**) illustrates the category of bioceramics.

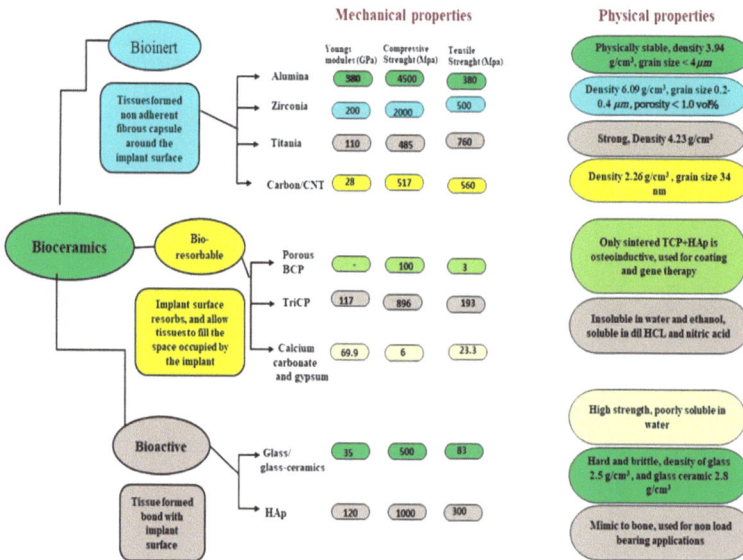

Fig. (1). Bioceramics category [2].

Biodegradable Implants

Biodegradable implants, as compared to their nondegradable counterparts, lead to a more patient-friendly treatment. The process of implant degradation facilitates the regeneration of tissue within the implanted site and does not impede radiological imaging in the absence of subsequent removal surgery.

Biodegradable materials are accessible in a variety of new forms, including biodegradable polymers, injectable *in situ* forming implants (ISFIs), bioresorbable ceramics, and biodegradable metal alloys. Different mechanical properties may be achieved, making it possible to tailor the implant to a specific use. The structural chains of biodegradable materials (often polymers or metal alloys) break down into smaller bits, then macrophages phagocytose the particles, and finally, the substance is dissolved chemically. Other types of biodegradable materials include bioceramics and metal alloys, which are made up of resorbable ingredients [3]. Their use is intended to enhance the performance of compromised biological frameworks like those seen in orthopedics and dentistry [4 - 6]. The following requirements should be met by the ideal biodegradable implant: (i). When viewed physiologically, it is biodegradable without harming the body. The pace of implant disintegration and the generation of debris particles should not be faster than the tissue's tolerance. (ii). The biocompatible implant surface is expected to facilitate favorable cellular proliferation in the adjacent tissue. (iii). Similar rates of implant degradation and healing of tissues are desirable. On the pace at which implants deteriorate, several variables have an impact, including implant shape, interaction with bodily fluids, implant position within the body, temperature, motion, component molecular weight, crystallinity, and formulation. To maintain a consistent drug release profile, the breakdown rate of biodegradable drug-loaded implants must stay constant [7].

Bioresorbable ceramics are special because they may degrade over time and be replaced by healthy tissue [8]. Ceramics that are both bioactive and resorbable find widespread usage in the field of bone repair, especially in the production of implants that form strong bonds with the bone (*e.g.*, in skull restorations after operations or trauma), tooth-root implants, biological tooth fillings, and the cure of periodontal disease (tissue around teeth).

They are also used in maxillofacial reconstruction, grafting and stabilizing skull bones, joint reconstruction, endoprosthesis of hearing aids, cosmetic eye prostheses, and so on. Resorbable ceramics can also be utilized to restore tendons, ligaments, tiny blood vessels, and nerve fibers [9, 10].

TCP ($Ca_3(PO_4)_2$), a kind of tricalcium phosphate [11] and calcite ($CaCO_3$) [12] are examples of bioresorbable ceramics. Because bone is a kind of calcium phosphate, different calcium phosphates are commonly employed in the production of bone replacements [11]. Typical monohydrate (MCPM) is the most acidic of the calcium phosphates, exhibiting exceptional solubility in an aqueous solution. Consequently, although MCPM cannot be employed in isolation, it can serve as a fundamental constituent in calcium phosphate cement when combined with -TCP. Calcium phosphate cement also employs dicalcium phosphate dihydrate (DCPD) and dicalcium phosphate anhydrous (DCPA) as raw materials [12]. Octacalcium phosphate (OCP) is a recognized HA precursor in teeth and bones. Despite exhibiting osteoconductivity, HA-sintered ceramics demonstrate limited bioresorbability, resulting in the retention of HA within the body for an extended duration following implantation [13].

It is also important for the initiation of *de novo* bone formation to have a stable surface, though a soluble phase is more conducive to osteoinduction [14]. Bioresorbable ceramics, such as calcium phosphate ceramics and silica-based glasses, are employed to facilitate, substitute, or stimulate the proliferation of healthy cellular tissue. Among the orthopedic applications for bioceramics is the treatment of significant bone thinning. The most widely explored bioresorbable ceramic materials are hydroxyapatite, dicalcium phosphate, dehydrate, and tricalcium phosphate [15, 16]. The tensile strength of bioceramics is low, and their compressibility is high. Many hard tissue degenerations develop as a result of conditions needing medication therapy, such as malignancies, infections, and osteoporotic fractures. There is an increasing interest in integrating pharmaceuticals onto or within the fabricated surfaces of bioceramics [17, 18]. In orthopedics and dentistry, bioceramic implants containing drug-loaded materials can be used to provide structural support as well as localized drug release. Bioceramics can attain greater structural stiffness and bioactivity as compared to polymer-based solutions. Thus, it is the better implant choice for use in hard tissues [14].

When it comes to the *in vivo* behavior of ceramic bone replacements, there are three critical factors to consider. The solubility of a molecule is an important initial characteristic since it determines whether or not it can be eliminated from the body. It is also important to note that the dissolution kinetics of the specific ceramic are linked to how fast it is eliminated from the body. Experiments indicate that even if a substance is soluble, it does not necessarily dissolve. In addition, the substance tends to disperse and precipitate into other substances. Apatites are the most stable phases of calcium phosphates, for instance, when subjected to physiological conditions. Furthermore, HA is a potential material for the creation of ceramic bioresorbable scaffolds to encourage new bone ingrowth [19]. It is also being utilized as an additive to create 3D-printed composite materials with good biocompatibility [20].

There exist diverse types of synthetic hydroxyapatite (HA) materials, ranging from highly dispersed powders to nonporous ceramics. Although these materials fail to fully emulate the crystal structure of natural HA found in bone, they are subject to resorption by the organism at varying rates, which are contingent upon factors such as their structure, chemical composition, and specific surface area

[21]); As a traditional resorbable bone replacement, -TCP is commonly employed. However, when quick reusability is required for TE, -TCP is preferred as a bone replacement and as a scaffold. The behavior of TCP ceramics within the human body is impacted by both their phase composition and porosity. Bioactive and resorbable ceramic porous granules and powders are often utilized to fill bone deficiencies [22, 23].

The most common components of bioresorbable ceramics are amorphous calcium phosphates, calcium sulfate, hydroxyapatite (HAP), and /-tricalcium phosphates. Currently, the aforementioned materials are predominantly utilized to restore and revitalize rigid tissues, specifically those of the dental and skeletal systems. Furthermore, it is possible to create thin coatings that can be applied to metallic implants. Several techniques, such as wet chemical precipitation, have been used in the production of these compounds [24]. Bioresorbable ceramic nanoparticles (*e.g.*, HAP and β-tricalcium phosphate) have recently been created using green synthesis to minimize or eliminate hazardous components [25]. Moreover, a stable surface is required for the initiation of *de novo* bone formation, despite a more soluble phase being favorable for osteoinduction [26].

Resorption Process

Endocytosis is a process by which cells absorb macromolecules and particle substances, including decaying polymeric material. The cell's plasma membrane tightens around the particle, eventually pinching it off to form a vesicle within the cell cytoplasm. Phagocytosis refers to the mechanism of endocytosis that involves the ingestion of relatively larger particulate matter, typically exceeding 250 nanometers in size. In the body, there are two types of specialized phagocytic cells: macrophages and neutrophils. The major role of macrophages and neutrophils is to devour injured cells and cell debris, as well as invading microbes. Antibodies that adhere to and induce phagocytosis can recognize exogenous particles originating from a biodegradable implant. A phagocytic vesicle finally unites with lysosomes to trigger particle degradation *via* a range of oxidative compounds and hydrolytic enzymes that speed up the material's destruction.

Macrophages can combine to generate multinucleated foreign body giant cells (FBGCs) and osteoclasts (in bone) to engulf and degrade bigger material pieces. Cells can internally distribute enzymatic proteins to biodegrade a degradable implant. They can also do so extracellularly. However, phagocytosis is more successful for enzymatic transfer and its effect on degradable substrates. Enzymes are relatively large in comparison to the gaps frequently observed among polymer chains. As a result, some amount of hydrolytic activity must first occur to produce bigger holes and fissures that enhance the implant's accessible surface area. These gaps in the bulk structure subsequently allow enzymes to enter and initiate enzymatic degradation of the substance. Researchers explored the effects of modifying surface roughness to increase enzyme activity in the process of degradation [27].

The intended medical purpose of an absorbable implant serves as its design purpose. Absorbable and/or degradable materials have a wide range of uses, including temporary scaffolds, temporary barriers, medication delivery systems, and multifunctional devices. An absorbable scaffold is used to give mechanical support that gradually shifts weight to the natural tissue that is recovering. An absorbable suture can be considered a scaffold as it initially serves to close a wound, but ultimately undergoes absorption as the skin undergoes recovery. Scar tissue between the abdominal wall and important internal organs is frequently formed during hernia treatment. A barrier that is capable of degrading is well-suited to impede the occurrence of unwanted tissue adhesion until the process of recuperation is fully realized. Absorbable drug delivery devices enable precise regulation of drug release within the specific vicinity of the intended tissue, eliminating the need for a follow-up procedure to remove the carrier implant. Some implants can necessitate surgery.

The integration of a device that possesses multiple functionalities. An absorbable orthopedic implant may be required to offer primary mechanical reinforcement while concurrently releasing therapeutic agents or growth factors into the adjacent osseous tissue. Whatever the primary purpose, the implant should have enough initial strength as well as a recorded, regulated rate of strength decrease. During the degradation of an implant, two forms of erosion happen. Surface erosion is a phenomenon that results in the reduction of an implant's size due to the gradual loss of its surface material. This process can be compared to the melting of an ice cube. Bulk deterioration is the second form of erosion. If the deterioration happens throughout the implant, this typically results in a surface-center disparity. The damage to the biomaterial surface may result in the production of a large number of acidic compounds, causing local tissue necrosis before the acidic compounds are removed from the site. When compared to bulk erosion, surface erosion is more predictable. Modeling the bulk erosion process and the resulting change in implant strength is a significant challenge.

The implant's chemical makeup should allow for total dissolution as an endpoint. The degradation of by-products and the implant itself during implant removal must adhere to a low toxicity level to avoid harm to the surrounding cells and tissue. To be manufactured in the required sizes and shapes, biodegradable materials should be malleable, as well as sterile. Chemical composition, shape, molecular weight, crystallinity, and molecular weight all impact the degradation rate of materials. Consequently, it is crucial to comprehend the intrinsic rate of deterioration of all utilized materials, along with any consequent decline in mechanical potency over time. This necessitates a particular emphasis on any alterations in physicochemical and/or mechanical characteristics that may arise due to processing, sterilization, or implantation impacts [27].

Mechanisms of In vivo Biodegradation

Bioresorption mechanisms differ depending on the type of material used. It is generally considered a dissolution process for ceramic materials, such as calcium phosphate. When dealing with bioresorbable materials, it is imperative to take into account the tissue response, wherein the degradation products (or bioactive additives) have an impact on the process of tissue repair. Biological responses to degradation products need to be predicted based on changes in mechanical properties over time. Scientists working in biomaterials face unique challenges, as materials may stay in the body for weeks, months, or even years at a time [27].

Hydrolysis and oxidation are two major chemical pathways used by the host to destroy biomaterials/foreign compounds. The principal chemical process by which the host achieves biodegradation of biomaterials is hydrolysis, which refers to the breaking of a chemical bond by the use of water. Free radicals can also break down covalent bonds, a process known as oxidative scission. It is unclear if enzymes have a role in catalyzing the chemical process responsible for the severing of covalent bonds. How well a biomaterial functions and where it is implanted in the body are two key factors. As a consequence of wear and water solubilization, biodegradation is affected by hydrolysis and oxidation. Over time, the biomaterial that was implanted shrinks until it disappears entirely [28].

Ceramic materials exhibit heightened susceptibility to pH fluctuations, with -TCP and hydroxyapatite ceramics being subject to degradation in acidic conditions [14]. The rate of deterioration of bioresorbable materials has been a prominent issue in the previous decade. The indent aims to identify the cells accountable *in vivo* for the disintegration of the inserted ceramic. The degradation of a ceramic implanted in a bone can be regulated through two mechanisms. Firstly, the material can be destroyed by phagocytosis and low-grade resorption, facilitated by a combination of inflammatory multinucleated giant cells and macrophages. Secondly, the bone tissue can be removed through resorption by osteoclasts [15]. Considerable debate has arisen regarding the classification of multinucleated large cells involved in the degradation of biomaterials as osteoclasts, given that they do not exhibit all of the characteristic features of osteoclasts. There is growing evidence that osteoclasts may resorb calcium phosphate (CaP) ceramics *in vitro* and *in vivo* [18-20]. Cultivated osteoclasts from human bone exhibit characteristic ultrastructural traits, including a polarized dome shape, a clear zone, and a ruffled border, that are unique to osteoclasts [19, 20]. The alteration of the morphology and compactness of calcium phosphate (CaP) crystals located beneath the ruffled border may involve the presence of an acidic microenvironment. Furthermore, it has been observed that osteoclasts are capable of degrading ceramics through the use of both resorption and phagocytic mechanisms. The progressive degradation of the ceramic after implantation is affected by several environmental factors, including physicochemical processes (dissolution-precipitation) and the activity of various cell types. Various cells, such as fibroblasts, osteoblasts, or monocytes and macrophages, can degrade ceramics, or an acidic mechanism employed by a proton pump can change the pH of the microenvironment and resorb the synthetic material [28].

Bioresorbable ceramics are employed in a variety of medical sectors as an alternative to metallic and polymeric biomaterials. Because biodegradable metals and polymeric materials create local inflammation in the human body by generating acidic byproducts [29]. Once implanted in a person, a bioresorbable material degrades and is ultimately replaced by growing tissues (bones). The most popular bioresorbable ceramics are calcium sulfate, calcium phosphates (CaP) and their salts, and porous HAP. These ceramics are frequently used to heal cracked bones without inducing inflammation [30]. Because of their tight association, bioactive and biodegradable compounds, such as CaP and HAp, are frequently researched together. Many writers consider both to be second-generation bioceramics, as opposed to first-generation bioinert and third-generation scaffolds utilized in bone tissue creation [2].

Bioresorbable Ceramic Types

Numerous improvements have been made in the handling and application of calcium phosphate (CaP) cement [1, 2]. Dicalcium phosphate dihydrate (DCPD; $CaHPO_42H_2O$), dicalcium phosphate (DCP; $CaHPO_4$), and octocalcium phosphate (OCP; $Ca_8H_2(PO_4)_65H_2O$) are all examples of low-temperature CaP bone replacements whose production has been improved with a wide range of precipitated CaP (PHA) [31].

The low-temperature CaPs hold great significance for *in vivo* application as opposed to the commonly implanted CaPs such as -tricalcium phosphate (-TCP; -$Ca_3(PO_4)_2$) and sintered hydroxyapatite (SHA) due to their presence *in vivo* [32]. Studies have also indicated the efficacy of DCPD [33], amorphous calcium phosphate (= ACP), DCP, OCP, and PHA [34]. Various studies have shown the significance of tracing this progression in parallel. Mineralization is regulated by ionized metals such as magnesium, silicon, and strontium [35, 36].

In vivo, investigations using bioresorbable ceramics have revealed significant variances in their *in vivo* characteristics [37]. As an example, calcium sulfate dihydrate (CSD; $CaSO_4 2H_2O$) dissolves rapidly from an implantation defect, whereas SHA will probably take decades to dissolve.

Alternatively, following *in vivo* implantation, the conversion of specific CaPs into a less soluble phase, such as apatite, results in a considerable reduction in their elimination rate. These distinctions are due in part to geometrical and chemical considerations.

As a result, any non-apatitic CaPs compound placed in the bone will eventually change into apatite. These three elements, namely solubility, dissolution, and conversion, are inextricably linked yet remain distinct. A given bioresorbable ceramic, such as -tricalcium phosphate (-TCP; $-Ca3(PO_4)_2$, may be soluble in the body but may take years to remove owing to *in vivo* conversion, such as into apatite.

OCP regulates bone solubility in living organisms, and its presence governs the body's CaP concentrations by simple physicochemical equilibrium.

Determining the solubility of a ceramic is necessary for classifying it as either more or less soluble than OCP. According to existing literature, biosoluble ceramics exhibit greater solubility in comparison to OCP, while bioresorbable ceramics demonstrate comparatively lower solubility. The initial approximation suggests that the elimination rate of Calcium Phosphate (CaP) from the human body is directly proportional to the solubility of CaP. In the given scenario, the rate of elimination ought to take place in the subsequent sequence. The order of the listed calcium phosphates, from the highest to the lowest in terms of solubility, is as follows: MCPM > TetcP > αTCP > DCPD > DCP > OCP > PA ≈ -TCP > SHA. The existing literature provides ample evidence to suggest that -TCP exhibits a higher degree of bioresorbability when compared to SHA. There is also evidence that DCPD has a higher *in vivo* clearance rate than –TCP. Solubility is affected by the crystals' stoichiometry and size, both of which have been neglected. DCPD and other similar compounds exhibit larger crystal diameters in comparison to apatite crystals, typically in the micrometer range. The bioresorption rate of implanted CaP materials appears to be influenced by solubility, which is influenced by chemical (= composition) and physical (= crystal size) factors.

Extensive *in vitro* exploration has been conducted on the impact of kinetics on both crystallization and dissolution. Based on thermodynamic data regarding solubility, it can be inferred that only two phases, namely DCP, and hydroxyapatite, have the potential to precipitate from an aqueous solution.

According to Tang *et al.*, the dissolution kinetics of CaP materials exhibit a decrease in the dissolving rate that is proportional to the dissolution time. Despite TCP being the primary constituent of the majority of CaP cement, Brunner *et al.* observed that -TCP nanopowders generated through the calcination of ACP may exhibit significantly low reactivity under certain circumstances.

Implant geometry has been shown to have a significant influence on resorption and hence, has attracted a lot of attention. The implant's dimensions, shape, and porosity have been found to play a major role in the outcome of treatment. The size and shape of the implant have an impact on the entire bioresorption period. Through large holes with diameters and connections greater than 50–100 m, blood vessels, and cells can be penetrated easier, and this causes a quicker rate of bioresorption. Additionally, it is believed that microporosity plays a substantial role in the process of bioresorption. The degree of microporosity holds greater significance in determining the rate of biodegradation in ceramics composed of whitlockite (also known as -TCP), as compared to the degree of macroporosity.

By simply dissolving and transporting biosoluble CaPs into solution *in vivo* [38], The removal of substances from the body can be achieved through a process of dissolution. The ions produced by the degradation of these compounds, on the other hand, might combine with bodily fluids and generate new, less-soluble molecules [39]. Driessens inferred from this and analogous findings with bioresorbable CaPs that the solubility is commonly not ascertained by the solubility product but by the surface layer's alteration. This phenomenon is evident in hydroxyapatite ceramics, -TCP ceramics, potentially -TCP ceramics, or even total transformation. As a result, the solubility product is not a strong predictor of CaP implant behavior *in vivo*. At physiological pH, the CaP compound hydroxyapatite has the highest stability. As a result, in such circumstances, it is widely accepted in the field that CaPs, except hydroxyapatite, exhibit thermodynamic instability and are expected to undergo conversion to either hydroxyapatite or an apatitic molecule. This is due to the ease with which hydroxyapatite can accommodate alien ions, including but not limited to carbonates, sulfates, magnesium, and sodium.

When -TCP is introduced in an aqueous solution, it easily changes to CDHA. Suzuki *et al.* also discovered that the DCP, OCP, and ACP all transformed into the apatitic phase after being transplanted into the subperiosteal area of the calvaria of BALB/c mice. In a recent work, Bohner *et al.* demonstrated that the central component of brushite calcium phosphate cement transformed into an apatite compound after 6 to 8 weeks after implantation. Concurrent with this transformation, the removal rate decreased significantly, implying that the transformation decreased removal rates. The quick transformation of brushite cement into an apatite compound was also reported in a separate study, resulting in an extremely sluggish resorption rate. It is surprising to find that the extent and location of the change are affected by the substance, in addition to the position and volume. Following a period of 24 weeks post-implantation in an alternative DCPD cement formulation, it was observed that the entire cement residue underwent a conversion process resulting in the formation of carbonated apatite. Surprisingly, the rate of conversion was affected by the site of the implant.

There is currently no knowledge of the parameters that determine the site of the conversion process. In addition to solubility, it is imperative to consider the role of kinetics and diffusion. In addition, it is currently unknown whether apatites can precipitate at a pH level below 6.5-7 in physiological conditions (at a pressure of 1 atm and a temperature of 37°C). This is likely due to the high interfacial energy between apatite and an aqueous solution, which inhibits apatite nucleation. This lack of experimental evidence has been noted. The anticipated occurrence of the transformation of TetcP, DCPD, DCP, and OCP is within or close to the range of pH values that are typical of physiological conditions [27].

Ceramics Based on α-TCP

TCP is a bioactive and bioresorbable ceramic utilized in oral and maxillofacial surgery, implantology, and periodontology as a bone graft alternative [40]. Additionally, TCP ceramics have the potential to serve as a viable option for bone substitution and scaffolding in tissue engineering applications that demand superior bioresorbability compared to β-TCP. Because -TCP has a greater solubility than β-TCP, a faster degradation rate in the body is to be expected. The production of TCP sintered ceramics can be achieved through a conventional sintering process, as this material is a thermodynamically stable phase when exposed to temperatures exceeding 1100°C. Zinc is considered an essential micronutrient that facilitates the process of bone development. According to reports, ceramics designed for bone repair that incorporate Zinc-containing Tricalcium Phosphate have been developed. Zinc-containing b-TCP (ZnTCP) was synthesized by Ito *et al.* through the combination of calcium oxide, phosphoric acid, and zinc nitrate hexahydrate [41]. Moreover, based on the findings of the β-TCP investigation, ceramics have been produced as bone replacements [11, 22]. Because β-TCP has a higher thermodynamic solubility than HA, β-TCP ceramics degrade faster in the body than HA ceramics. Meanwhile, the bioresorbability of calcium phosphate ceramics is influenced by both the solubility of its constituents and its morphology, encompassing factors such as porosity and pore structures.

Tissue engineering techniques have also been used using β-TCP porous ceramics. Bone morphogenetic protein (BMP) and -TCP have been shown to enhance bone repair [42 - 44]. The proliferation of cells depends on both the scaffold and the growth factor. The potential for -TCP porous ceramics to attract cells involved in bone healing makes them a promising scaffolding material. When exposed to bodily conditions, the ceramics themselves would disintegrate due to bone repair.

Doped Calcium Phosphates

To dope CaPs, two approaches have been used: a chemical approach and a biological one. The utilization of ions in chemical methodology has been implemented to modify the solubility properties of molecules, thereby affecting their bioresorption rate. An instance can be observed in the incorporation of carbonate ions into apatites, which leads to an increase in solubility and consequently enhances the bioresorption rate [34]. Fluoride utilization in toothpaste is considered another indicator of reducing enamel solubility and mitigating the risk of caries. In the biological approach, bioresorbable ceramics have been modified by incorporating compounds such as Mg, Si, Sr, and Zn to alter their biological activity [45]. It is reasonable to assume that the latter approach should be employed for a molecule that exhibits sufficient solubility to be efficacious. Nevertheless, despite the doping of insoluble compounds such as sintered HA, there seems to be a modification in the biological behavior [27].

Calcium Carbonates

The discovery that corals could replace bones in an efficient manner led to the use of calcium carbonate as a bone. There are three different forms of calcium carbonate, or polymorphs; aragonite, calcite, and vaterite make up coral. The following is a list of the soluble polymorphs: calcite aragonite vaterite [46]. When Vaterite is exposed to water, it quickly changes into calcite. The different polymorphs' *in vivo* behavior was found to be comparable.: (i) Bioresorbability was discovered in all polymorphs.; [47] (ii) Calcite and bone were shown to have a direct connection.; (iii) Similar to hydroxyapatite, calcium carbonate (aragonite) mined from marine corals also contributes to bone formation.; [47] (iv) The particular cytocompatibility of coral crystallized in either aragonite or calcite is the same. The interfacial reaction appears to be the main difference between aragonite and calcite. A calcium phosphate layer was found to form on aragonite by many authors, [46] but not on calcite [48]. As a result, unlike calcite, aragonite seems to be biosoluble. Combes *et al.* [49], for example, designed calcium carbonate hydraulic cement using the polymorphism nature of calcium carbonate. Combes *et al.*, for example, designed calcium carbonate hydraulic cement using the crystalline polymorphs of calcium carbonate.

In fact, the rate of bioresorption may be modulated by varying the degree of crystallinity. Calcium pyrophosphate (CPP; $Ca_2P_2O_7$), a pyrophosphate of calcium as a bone replacement, has been proposed. Importantly, *in vivo* findings imply that CPP may have an anti-osteoporosis effect, owing to its molecular similarities to bisphosphonates. Consistent findings showed the following degrading order: -TCP > doped CPP > undoped CPP silicate of calcium Wollastonite ($CaSiO_3$), pseudowollastonite ($CaSiO_3$), larnite (-Ca_2SiO_4), and calcium olivine (Ca_2SiO_4) are all types of calcium silicate.

As a result, it is anticipated that implant removal take place through chemical events rather than physiological processes. Nevertheless, akin to the more soluble calcium phosphates, calcium sulfates possess the capability to transform into calcium phosphates after being implanted. The calcium ions released during calcium sulfate decomposition and phosphate ions are present in the physiological fluid. *In vivo*, gypsum disappears significantly faster than brushite cements. Calcium sulfates are frequently utilized in the form of compact pellets or in combination with HA granules to decrease the rate of clearance and enhance biological efficacy [27].

Application of Bioresorbable Ceramics

A bioresorbable scaffold in tissue engineering provides the initial form of regenerating tissue. Moreover, the use of ceramics in orthopedics, along with conventional artificial joints, dates back many years.

Implants can be advantageous in various applications, as a substitute for bone void fillers, until the point where osteoclasts are involved. Bioresorbing cells present in bone tissue interact with bioresorbable ceramics, resulting in the creation of a durable non-soluble foundation for the purpose of bone regeneration. Osteoblasts and osteoclasts are responsible for the continuous pre-fabrication of the mineralized and organic matrix of bones, which occurs in response to the application of loads in healthy bone. Osteoclasts are responsible for the continuous uptake of calcium phosphate by bones, whereas osteoblasts are involved in the gradual formation of new bone, which is contingent upon the degree of stress experienced by the skeletal system. In the short term, bioresorbable ceramic implants can help stabilize an injured or insufficient bone region until the body's natural resorption mechanism, mediated by osteoclasts, can take over [27].

Drugs can be loaded into bioceramics in two different ways: either inside the pores of the material or as polymerized attachments to the functional groups, both approaches are capable of achieving the regulated drug release. Calcium phosphates and bioactive glasses, which are porous bioceramics called mesoporous bioceramics, can act as drug reservoirs. (*e.g.*, antibiotics, anticancer drugs, anti-inflammatory drugs) [17, 18]. The porosity of the material can aid tissue growth by creating a situation that is conducive to the diffusion of nutrients, oxygen, and metabolic byproducts [16, 50]. The ability to load more drugs onto a surface is enhanced by increased porosity, however, this comes at the expense of mechanical strength [16]. As a result, when used in biological contexts, materials must strike a balance between porosity and mechanical strength. An alternative approach to drug loading involves the conjugation of active therapeutic agents onto bioceramics-functionalized moieties [51]. This kind of dosing can be used to slow down the drug's metabolism and slow down its removal from the body. As a result, material arrangements are often more compact and mechanically stronger. Although bioceramics have been approved for their appropriate compatibility with living organisms, different aspects of drug-loaded bioceramics should be taken into account during their design and development. The interphase of an implant can establish interaction and form a bond with the surrounding tissue. Bioactive ceramics can enhance the development of cells through the promotion of protein synthesis, cell adhesion, and tissue repair [51]. The degradation of a biodegradable material occurs at the same time as the replacement and regeneration of the natural host tissue occur. Loading is shifted to tissue during implant deterioration, avoiding the stress-shielding effect [52]. The shape of implant is frequently tailored to resemble natural body elements. To improve delivery and interactions with cells, drugs and proteins can be loaded on the surface of bioceramics [51].

The implants can withstand mechanical loads and strains. This is frequently extremely dependent on the material's pore size and shape [16, 53]. The size of pores may range from hundreds of micrometers to nanometers. For cells to grow and transport nutrients, oxygen, and metabolic products, pores must be at least 100 m m [16, 53].

Ceramic implants that are loaded with drugs before they are bioresorbable are a great replacement for polymer-based implants. Supporting biological structures and allowing for bioactivity, are crucial parts of the many biomaterial engineering techniques. At present, the utilization of bioceramics is restricted to addressing minor bone ailments and for the purpose of dental implantation. However, because of the restricted number of applications, progress in developing bioceramics loaded with medicinal drugs is slowed [7].

Understanding the In vivo Environment

Absorbable implants require careful consideration of the *in vivo* setting in which they will be used. The deterioration rate of a certain implant might vary greatly.

The variation in the response of a substance when introduced into distinct tissue types can be attributed to alterations in the biochemical and biomechanical factors present in the surrounding atmosphere. Because enzyme activity varies inside the human body, the location of an enzymatic-sensitive substance dictates its breakdown rate. The issue of toxicity is also crucial in the development of implants. The bulk material and accompanying degradation substances that permeate the adjacent tissue ought to exhibit a degree of inertness. It is necessary to evaluate the cytotoxicity of monomeric species and other degradants that arise from the degradation of polymers. This evaluation should be carried out using cell and tissue types that are suitable for the intended use. The occurrence of toxic reactions can elicit an inflammatory or immunological reaction, potentially resulting in alterations to the pH levels in the affected area. Physiological factors can induce alterations in pH levels, which can accelerate the process of degradation and lead to a reduction in mechanical resilience. A necessity for a provisional scaffold or barricade is its ability to sustain its structural integrity for a duration that is conducive to the ideal regeneration and maturation of tissue. In Fig. (**2**), a schematic of the mechanism of degradation, erosion of surface, and erosion of bulk are shown [27].

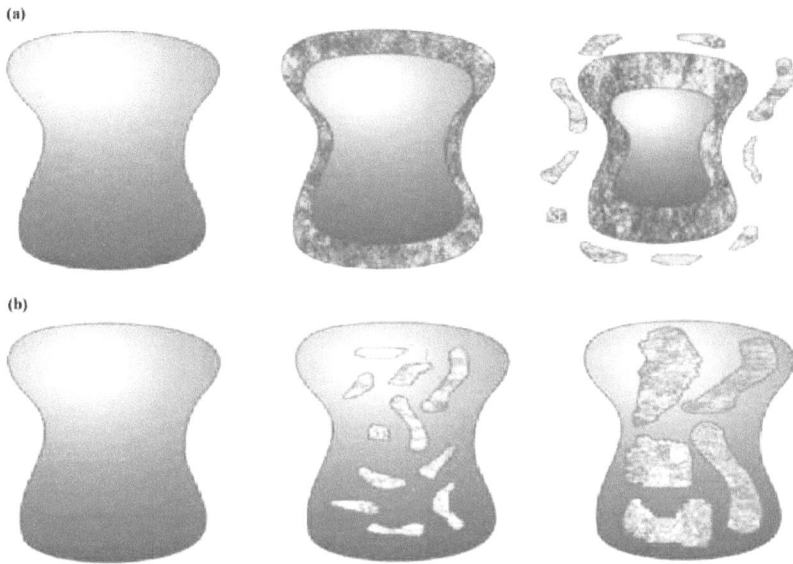

Fig. (2). Mechanism of degradation. **a)** Erosion of surface **b)** Erosion of bulk [27].

If they are sufficiently massive and/or incapable of being further digested, they can remain in the body eternally. It is critical to identify these non-biodegradable byproducts to minimize future issues from the leftover material. Degradation products that are too big to be removed by regular physiological routes may nevertheless be tiny enough to enter the body and cause damage. The development of a fibrous capsule composed of collagen in reacting to the presence of a foreign object is a common physiological reaction. The entrapment of biodegradable implants through a capsule may lead to the accumulation of leachable products, thereby jeopardizing the structural stability of the implant due to a concomitant alteration in pH. Moreover, the infiltration of tissue nearby by by-products separate from the fibrous layer may result in tissue damage. The presence of a fibrous capsule has the potential to impede the functionality of a delivery apparatus and the intended sustained discharge of therapeutic agents. To prevent any potential hindrances to the inherent process of tissue regeneration, it is imperative to conduct a comprehensive assessment of the properties of the materials involved [27].

The Primary Role of an Osteoclast

CaP ceramics are resorbed by osteoclasts using the same three phagocytic stages seen in normal bone resorption: (i) Crystal phagocytosis, (ii) shedding of the membrane that surrounds the endophagosome, and (iii) the process of crystal fragmentation within the cytoplasm after phagocytosis. Moreover, the physicochemical characteristics of ceramics, specifically their solubility and calcium composition, have an impact on the aforementioned degradation mechanism. Ceramic materials exhibiting rapid dissolution rates have been observed to induce an increase in intracellular calcium concentrations within osteoclasts. This, in turn, leads to the disruption of the intracellular actin network located in the osteoclast podosomes, ultimately resulting in the detachment of the osteoclast from the material surface.

Material Considerations Influencing Biodegradation

The characteristics of the biomaterial and host responses regulate implant *in vivo* biodegradation. Understanding these material characteristics is therefore critical to recognizing the impact of host reactions on biodegradation. This is especially significant because biomaterial scientists may modify the characteristics of biomaterials to correspond to certain host reactions. This section discusses some of the important material characteristics that influence a biomaterial's susceptibility to deterioration. In practice, a biodegradable material is exposed to a variety of conditions at the same time, each of which eventually influences its total *in vivo* response. Even if each aspect of the biomaterial can be controlled before implant, the overall *in vivo* response cannot be understood in isolation [27].

Particle Size

The size of the particles is an important determinant in their degradation *in vivo*. Only particles with submicron diameters may be mobilized and eliminated by the host. The lymphatic system is responsible for absorbing particles larger than a few nanometers, whereas the circulatory system is responsible for exchanging smaller particles (less than a few nanometers, particles between 50 and 200 nm in size are transferred).

The impact of clinical use: Lymph nodes play a significant role in the host's immune response by trapping and phagocytizing lymph. APCs can also carry foreign or trash particles within the cell *via* a process called phagocytosis. After taking in an antigen by phagocytosis, APCs go to the closest lymph node through the lymphatic channels to offer the antigen to MHC Class II-bearing T-helper cells for further processing inside the cell. As a result, the biomaterial must be broken down by the host to a size manageable by the lymphatic and circulatory systems, where it can be taken up by metabolic pathways or removed immediately [27].

Hydrophilicity

Due to the importance of water as a necessary component of all living activities, hydrophilicity plays a crucial role in the biodegradation process. Both a material's polymer content and shape affect its hydrophilicity, which is a measure of how well it absorbs, swells, or dissolves in water. The extent to which a surface attracts and retains water is referred to as its hydrophilicity. Whereas bulk hydrophilicity refers to how much water enters the substance's mass. Hydrophilicity has a significant influence on the rate at which polymers degrade by hydrolysis. To control the rate at which fluid and nutrients may penetrate a tissue engineering scaffold, hydrophilicity might be crucial, hence supporting cellular processes and scaffold remodeling. It has the potential to modify drug delivery pharmacokinetics, including onset, rate, and duration [27].

Surface Chemistry

The majority of biomaterials studies have focused on improving the compatibility of surfaces with living organisms. This is because it has been shown that surface chemistry controls how the host reacts. Hydrophobic surfaces, on the other hand, have been shown to increase protein adsorption, degradation rates of bioresorbable materials, and cell adhesion, all of which contribute to an inflammatory response. Lymphocyte proliferation and initial monocyte adherence may both be dramatically reduced by using hydrophilic surfaces. The adherence of monocytes was somewhat reduced, while the fusion of macrophages was sped up, on both anionic and cationic surfaces, whereas cationic surfaces slowed lymphocyte growth and stopped monocytes from doing their thing. It was shown that anionic surfaces were more effective than hydrophilic or cationic chemistries at inducing apoptosis of adherent macrophages. Apoptosis was also discovered to have a negative effect on the development of macrophage fusion into giant cells on various surfaces. Because of this, modifying the surface chemistries may offer a way to prevent the detrimental effects of adhering macrophages on biomaterial degradation. Researchers have also looked at how different surface chemistries affect the *in vitro* cytokine expression of monocytes and macrophages [27].

Host Response as a Measure of Clinical Use

Biodegradation causes more issues and complexities for the host organism than do non-degradable biomaterials. As long as the implant region is not completely destroyed and the biomaterial is not completely destroyed, the host must cope with environmental changes caused by deteriorating biomaterials' changes in shape, size, and surface topography. Also, degradation products are released continuously and must be removed at the earliest opportunity. Additionally, these particles or partially degraded polymers can trigger the host's reaction. The concerns described may create unnecessary complications for the intended use of biomaterials, however, degradable biomaterials offer several advantages, including the potential for tissue remodeling and regeneration, as well as the absence of long-term irritation to the host. As a result, biodegradable substances have achieved widespread acceptance and show great potential for use in tissue engineering as carriers for cells, medications, bioactive chemicals, or gelatin, or as short-term scaffolds for tissue replacement. Several host and biomaterial factors influence the pace, amount, and mode of biodegradation of a foreign material (biomaterial). In addition, the biomaterial's properties are carefully considered since they may be altered by the designer to affect the host's response pathways [27].

Host Response

Surgery to implant a biomaterial (including minimally invasive injections) will cause some physical harm. Damage of this sort triggers the host organism's wound-healing mechanisms, such as hemostasis, inflammation, proliferation, and maturation (fibroplasia/regeneration). These processes try to restore tissue integrity and function. The presence of a foreign substance inserted, on the other hand, affects the usual recovery procedure. In addition, biomaterial properties are given special attention, as they can significantly affect host response pathways and are within the designer's control. This leads to intricate host response networks in which cellular and metabolic activity are tightly regulated. Hemostasis initiates the host response process, the implantation of biomaterials can lead to native vascular tissue damage, resulting in hemorrhage. This, in turn, triggers blood coagulation, which initiates the procedure of wound recovery. The outcome of this process is the generation of a platelet plug and provisional fibrin matrix. The aforementioned items serve to arrest hemorrhage at the site of trauma and restore the damaged region or tissue deficit resulting from said trauma. Additionally, this function serves to regulate the subsequent immune and tissue repair reactions [28].

There are a large number of components and processes that compose the biosphere, but we become aware of their complicated interactions when we experience tissue damage [28].

Inflammatory Response

Inflammation is defined as the heat and redness caused by a vascular response to tissue injury. It is derived from the Latin word for fire. In response to the injury, a variety of cells, plasma proteins, and signaling molecules coordinate their actions to remove the injurious substance. As opposed to 'adapted and acquired immunity,' which explains the generation of long-term immunity through antibody production, the inflammatory reaction is often referred to as 'innate immunity.

The phase of inflammation ensues after tissue injury, thereby facilitating the immune system's access to the changed site. The initial reaction, referred to as chronic inflammation, is marked by the presence of edema and penetration of inflammatory cells, irrespective of the ultimate outcome. The principal function of the severe inflammatory procedure is to safeguard the wounded area against potential infection by initiating a prompt yet moderately unspecific phagocytic reaction that entails neutrophils. Neutrophils eat and eliminate infected cells, the cell remains, and substances with low molecular weights that can be released from biomaterials that have been implanted. The initiation of recovery is contingent upon the removal of the inflammatory stimulus, which in this case is the biomaterial. If the implant-induced stimulation or mechanical trauma endures, the initial acute inflammatory response will transition into a state of chronic inflammation. In the context of chronic inflammation, the host endeavors to maintain a delicate equilibrium between the infliction of tissue damage and the elicitation of the requisite inflammatory response to the cure [28].

Mediators of Inflammation

Communication between cells is required to coordinate the host's reaction to harm with secretory granules inside leukocytes and mast cells. These molecules are either retained inside the cells for release in response to unpleasant stimuli, or they are made and released in response to unpleasant stimuli. These molecules affect the cells differently depending on whether they influence the cells they were created in (autocrine) or the cells in the surrounding tissues (paracrine) [28].

Acute Inflammation

Acute inflammation is characterized as a vascularized tissue's response to damage or infection. Celsus defined the four cardinal indications of inflammation in the first century AD: redness (rubor), swelling (tumor), heat (calor), and pain. Virchow added loss of function (functio laesa) to his list of pathological factors after observing that inflamed tissue frequently does not function properly. Injury to a vascularized tissue, regardless of the origin, will result in an immediate, acute inflammatory response that is limited to the wounded region and has a relatively brief duration, on the order of a few days. There are two stages to this process: alterations to the vascular flow and permeability, followed by leukocyte expulsion from the circulation [28].

Chronic Inflammation

Acute inflammation usually goes away within a few days without causing long-term complications. Acute inflammation will progress to chronic inflammation if the initiating stimulus persists, as in chronic exposure to chemical, physical, or biological agents. There is persistent activation and recruitment of macrophages and FBGC in chronic inflammation, which results in a stalled cycle of tissue damage and repair. An example of this harm may result from a recapitulation of events that occurred during the acute immune response, which reinforces the idea that acute and chronic are oversimplified descriptions. In most cases, though, prolonged inflammation results in the loss of the parenchyma and the stroma, with a fibrous scar healing the injured area [28].

CONCLUSION

Bioceramics are also divided into three types: oxide- and nitride-based bioinert ceramics, bioresorbable calcium phosphate-based materials, and bioactive glasses/glass-ceramics. For biomedical purposes, many ceramic materials are employed. The categorization of bioceramics can be delineated into three distinct classifications: namely bioinert, bioactive, and bioresorbable.

In general, bioresorbable means able to dissolve or absorb using cellular activity in a biological environment, which is often regarded as synonymous with absorbable. Biodegradation refers to the degradation of a substance that has been degraded through the action of a biological agent, typically an enzyme or microbe. The process may lead to the remodeling of the altered substance within an individual's body.

Bioresorbable ceramics possess the unique advantage of undergoing *in vivo* resorption, leading to their replacement by natural bone within bone tissue.

Calcium phosphates (CaP) are a common bioactive and bioresorbable ceramic used for bone substitutes [54, 55]. The most commonly used CaP phases are hydroxyapatite (HA, $Ca_{10}(PO_4)6(OH)_2$) and tricalcium phosphate (-TCP, $Ca_3(PO_4)_2$) due to their chemical compositions close to the mineral component of bone. They are both biocompatible and osteoconductive. It is well known that HA is more effective at bonding directly with bone [56], and TCP has a higher *in vivo* bioresorbability. Biphasic Calcium Phosphate (BCP) materials are made up of HA and –TCP. Because of the connection between the two CaP phases, the accepted HA/-TCP rate permits control of the dissolution rate: the lower the rate, the higher the resorbability.

Additionally, *in-vivo* investigations demonstrate that -TCP is more resorbable than -TCP during bone development, which is important for tissue engineering [57]. TCP is more bioresorbable in both stages than Hap [2].

The *in vivo* fate of ceramics is a complex process that is influenced by various biological factors, including the flow of bodily fluids and the surrounding pH levels, which can be generated by osteoclasts, among other sources. (ii) Ceramic solubility and dissolution kinetics are two examples of chemical variables, and (iii) crystal size and shape are two examples of physical variables. If the ceramic material exhibits solubility within an extracellular milieu, such as blood, it may be subject to direct elimination; if it is soluble in an intracellular environment, it can be eliminated indirectly; or by *in vivo* dissolution, precipitation in a new phase, and bioresorption.

Ceramics have become essential for biological applications due to their three primary types: bioactive, bioinert, and bioresorbable ceramics [2]. While the body regenerates new tissues, bioresorbable components degrade, are ingested, and are freed by metabolic activity. And, in the case of new-generation ceramics, the concepts of bioactivity and biodegradability are integrated, and material qualities are mixed with their ability to signal and trigger certain cellular activity and behavior.

REFERENCES

[1]   Mallick KK, Cox SC. Biomaterial scaffolds for tissue engineering. Front Biosci 2013; 5(1): 341-60.
        [http://dx.doi.org/10.2741/E620] [PMID: 23276994]

[2]   Punj S, Singh J, Singh K. Ceramic biomaterials: Properties, state of the art and future prospectives.
        Ceram Int 2021; 47(20): 28059-74.
        [http://dx.doi.org/10.1016/j.ceramint.2021.06.238]

[3]   Kang CW, Fang FZ. State of the art of bioimplants manufacturing: part II. Advances in Manufacturing
        2018; 6(2): 137-54.
        [http://dx.doi.org/10.1007/s40436-018-0218-9]

[4]   Nabiyouni M, Brückner T, Zhou H, Gbureck U, Bhaduri SB. Magnesium-based bioceramics in

orthopedic applications. Acta Biomater 2018; 66: 23-43.
[http://dx.doi.org/10.1016/j.actbio.2017.11.033] [PMID: 29197578]

[5]   Garrido CA, Lobo SE, Turíbio FM, LeGeros RZ. Biphasic calcium phosphate bioceramics for orthopaedic reconstructions: Clinical outcomes. Int J Biomater 2011; 2011: 1-9.
[http://dx.doi.org/10.1155/2011/129727] [PMID: 21760793]

[6]   Kirtane AR, Abouzid O, Minahan D, *et al.* Development of an oral once-weekly drug delivery system for HIV antiretroviral therapy. Nat Commun 2018; 9(1): 2.
[http://dx.doi.org/10.1038/s41467-017-02294-6] [PMID: 29317618]

[7]   Rajgor N, Patel M, Bhaskar V H. Implantable drug delivery systems: An overview. Systematic Reviews in Pharmacy 2011; 2(2).

[8]   Palmero P, Montanaro L, Reveron H, Chevalier J. Surface coating of oxide powders: A new synthesis method to process biomedical grade nano-composites. Materials 2014; 7(7): 5012-37.
[http://dx.doi.org/10.3390/ma7075012] [PMID: 28788117]

[9]   Rahaman MN, Day DE, Sonny Bal B, *et al.* Bioactive glass in tissue engineering. Acta Biomater 2011; 7(6): 2355-73.
[http://dx.doi.org/10.1016/j.actbio.2011.03.016] [PMID: 21421084]

[10]   Prakasam M, Locs J, Salma-Ancane K, Loca D, Largeteau A, Berzina-Cimdina L. Fabrication, properties and applications of dense hydroxyapatite: A review. J Funct Biomater 2015; 6(4): 1099-140.
[http://dx.doi.org/10.3390/jfb6041099] [PMID: 26703750]

[11]   Dorozhkin SV. Calcium orthophosphates. J Mater Sci 2007; 42(4): 1061-95.
[http://dx.doi.org/10.1007/s10853-006-1467-8]

[12]   Liu C, Shao H, Chen F, Zheng H. Effects of the granularity of raw materials on the hydration and hardening process of calcium phosphate cement. Biomaterials 2003; 24(23): 4103-13.
[http://dx.doi.org/10.1016/S0142-9612(03)00238-2] [PMID: 12853240]

[13]   Kamitakahara M, Ohtsuki C, Miyazaki T. Review paper: Behavior of ceramic biomaterials derived from tricalcium phosphate in physiological condition. J Biomater Appl 2008; 23(3): 197-212.
[http://dx.doi.org/10.1177/0885328208096798] [PMID: 18996965]

[14]   Major I, Lastakchi S, Dalton M, McConville C. Implantable drug delivery systems. Engineering Drug Delivery Systems. Woodhead Publishing 2020; pp. 111-46.
[http://dx.doi.org/10.1016/B978-0-08-102548-2.00005-6]

[15]   LeGeros RZ, Lin S, Rohanizadeh R, Mijares D, LeGeros JP. Biphasic calcium phosphate bioceramics: Preparation, properties and applications. J Mater Sci Mater Med 2003; 14(3): 201-9.
[http://dx.doi.org/10.1023/A:1022872421333] [PMID: 15348465]

[16]   Prakasam M, Locs J, Salma-Ancane K, Loca D, Largeteau A, Berzina-Cimdina L. Biodegradable materials and metallic implants—a review. J Funct Biomater 2017; 8(4): 44.
[http://dx.doi.org/10.3390/jfb8040044] [PMID: 28954399]

[17]   Arcos D, Vallet-Regí M. Bioceramics for drug delivery. Acta Mater 2013; 61(3): 890-911.
[http://dx.doi.org/10.1016/j.actamat.2012.10.039]

[18]   Wang X, Li W. Biodegradable mesoporous bioactive glass nanospheres for drug delivery and bone tissue regeneration. Nanotechnology 2016; 27(22): 225102.
[http://dx.doi.org/10.1088/0957-4484/27/22/225102] [PMID: 27102805]

[19]   Jazayeri HE, Rodriguez-Romero M, Razavi M, *et al.* The cross-disciplinary emergence of 3D printed bioceramic scaffolds in orthopedic bioengineering. Ceram Int 2018; 44(1): 1-9.
[http://dx.doi.org/10.1016/j.ceramint.2017.09.095]

[20]   Bulina NV, Baev SG, Makarova SV, *et al.* Selective laser melting of hydroxyapatite: perspectives for 3D printing of bioresorbable ceramic implants. Materials 2021; 14(18): 5425.

[http://dx.doi.org/10.3390/ma14185425] [PMID: 34576648]

[21]   Ben-Nissan B, Choi AH, Macha IJ, Cazalbou S. Sol-gel nanocoatings of bioceramics. Handbook of Bioceramics and Biocomposites. 2016.

[22]   Uchino T, Yamaguchi K, Suzuki I, Kamitakahara M, Otsuka M, Ohtsuki C. Hydroxyapatite formation on porous ceramics of alpha-tricalcium phosphate in a simulated body fluid. J Mater Sci Mater Med 2010; 21(6): 1921-6.
[http://dx.doi.org/10.1007/s10856-010-4042-4] [PMID: 20224935]

[23]   Sheikh Z, Abdallah MN, Hanafi A, Misbahuddin S, Rashid H, Glogauer M. Mechanisms of *in vivo* degradation and resorption of calcium phosphate based biomaterials. Materials 2015; 8(11): 7913-25.
[http://dx.doi.org/10.3390/ma8115430] [PMID: 28793687]

[24]   Yelten-Yilmaz A, Yilmaz S. Wet chemical precipitation synthesis of hydroxyapatite (HA) powders. Ceram Int 2018; 44(8): 9703-10.
[http://dx.doi.org/10.1016/j.ceramint.2018.02.201]

[25]   Jahangirian H, Ghasemian Lemraski E, Rafiee-Moghaddam R, Webster T. A review of using green chemistry methods for biomaterials in tissue engineering. Int J Nanomedicine 2018; 13: 5953-69.
[http://dx.doi.org/10.2147/IJN.S163399] [PMID: 30323585]

[26]   Habraken W, Habibovic P, Epple M, Bohner M. Calcium phosphates in biomedical applications: Materials for the future? Mater Today 2016; 19(2): 69-87.
[http://dx.doi.org/10.1016/j.mattod.2015.10.008]

[27]   Buchanan F, Leonard D. Influence of processing, sterilisation and storage on bioresorbability. Degradation Rate of Bioresorbable Materials. Woodhead Publishing 2008; pp. 209-33.
[http://dx.doi.org/10.1533/9781845695033.4.209]

[28]   Loughran MJ. The degradation and drug-eluting properties of biodegradable polymers and their potential use as coatings for coronary stems. United Kingdom: The University of Liverpool 2009.

[29]   Khan F, Tanaka M, Ahmad SR. Fabrication of polymeric biomaterials: A strategy for tissue engineering and medical devices. J Mater Chem B Mater Biol Med 2015; 3(42): 8224-49.
[http://dx.doi.org/10.1039/C5TB01370D] [PMID: 32262880]

[30]   Sheikh Z, Najeeb S, Khurshid Z, Verma V, Rashid H, Glogauer M. Biodegradable materials for bone repair and tissue engineering applications. Materials 2015; 8(9): 5744-94.
[http://dx.doi.org/10.3390/ma8095273] [PMID: 28793533]

[31]   Benoit JP, Faisant N, Venier-Julienne MC, Menei P. Development of microspheres for neurological disorders: From basics to clinical applications. J Control Release 2000; 65(1-2): 285-96.
[http://dx.doi.org/10.1016/S0168-3659(99)00250-3] [PMID: 10699288]

[32]   Bourges JL, Touchard E, Kowalczuk L, *et al.* [Drug delivery systems for intraocular applications]. J Fr Ophtalmol 2007; 30(10): 1070-88.
[http://dx.doi.org/10.1016/S0181-5512(07)79290-2] [PMID: 18268450]

[33]   Chu FM, Jayson M, Dineen MK, Perez R, Harkaway R, Tyler RC. A clinical study of 22.5 mg. La-2550: A new subcutaneous depot delivery system for leuprolide acetate for the treatment of prostate cancer. J Urol 2002; 168(3): 1199-203.
[http://dx.doi.org/10.1016/S0022-5347(05)64625-3] [PMID: 12187267]

[34]   De Jong SJ, De Smedt SC, Demeester J, van Nostrum CF, Kettenes-van den Bosch JJ, Hennink WE. Biodegradable hydrogels based on stereocomplex formation between lactic acid oligomers grafted to dextran. J Control Release 2001; 72(1-3): 47-56.
[http://dx.doi.org/10.1016/S0168-3659(01)00261-9] [PMID: 11389984]

[35]   Edlund U, Albertsson AC. Polyesters based on diacid monomers. Adv Drug Deliv Rev 2003; 55(4): 585-609.
[http://dx.doi.org/10.1016/S0169-409X(03)00036-X] [PMID: 12706051]

[36]   Einmahl S, Capancioni S, Schwach-Abdellaoui K, Moeller M, Behar-Cohen F, Gurny R. Therapeutic applications of viscous and injectable poly(ortho esters). Adv Drug Deliv Rev 2001; 53(1): 45-73.
[http://dx.doi.org/10.1016/S0169-409X(01)00220-4] [PMID: 11733117]

[37]   Guse C, Koennings S, Maschke A, *et al.* Biocompatibility and erosion behavior of implants made of triglycerides and blends with cholesterol and phospholipids. Int J Pharm 2006; 314(2): 153-60.
[http://dx.doi.org/10.1016/j.ijpharm.2005.12.050] [PMID: 16517106]

[38]   Lee KY, Yuk SH. Polymeric protein delivery systems. Prog Polym Sci 2007; 32(7): 669-97.
[http://dx.doi.org/10.1016/j.progpolymsci.2007.04.001]

[39]   Nair LS, Laurencin CT. Biodegradable polymers as biomaterials. Prog Polym Sci 2007; 32(8-9): 762-98.
[http://dx.doi.org/10.1016/j.progpolymsci.2007.05.017]

[40]   Bomze D, Ioannidis A. 3D-printing of high-strength and bioresorbable ceramics for dental and maxillofacial surgery applications-the LCM process. Ceramic Applications 2019; 7(1): 38-43.

[41]   Ito A, Otsuka M, Kawamura H, *et al.* Zinc-containing tricalcium phosphate and related materials for promoting bone formation. Curr Appl Phys 2005; 5(5): 402-6.
[http://dx.doi.org/10.1016/j.cap.2004.10.006]

[42]   Matsushita N, Terai H, Okada T, *et al.* A new bone-inducing biodegradable porous β-tricalcium phosphate. J Biomed Mater Res A 2004; 70A(3): 450-8.
[http://dx.doi.org/10.1002/jbm.a.30102] [PMID: 15293319]

[43]   Hoshino M, Egi T, Terai H, Namikawa T, Takaoka K. Repair of long intercalated rib defects using porous beta-tricalcium phosphate cylinders containing recombinant human bone morphogenetic protein-2 in dogs. Biomaterials 2006; 27(28): 4934-40.
[http://dx.doi.org/10.1016/j.biomaterials.2006.04.044] [PMID: 16759693]

[44]   Dong X, Shi W, Yuan G, Xie L, Wang S, Lin P. Intravitreal implantation of the biodegradable cyclosporin A drug delivery system for experimental chronic uveitis. Graefes Arch Clin Exp Ophthalmol 2006; 244(4): 492-7.
[http://dx.doi.org/10.1007/s00417-005-0109-1] [PMID: 16163496]

[45]   Grube E, Sonoda S, Ikeno F, *et al.* Six- and twelve-month results from first human experience using everolimus-eluting stents with bioabsorbable polymer. Circulation 2004; 109(18): 2168-71.
[http://dx.doi.org/10.1161/01.CIR.0000128850.84227.FD] [PMID: 15123533]

[46]   Sartor O. Eligard: Leuprolide acetate in a novel sustained-release delivery system. Urology 2003; 61(2) (Suppl. 1): 25-31.
[http://dx.doi.org/10.1016/S0090-4295(02)02396-8] [PMID: 12667884]

[47]   Sastre RL, Blanco MD, Teijón C, Olmo R, Teijón JM. Preparation and characterization of 5-fluorouracil-loaded poly(ε-caprolactone) microspheres for drug administration. Drug Dev Res 2004; 63(2): 41-53.
[http://dx.doi.org/10.1002/ddr.10396]

[48]   Kim MR, Park TG. Temperature-responsive and degradable hyaluronic acid/Pluronic composite hydrogels for controlled release of human growth hormone. J Control Release 2002; 80(1-3): 69-77.
[http://dx.doi.org/10.1016/S0168-3659(01)00557-0] [PMID: 11943388]

[49]   Serruys PW, Sianos G, Abizaid A, *et al.* The effect of variable dose and release kinetics on neointimal hyperplasia using a novel paclitaxel-eluting stent platform: the Paclitaxel In-Stent Controlled Elution Study (PISCES). J Am Coll Cardiol 2005; 46(2): 253-60.
[http://dx.doi.org/10.1016/j.jacc.2005.03.069] [PMID: 16022951]

[50]   Barrère F, Mahmood TA, de Groot K, van Blitterswijk CA. Advanced biomaterials for skeletal tissue regeneration: Instructive and smart functions. Mater Sci Eng Rep 2008; 59(1-6): 38-71.
[http://dx.doi.org/10.1016/j.mser.2007.12.001]

[51]   Lee WH, Loo CY, Rohanizadeh R. A review of chemical surface modification of bioceramics: Effects on protein adsorption and cellular response. Colloids Surf B Biointerfaces 2014; 122: 823-34.
[http://dx.doi.org/10.1016/j.colsurfb.2014.07.029] [PMID: 25092582]

[52]   Thomas S, Balakrishnan P, Sreekala MS, Eds. Fundamental biomaterials: ceramics. Woodhead Publishing 2018.

[53]   Schuessele A, Mayr H, Tessmar J, Goepferich A. Enhanced bone morphogenetic protein-2 performance on hydroxyapatite ceramic surfaces. J Biomed Mater Res A 2009; 90A(4): 959-71.
[http://dx.doi.org/10.1002/jbm.a.31745] [PMID: 18655137]

[54]   Bohner M. Calcium orthophosphates in medicine: from ceramics to calcium phosphate cements. Injury 2000; 31 (Suppl. 4): D37-47.
[http://dx.doi.org/10.1016/S0020-1383(00)80022-4] [PMID: 11270080]

[55]   Piconi C, Porporati AA. Bioinert ceramics: Zirconia and alumina. Handbook of Bioceramics and Biocomposites. Cham: Springer 2016; pp. 59-89.
[http://dx.doi.org/10.1007/978-3-319-12460-5_4]

[56]   Chetty A, Wepener I, Marei MK, Kamary YE, Moussa RM. Synthesis, properties, and applications of hydroxyapatite. Nova Science Publishers 2012.

[57]   Pripatnanont P, Praserttham P, Suttapreyasri S, Leepong N, Monmaturapoj N. Bone regeneration potential of biphasic nanocalcium phosphate with high hydroxyapatite/tricalcium phosphate ratios in rabbit calvarial defects. Int J Oral Maxillofac Implants 2016; 31(2): 294-303.
[http://dx.doi.org/10.11607/jomi.4531] [PMID: 27004276]

CHAPTER 8

Calcium Orthophosphates in Tissue Engineering

Sergey V. Dorozhkin[1,*]

[1] *Department of Physics, Moscow State University, Vorobievy Gory, Moscow, Russia*

Abstract: $CaPO_4$ (calcium orthophosphate) is an ideal class of materials for bone tissue engineering applications due to the similarity of its set of chemical compositions and structures with mammalian bones and teeth. The use of $CaPO_4$-based biomaterials in dental and orthopedic applications has become widespread in recent years. The biocompatibility, biodegradability, and varying stoichiometry of $CaPO_4$ scaffolds make them suitable candidates for drug loading and tissue engineering strategies. Therefore, calcium phosphate compounds, particularly hydroxyapatite (HA) and tricalcium phosphates (TCP) are highly attractive as bone grafts or drug delivery agents. Specifically, three-dimensional (3D) scaffolds and carriers made from calcium phosphate are created to promote osteogenesis and angiogenesis. These scaffolds are typically porous and can accommodate a range of drugs, bioactive molecules, and cells. In recent years, stem cells and calcium phosphate compounds have been used increasingly as bone grafts. This chapter explores the advantages, sources, and fabrication methods of $CaPO_4$ scaffolds for possible usage in tissue engineering.

Keywords: Calcium orthophosphates ($CaPO_4$), Hydroxyapatite (HA), Tricalcium phosphate (TCP), Scaffolds, Tissue engineering.

INTRODUCTION

Bones represent a supportive organ for living structures and give shape and form to the entire body. In the musculoskeletal system, bones are pivots and levers that enable movement direction and a range of motion to be controlled. Furthermore, bones protect vital organs and store vitamins and nutrients (*e.g.*, calcium). The bone cannot fully heal on its own when its repairing process is ignored when the defect is large or when the usual repair process is interrupted [1].

Commonly, the standard treatment for complex bone fractures is to use various types of fixation devices or implants combined with autografts or artificial bone replacement materials. The advantages of using autogenous bone grafts are clear.

* **Corresponding author Sergey V. Dorozhkin:** Department of Physics, Moscow State University, Vorobievy Gory, Moscow, Russia; E-mail: sedorozhkin@yandex.ru

Saeid Kargozar and Francesco Baino (Eds.)
All rights reserved-© 2024 Bentham Science Publishers

In brief, calcium phosphate compounds serve as a matrix that facilitates cell attachment and migration, resulting in bone formation (osteoconductive properties). They also may serve as a source of therapeutic proteins like growth factors to boost osteogenic differentiation (osteoinductive properties). A similar scenario applies to living cells with osteogenic properties. However, autografts have limitations such as limited tissue availability, the need for another operation (*e.g.*, iliac crest harvest), and the possibility of donor site morbidity. Therefore, allografts and xenografts have been studied and applied as alternatives to autografts. Accessibility to allografts in different shapes and sizes is easier; they provide substances that act as osteoconductive and osteoinductive substances as part of the healing process (if growth factors are kept intact). However, allografts do not show osteogenic properties due to the absence of living cells. As a result of the poor remodeling capacity of allografts and the risk of disease transmission and immune reactions, bone grafts have a significantly higher complication rate and need for reoperation than autografts [2].

Thus, medical professionals are exploring innovative approaches to address the restrictions of existing intervention approaches for complex bone defects. The goal of tissue engineering and regenerative medicine is to repair tissues by using scaffolds, biologically active compounds, and cells. Bone tissue engineering represents an advanced and widely studied field in this area of science [3].

This chapter aims to assess the role and impact of calcium orthophosphate ($CaPO_4$) materials in the repair and regeneration of injured hard tissues. The focus of this chapter is on the development of new formulations that can be translated into scaffolds with the required shape and structure. A variety of techniques can be used to influence the structure of materials, and factors affecting their effectiveness are discussed. In Table **1**, you will find a list of $CaPO_4$ products available, along with their standard abbreviations and key properties [4, 5].

Table 1. Existing calcium orthophosphates and their main properties [4, 5].

Ca/P Molar Ratio	Compounds and Abbreviations	Chemical Formula	Solubility (25 °C, -log(K_s))	Solubility (25 °C, g/L)	Stability Range pH (Aqueous Solutions) (25°C)
0.5	Monocalcium phosphate monohydrate (MCPM)	$Ca(H_2PO_4)_2 \cdot H_2O$	1.14	~ 18	0.0 – 2.0
0.5	Monocalcium phosphate anhydrous (MCPA or MCP)	$Ca(H_2PO_4)_2$	1.14	~ 17	[c]
1.0	Dicalcium phosphate dihydrate (DCPD), mineral brushite	$CaHPO_4 \cdot 2H_2O$	6.59	~ 0.088	2.0 – 6.0

(Table 1) cont.....

Ca/P Molar Ratio	Compounds and Abbreviations	Chemical Formula	Solubility (25 °C, -log(K_s))	Solubility (25 °C, g/L)	Stability Range pH (Aqueous Solutions) (25°C)
1.0	Dicalcium phosphate anhydrous (DCPA or DCP), mineral monetite	$CaHPO_4$	6.90	~ 0.048	[c]
1.33	Octacalcium phosphate (OCP)	$Ca_8(HPO_4)_2(PO_4)_4 \cdot 5H_2O$	96.6	~ 0.0081	5.5 – 7.0
1.5	α-Tricalcium phosphate (α-TCP)	$α\text{-}Ca_3(PO_4)_2$	25.5	~ 0.0025	[a]
1.5	β-Tricalcium phosphate (β-TCP)	$β\text{-}Ca_3(PO_4)_2$	28.9	~ 0.0005	[a]
1.2 – 2.2	Amorphous calcium phosphates (ACP)	$Ca_xH_y(PO_4)_z \cdot nH_2O$, $n = 3 – 4.5$; 15 – 20% H_2O	[b]	[b]	~ 5 – 12 [d]
1.5 – 1.67	Calcium-deficient hydroxyapatite (CDHA or Ca-def HA)[e]	$Ca_{10-x}(HPO_4)_x(PO_4)_{6-x}(OH)_{2-x}$ ($0<x<1$)	~ 85	~ 0.0094	6.5 – 9.5
1.67	Hydroxyapatite (HA, HAp or OHAp)	$Ca_{10}(PO_4)_6(OH)_2$	116.8	~ 0.0003	9.5 – 12
1.67	Fluorapatite (FA or FAp)	$Ca_{10}(PO_4)_6F_2$	120.0	~ 0.0002	7 – 12
1.67	Oxyapatite (OA, OAp or OXA)[f], mineral voelckerite	$Ca_{10}(PO_4)_6O$	~ 69	~ 0.087	[a]
2.0	Tetracalcium phosphate (TTCP or TetcP), mineral hilgenstockite	$Ca_4(PO_4)_2O$	38 – 44	~ 0.0007	[a]

[a] There is no precipitation of this type of compound from aqueous solutions.
[b] It is not possible to measure precisely. Still, the following values were reported: 25.7±0.1 (pH = 7.40), 29.9±0.1 (pH = 6.00), 32.7±0.1 (pH = 5.28). In the acidic buffer, ACP dissolves more readily than α-TCP >> β-TCP > CDHA >> HA > FA.
[c] Stable at temperatures over 100°C.
[d] Metastable at all times.
[e] The precipitated HA may also be called PHA (precipitated HA).
[f] There is some doubt regarding the existence of OA.

TISSUE ENGINEERING

Repairing tissues and organs has been the goal of surgery from antiquity to the present [6, 7]. This repair has traditionally taken place in two main ways: organ transplantation followed by tissue transplantation and replacement with allogeneic or synthetic materials. It is clear that both approaches have some limitations. Transplantation requires a second surgical site that can lead to morbidity, and organ transplants are in particular constrained by the limited amount of available material. The poor integration of synthetic materials with host tissues may lead to failure after implantation because of wear, fatigue, or adverse reactions inside the body [8]. In addition, modern prosthetic implants for orthopedic surgery lack three very important characteristics, including (i) self-healing properties, (ii) blood compatibility, and (iii) the quick response to external cues like mechanical loading by changing structure and properties [9]. It goes without saying that bone has all these properties along with the self-generation capacity, hierarchical architecture, multifunctional properties, and biodegradability. Therefore, an ideal artificial bone graft should have similar properties [10].

In recent years, there has been a surge of innovative ideas and technologies aimed at addressing the challenge of repairing and replacing damaged or diseased tissues. Consequently, tissue engineering has been growing as a new medical technology. An individual's tissues and organs can be healed, repaired, replaced, protected, and enhanced with the aid of tissue engineering, which involves combining cells, advanced materials, and appropriate growth factors [11 - 14]. Despite the importance of cells and growth factors in tissue repair, this discussion is exclusively focused on $CaPO_4$-engineered materials.

The goal of tissue engineering is to construct tissues and organs *de novo* instead of repairing them [13]. This research discipline started over 20 years ago [15, 16] and Langer and Vacanti's famous paper [17] helped promote tissue engineering worldwide.

Tissue engineering requires collaboration between cellular and molecular biology, biochemistry, materials science, bioengineering, and clinical research fields. It is particularly relevant in the case of bone substitutes since these tissues typically operate in a mechanically demanding environment [18 - 21]. Success requires researchers with expertise in one field to understand the knowledge and challenges in other fields. Due to significant technical, regulatory, and commercial challenges, new products are likely to be slow to develop and introduce [13].

Tissue engineering is currently realizing its full research potential with respect to the features mentioned as follows. It is long-term, safer, and more cost-effective than other methods. (ii) It requires fewer amounts of donor tissues and there are no problems of immunosuppression. (iii) It also offers a solution without residual foreign bodies [22, 23].

SCAFFOLDS AND THEIR PROPERTIES

Patients and physicians would be pleased if devastated patient tissues and organs were regenerated simply by injecting cells into the target area; however, such cases are not very common. Many large, three-dimensional tissues and organs need support (scaffold) to be created from cells [15, 18 - 27]. Like natural extracellular matrices, scaffolds are expected to aid cell growth, differentiation, and biosynthesis. Scaffolds placed at the injured location can also prevent disturbed cells from entering the domain [26, 27]. A Spanish classical guitarist Andrés Segovia (1893 - 1987) clearly described the role of scaffolds: when constructing a building, scaffolding is used to place everything in its proper place. But when tearing down the scaffolding, the building must stand alone without any trace of the scaffold." That is how a musician should work. But the term 'template' may be more appropriate for tissue engineering, because Prof. David F. Williams claims that the term 'scaffold' conveys an old-fashioned meaning of an inert external structure that helps construct or repair inanimate objects like buildings, but does not contribute to the final product's characteristics [28, 29].

Basically, tissue engineering involves expanding autologous cells on scaffolds *in vitro* or implanting cellular templates *in vivo*, and then repairing the tissue driven by the scaffold with the patient's own cells. A bioreactor is used to create tissue constructs *in vitro*; a nutrient medium and a metabolically and mechanically supported environment allow the cells and scaffolds to grow in a controlled environment. This supports cells proliferation, migration, infiltration, and differentiation, ideally leading to generating the extracellular matrix [29, 30]. The second stage involves implanting constructs at the appropriate anatomical location to re-create organs and tissues *in vivo* [31, 32]. It is crucial for tissue formation and maturation to take place *in vitro* and *in vivo*, which involves (i) proliferation, sorting, and differentiation of cells, (ii) synthesis and organization of extracellular matrix, (iii) degradation of scaffolds, and (iv) growth and remodeling of tissues [33].

To obtain a successful tissue rebuilding, scaffolds need to fulfill a set of specific criteria including (i) macroscopic morphology; (ii) porous texture; (iii) pores' size and morphology; (iv) surface topography; and (v) micro-, submicro-, and nanoporosities. It is also important that scaffolds be biodegradable [24 - 28]. Additionally, the seeding and fixing of cells depend on sufficient surface roughness [34 - 39]. For cell migration, neovascularization, and nutrient transport, sufficient porosity and void dimensions are essential (Table **2**) [41, 42]. In the words of French architect Robert le Ricolais (1894-1977): "The art of structure is where to put the holes". Therefore, a scaffold with a highly interconnected porous structure ensures tissue viability, neovascularization, and nutrient delivery. This network should consist of both macro- and micropores, with at least 60% of the pores being ~150 μm to ~400 μm in size and at least ~20% smaller than 20 μm [41, 43 - 49]. To resist contractile forces and remodel the damaged tissue, appropriate mechanical properties (strength and stiffness) are also required [50, 51]. Moreover, scaffolds should be manufactured from materials that are biodegradable and bioresorbable, such as $CaPO_4$, which will eventually replace the scaffold with new bone [18, 44, 52]. Besides, the degradation by-products of the scaffold should not be cytotoxic. Moreover, the resorption rate should match the bone formation rate as much as possible (*i.e.*, from a few months up to two years) [53]. In other words, this means the scaffold can maintain structural integrity in the body while the cells start forming an extracellular matrix for subsequent replacement of the implanted scaffold. However, it should be borne in mind that the process of scaffold degradation changes its structure and by-products can influence the physiological phenomena. In addition, scaffolds need to be easily machinable in different forms and dimensions [54], moldable to fit defects with irregular shapes, and for mass production, the machining process must be easily scalable. As in the case of self-hardening $CaPO_4$ formulations, shape, injection ease, and processing dictate the choice of biomaterials. Finally, both laboratory and industrial scaffold production require sterilization without compromising properties [18 - 20]. Therefore, each scaffold (template) must fulfill a series of functions in all steps from the processing to post-implantation.

Table 2. The dimensions and the possible function of 3D pores scaffolds in tissue engineering [40].

Dimensions of a 3D Scaffold's Pores	Effects or Functions of Biochemistry
< 1 μm	Protein interaction
	Bioactive
1 – 20 μm	Cell adhesion
	Development of cells
	Cellular ingrowth orientation and directionality
100 – 1000 μm	Growth of cells
	Ingrowth of bone
	Mechanical strength's predominant function
> 1000 μm	Functionality of implants
	Implant shape
	Ethic of implants

The production of $CaPO_4$-based porous scaffolds with diverse architectural properties is accessible through various manufacturing techniques. The manufacturing process needs to be optimized to achieve the properties of interest at a minimum cost [55]. Smart scaffolds for bone replacement are being developed to improve both osteoconduction [56] and osteoinduction [57]. Regarding $CaPO_4$, the smart scaffolds are a biphasic formulation (the ratio of HA/β-TCP equals 20/80) with macropores, mesopores, and micropores (>100, 10-100, and <10 μm, respectively) porosities of about 73%. Additionally, this composition contains 40% of micropores (<10 μm), a crystal size of <0.5-1 μm, and a specific surface area (S_{BET}) of ~6 m^2/g [57]. The discovery of CaPO$_4$ has opened up the possibility of fabricating synthetic scaffolds containing cells and growth factors for tissue engineering. Fig. (**1**) shows a schematic illustration of the fabrication process for scaffolds that are used for bone regeneration [58].

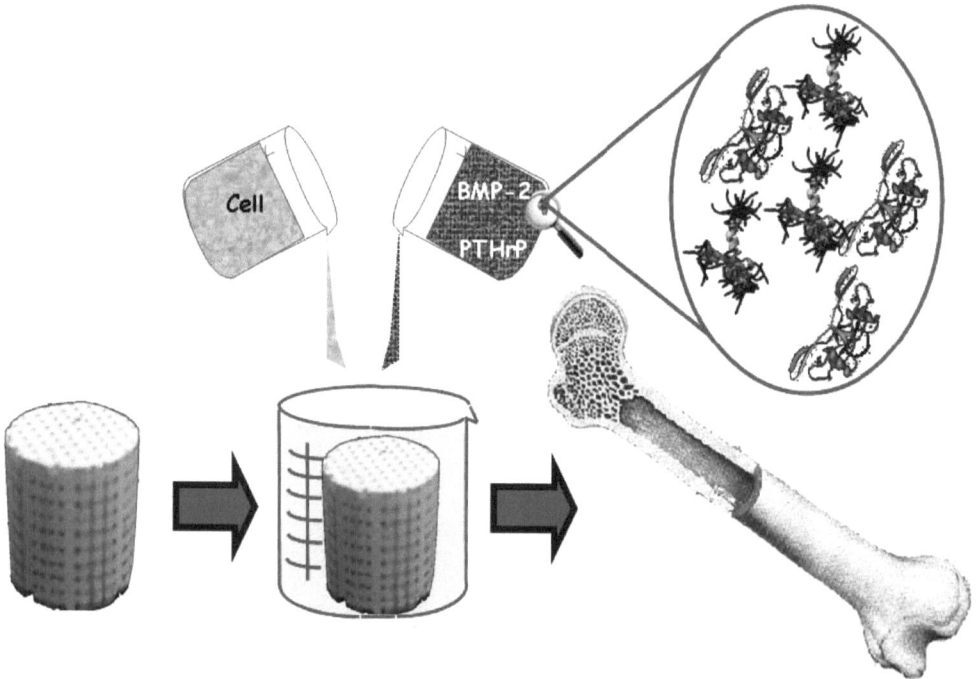

Fig. (1). Schematic representation of a third-generation biomaterial, which uses porous CaPO$_4$ ceramics to transport cells, growth factors, *etc.* Reprinted with permission from [58].

Finally, it should be noted that it is essentially impossible to create an ideal bone scaffold. Since the dimensions, shape, and structure of the human skeleton vary greatly according to its function and site, it is likely that artificial bones will be required with diverse compositions, dimensions, morphology, textural and mechanical properties, as well as resorbability. In order to achieve high strength, HA biomaterials with 0-15% porosity are suitable for iliac and intervertebral spaces. In laminoplasty requiring osteogenesis and medium strength, HA biomaterials with 30-40% porosity can be used as spinous process spacers, while 40-60% porosity can be used as calvaria plates for rapid bone formation (Fig. **2**) [59]. Moreover, defining the optimal parameters of an artificial scaffold is essentially an endeavor to provide a reasonable balance between different conflicting functional requirements. That is, a strong and dense structure is required to increase the mechanical strength of the bone graft, while an interconnected porous structure is required for the formation of colonies by cells on the bone graft surface. The details and discussions on this topic are well documented in the literature [60], as stated by the authors: "There is enough evidence to postulate that ideal scaffold architecture does not exist." (p. 478).

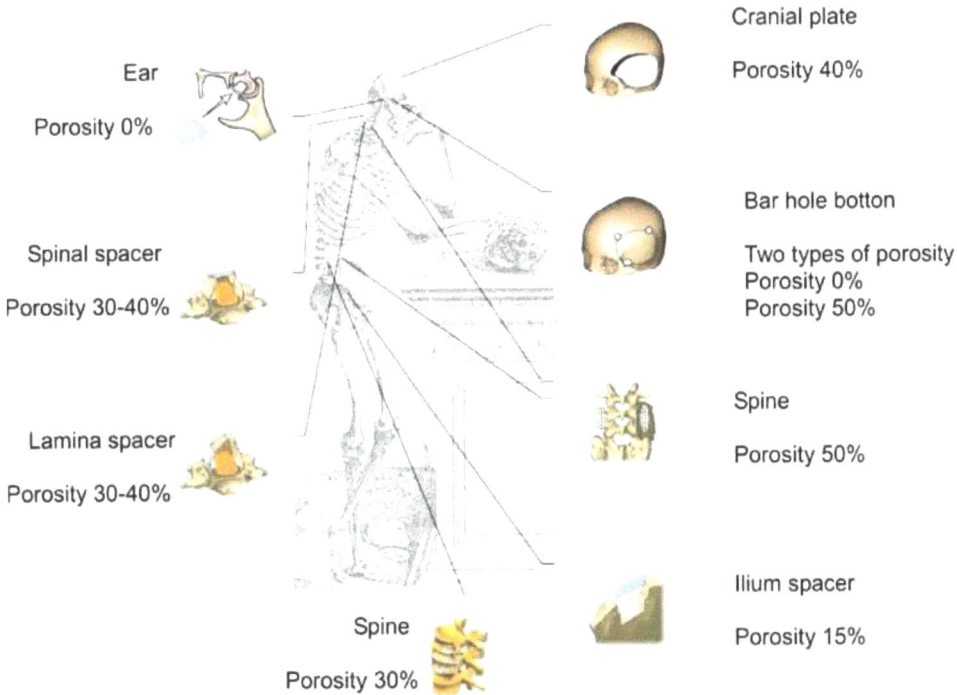

Fig. (2). Schematic representation of the possible applications for HA with different porosities. Reprinted with permission from [59].

CAPO$_4$

Today, CaPO$_4$ can be prepared from a variety of raw materials [61 - 70]. However, scientists have only been relatively successful in synthesizing bone replacement materials that are physiologically safe, biocompatible, and stable for clinical use. There is no doubt that, natural structures are superior and complex [10]. It is generally required to characterize different aspects of CaPO$_4$ biomaterials, such as their physical, chemical, mechanical, and biological characteristics. From a chemical point of view, most CaPO$_4$ biomaterials are based on HA [71 - 75], both types of TCPs [76 - 86], and their various multiphasic formulations [87]. Biphasic formulations, also known as BCP-biphasic calcium phosphates, comprise β-TCP + HA [88 - 96], α-TCP + HA [97 -

99], and biphasic TCP (BTCP) consisting of α-TCP and β-TCP [100 - 105]. Triphasic preparations (HA + α-TCP + β-TCP) were prepared as well [106 - 109]. An in-depth review of this topic is available [87]. It should be noted that besides the main topic of self-curing formulations of DCPD formation [110, 111], other types of $CaPO_4$ bioceramics have only been described in a few publications [112 - 120].

Various $CaPO_4$ preparation techniques are extensively discussed in the literature [121 - 125] and interested readers are directed there. The lower solubility (Table 1) and slower absorption rates of HA compared to both α- and β-TCP can be attributed to its more stable phase after exposure to physiological solutions [126 - 128]. Accordingly, BCP characteristics depend on the optimal balance between HA and TCP phases. BCP with a high ratio of TCP to HA shows higher reactivity in solution because of having more biodegradable α- or β-TCP components [89]. The same results apply to biphasic, triphasic, and more complex compositions [87].

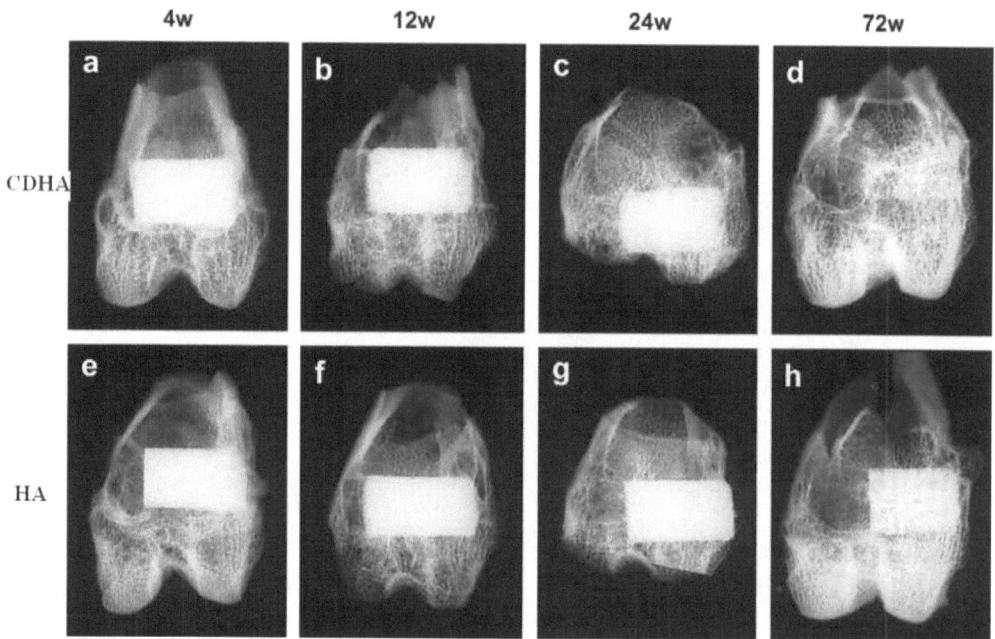

Fig. (3). Soft X-ray images of the rabbit femur treated with CDHA and sintered HA after 1, 3, 6, and 12 months (a-d and e-f, respectively). Reprinted with permission from [129].

Implants made of sintered HA will remain at damaged locations for a long time (Fig. **3** bottom), so biomaterials fabricated using $CaPO_4$ are better choices (Fig. **3** top) for biomedical purposes [76 - 120, 129, 130]. Moreover, BCP is more effective at adsorbing fibrinogen, insulin, and collagen (type I) than HA [131]. Therefore, $CaPO_4$ biomaterials are ranked according to osteogenic parameters as follows; low, medium, and high sintering temperature BCP (rough and smooth), TCP, sintered (low and high temperatures), and unsintered HA [132]. In 2000, this sequence was developed and did not include multiphase mixtures or other types of $CaPO_4$.

SCAFFOLDS FROM CAPO$_4$

Since the late 1990s, biomaterials researchers and clinicians decided to pay attention to tissue regeneration rather than tissue replacement [133, 134]. *In vitro* regeneration of bio-cellular constructs for transplantation can be performed using hierarchical bioactive scaffolds or by using bioabsorbable bioactive microparticles and porous networks *in vivo* [135, 136]. The aim of $CaPO_4$ is therefore to act as a constructive material for porous scaffolds that can provide physicochemical cues to direct cell behaviors (*e.g.*, adhesion and differentiation) and thereby the assembly of newly formed bone into 3D tissue. Scaffold characteristics including particle size, morphology, surface area, and surface roughness have been demonstrated as the main influencers of cellular behaviors like adhesion, migration, and proliferation [34 - 39]. Furthermore, surface morphology and its energy play a role in the attraction of certain proteins to the $CaPO_4$ surface and thus affect the affinity of cells for the material. Moreover, cell behaviors are greatly determined through chemical composition and their osteogenic function may depend on the granular morphology of the scaffold. For example, osteoblasts have been found to function better in nano-sized fibers than in nano-sized spheres because of more similarity in shape with the biological apatite in bone [137]. Additionally, a notable increase in osteoblast proliferation was observed for sintered HA at 1200°C in comparison to its sintered counterparts at 800°C and 1000°C [138]. Furthermore, $CaPO_4$ is a drug functionality candidate since calcium and orthophosphate ions modulate bone metabolism. Fig. (**4**) illustrates the key characteristics of a scaffold that can affect the sequence of biological events that take place following the implantation of a $CaPO_4$ scaffold [139].

Therefore, great emphasis has been placed on the further development of $CaPO_4$ biomaterials in order to meet tissue engineering requirements [140 - 142]. In terms of chemistry, new types of ion-substituted $CaPO_4$ are developed. For example, the supplementation of $CaPO_4$-based bone replacement materials with bioinorganic materials, especially strontium, silicon, and magnesium, have had a significant impact on improving physicochemical properties and increasing new bone formation [143]. The positive effect of doping $CaPO_4$ with carbonate has also been noted [144]. In terms of materials, researchers are interested in the following topics: nano-sized and nano-crystalline structures [145 - 148], amorphous compounds [149, 150], (bio)organic/$CaPO_4$ formulation and hybrid compositions [151, 152], biphasic, triphasic and polyphasic formulations [87] with different structures. The latter include fibers, wires, whiskers, filaments [153 - 166], macro-, micro-, and nano-dimensional spheres, beads, granules [165 - 178], quads, pyramids [179], micro- and nano-sized tubes [180 - 184], aerogels [185], 'flowers' [186], ACP [187], DCPD/DCPA [188 - 191], OCP [192, 193], TCP [194 - 197], HA [198 - 202], TTCP [203] and biphasic formulations [197, 204 - 209], porous 3D scaffolds with graded porosity [210 - 213] and hierarchically organized [214, 215]. It is possible to improve the properties of $CaPO_4$ scaffolds using advanced composite techniques. One example of such a technique involves modifying the surface of porous $CaPO_4$ scaffolds with $CaPO_4$ whiskers to increase the degrees of ossification and homogenization [216, 217] and then reinforcing with multilayer free nanosized CDHA particles [217]. $CaPO_4$-based biomaterials with significant softness, flexibility, and/or good elasticity have recently been fabricated using HA wires with nanosized ultra-lon--range structures [218, 219]. Besides, the introduction of defects by grinding [220, 221] or the removal of defects by heat treatment [222] is considered for modifying $CaPO_4$ chemical reactivity. Furthermore, there are crystallographically aligned $CaPO_4$ [223].

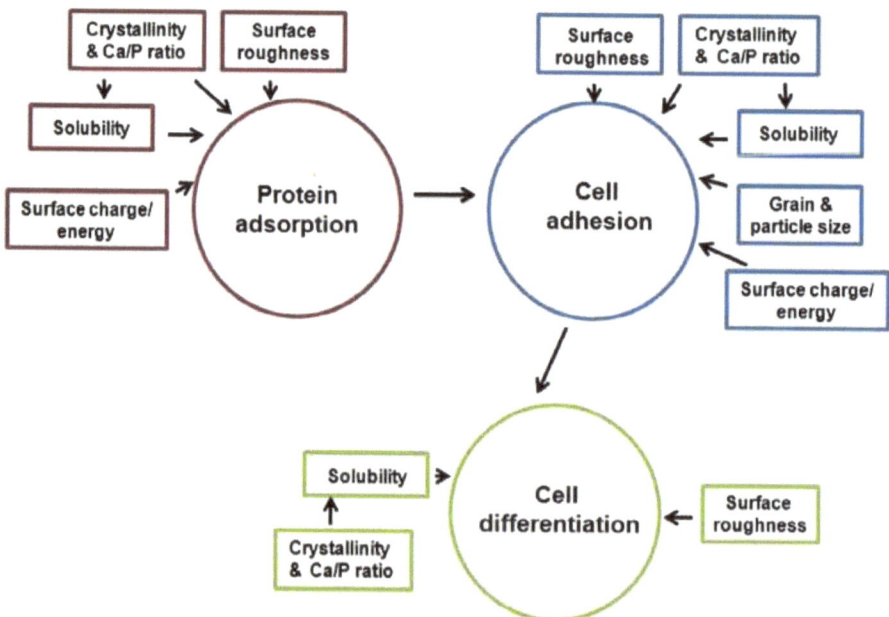

Fig. (4). A representative illustration of the main properties of a scaffold that affect biological functions occurring after $CaPO_4$ implantation. Reprinted with permission from [139].

There are typically three primary approaches for treating damaged or diseased tissues including (i) transplanting fresh cultured or isolated cells, (ii) transplanting tissue that has been assembled from scaffolds or cells *in vitro*, and (iii) regenerating tissue *in situ*. In cell transplantation, individual cells or small clusters of cells are injected directly into the damaged area by a patient or donor. Alternatively, the cells can be combined with biodegradable scaffolds *in vitro* before transplantation. In the case of tissue transplantation, the patient's or donor's cells along with a bioabsorbable scaffold are used to generate a complete 3D tissue *in vitro*. In order to replace diseased or damaged tissue, transplantation is performed on the patient. To facilitate local tissue repair, scaffolds are implanted directly into damaged tissue to stimulate body cells. In either case, it is not enough to simply capture cells in a specific location on the surface; cell differentiation must be stimulated, which is not possible in the absence of appropriate biochemical factors [224]. In other words, they serve as suitable materials to produce suitable 3D substrates for cell colonization before being implanted [225 - 227]. $CaPO_4$ scaffolds have been evaluated *in vitro* for tissue engineering in other publications [228]. $CaPO_4$'s mechanical properties in tissue engineering are also available [229 - 231]. HA-based biomaterials have also been reported to positively affect gene expression in osteoblast-like cells [232].

In conclusion, the good biocompatibility of $CaPO_4$, its possible osteoinductive properties, and its high loading of drugs [233 - 237], cells, and proteins [234, 238, 293]. The future of $CaPO_4$ is wide open by making feasible scaffolds with tailored physicochemical, mechanical, and biological properties [232 - 240].

A CLINICAL EXPERIENCE

In the human bone tissue engineering field, only a few papers have been published utilizing $CaPO_4$ bioceramics laden with cells. Historically, a team of researchers reported the first successful treatment of large bone defects (4-7 cm) in three patients aged 16 to 41 in 2001. Conventional surgical treatments had failed in these cases. Researchers loaded autologous bone marrow stromal cells into non-resorbable porous HA scaffolds that had been expanded *in vitro*. Within two months after surgery, radiographs and CT scans showed abundant callus formation around the implant. Additionally, the implant integrated well with the host bone in all three patients [241]. Another report described a man with traumatic avulsion of the distal phalanx of the thumb in the same year. Replacement of damaged phalanx with unique porous HA implants made from natural coral (ProOsteon 500; pore diameter ~ 500 micrometers) that had been specially treated. The implant was then colonized with autologous periosteal cells that had been cultured *in vitro* before transplantation. The operation enabled the thumb to be restored to its normal length and function properly, without the pain and complications that typically accompany bone grafting. The biomechanical stability of the thumb was also improved [242].

Japanese researchers repaired defects caused by benign bone tumor surgeries using HA scaffolds seeded with autologous bone marrow cells. There were two bone defects in the tibia and one in the femur that were treated. The osteogenic potential of the cells was demonstrated by heterotopic implantation in nude mice, but no details were provided about the bone content or volume of the implants [243]. Later, cell-seeded $CaPO_4$ scaffolds showed superior bone formation at the xenograft site compared to autografts, allografts, and cell-seeded allografts [244]. Furthermore, periodontal ligament cells combined with β-TCP are considered to be a potential new therapeutic strategy for restoring periodontal defects [245]. In a further study, human periodontal ligament stem cells grown on HA-coated genipin-chitosan scaffolds demonstrated bone regeneration potential *in vitro* and *in vivo* [246]. In a study, this formulation was shown to be capable of regenerating and repairing bone tissue.

Researchers from the Netherlands evaluated the vascularization and osteogenic potential of adipose stem cell-containing adipose tissue seeded on two different $CaPO_4$ carriers in a phase I study within a human maxillary sinus floor elevation model using the stromal vascular fraction (SVF) [247]. The maxillary sinus floor was elevated using autologous neural SVF from 10 patients loaded into TCP scaffolds or BCP scaffolds (HA + TCP scaffolds). Histomorphometry analysis was performed on biopsies taken at implant placement after 6 months. Blood vessel numbers were quantified using immunohistochemical staining of markers. The bone fraction was correlated with vessel formation, particularly in cranial biopsies. This was higher in patients treated with β-TCP and in controls. $CaPO_4$ scaffolds containing adipose stem cells were found to be safe, feasible, and efficient for elevating the maxillary sinus floor in human volunteers and suggested that SVF may promote angiogenesis [247]. Brief descriptions of several other cases can be found in the literature [248]. As a final note, $CaPO_4$ is also used in veterinary orthopedics to promote bone healing in areas where bone defects exist in animals [249, 250].

CONCLUSION

Over the past 30 years, bone tissue engineering has seen rapid growth, but there is a disconnect between research and clinical practice, with academia tending to be complex and clinical practice tending to be simple, inhibiting its further advancement. There are a number of causes for bone defects and fractures from a clinical perspective, including fractures (single or bicondylar), nonunions, and tumors, while each patient has a naturally individualized intrinsic healing pattern that is determined by their physical condition, age, associated comorbidities, compliance and loading patterns, and lifestyle choices. Consequently, a one-size-fits-all approach is unlikely to succeed in any regenerative medicine treatment concept. Additionally, researchers should consult surgeons in order to consider the end users of the product. As new technologies and therapies are adopted, the emphasis must be on ease of use and technical feasibility. Overly complex engineering designs may hinder widespread adoption. The induction of the host's endogenous regenerative capacity should also be promoted and utilized since bones possess an innate regenerative capacity [251].

Despite significant scientific advances over the last few decades, we are still far from "pulling a newly engineered organ out of a Petri dish". Thus, bone tissue engineering has already developed scaffolds that can be used for space-making, biodegradable substitutes for implantation into new bones, drug delivery systems, and even support structures that enable seeded cells to attach, multiply, and produce, depending on the bone defect situation. Despite this, clinically fully satisfactory scaffolds have yet to be developed. The development of more functional scaffolds is therefore necessary.

REFERENCES

[1]   Seeman E, Delmas PD. Bone quality--the material and structural basis of bone strength and fragility. N Engl J Med 2006; 354(21): 2250-61.
        [http://dx.doi.org/10.1056/NEJMra053077] [PMID: 16723616]

[2]   Giannoudis PV, Dinopoulos H, Tsiridis E. Bone substitutes: An update. Injury 2005; 36(3): S20-7.
        [http://dx.doi.org/10.1016/j.injury.2005.07.029] [PMID: 16188545]

[3]   Lutolf MP, Hubbell JA. Synthetic biomaterials as instructive extracellular microenvironments for morphogenesis in tissue engineering. Nat Biotechnol 2005; 23(1): 47-55.
        [http://dx.doi.org/10.1038/nbt1055] [PMID: 15637621]

[4]   Dorozhkin SV. Calcium orthophosphates: applications in nature, biology, and medicine. Singapore: Pan Stanford 2012.
        [http://dx.doi.org/10.1201/b12312]

[5]   Dorozhkin SV. Calcium orthophosphate-based bioceramics and biocomposites. Weinheim, Germany: Wiley-VCH 2016.
        [http://dx.doi.org/10.1002/9783527699315]

[6]   Bose S, Tarafder S. Calcium phosphate ceramic systems in growth factor and drug delivery for bone tissue engineering: A review. Acta Biomater 2012; 8(4): 1401-21.
        [http://dx.doi.org/10.1016/j.actbio.2011.11.017] [PMID: 22127225]

[7]   Arcos D, Vallet-Regí M. Bioceramics for drug delivery. Acta Mater 2013; 61(3): 890-911.
        [http://dx.doi.org/10.1016/j.actamat.2012.10.039]

[8]   Hollister SJ. Porous scaffold design for tissue engineering. Nat Mater 2005; 4(7): 518-24.
        [http://dx.doi.org/10.1038/nmat1421] [PMID: 16003400]

[9]   Jones JR, Hench LL. Regeneration of trabecular bone using porous ceramics. Curr Opin Solid State Mater Sci 2003; 7(4-5): 301-7.
        [http://dx.doi.org/10.1016/j.cossms.2003.09.012]

[10] Valletregí M, González-Calbet JM. Calcium phosphates as substitution of bone tissues. Prog Solid State Chem 2004; 32(1-2): 1-31.
[http://dx.doi.org/10.1016/j.progsolidstchem.2004.07.001]

[11] Williams DF. To engineer is to create: The link between engineering and regeneration. Trends Biotechnol 2006; 24(1): 4-8.
[http://dx.doi.org/10.1016/j.tibtech.2005.10.006] [PMID: 16289395]

[12] Griffith LG, Naughton G. Tissue engineering--current challenges and expanding opportunities. Science 2002; 295(5557): 1009-14.
[http://dx.doi.org/10.1126/science.1069210] [PMID: 11834815]

[13] Goldberg VM, Caplan AI. Orthopedic tissue engineering basic science and practice. New York, USA: Marcel Dekker 2004.

[14] Van Blitterswijk CA, Thomsen P, Hubbell J, Eds., *et al.* Tissue engineering. Burlington, MA, USA: Academic Press 2008.

[15] Ikada Y. Challenges in tissue engineering. J R Soc Interface 2006; 3(10): 589-601.
[http://dx.doi.org/10.1098/rsif.2006.0124] [PMID: 16971328]

[16] Cima LG, Langer R. Engineering human tissue. Chem Eng Prog 1993; 89: 46-54.

[17] Langer R, Vacanti JP. Tissue engineering. Science 1993; 260(5110): 920-6.
[http://dx.doi.org/10.1126/science.8493529] [PMID: 8493529]

[18] El-Ghannam A. Bone reconstruction: From bioceramics to tissue engineering. Expert Rev Med Devices 2005; 2(1): 87-101.
[http://dx.doi.org/10.1586/17434440.2.1.87] [PMID: 16293032]

[19] Kneser U, Schaefer DJ, Polykandriotis E, Horch RE. Tissue engineering of bone: The reconstructive surgeon's point of view. J Cell Mol Med 2006; 10(1): 7-19.
[http://dx.doi.org/10.1111/j.1582-4934.2006.tb00287.x] [PMID: 16563218]

[20] Boyan BD, Cohen DJ, Schwartz Z. Bone tissue grafting and tissue engineering concepts. In: Ducheyne P, Ed. Comprehensive Biomaterials II. Oxford: Elsevier 2017; pp. 298-313.
[http://dx.doi.org/10.1016/B978-0-12-803581-8.10240-1]

[21] Koons GL, Diba M, Mikos AG. Materials design for bone-tissue engineering. Nat Rev Mater 2020; 5(8): 584-603.
[http://dx.doi.org/10.1038/s41578-020-0204-2]

[22] Lutolf MP, Hubbell JA. Synthetic biomaterials as instructive extracellular microenvironments for morphogenesis in tissue engineering. Nat Biotechnol 2005; 23(1): 47-55.
[http://dx.doi.org/10.1038/nbt1055] [PMID: 15637621]

[23] Ma PX. Biomimetic materials for tissue engineering. Adv Drug Deliv Rev 2008; 60(2): 184-98.
[http://dx.doi.org/10.1016/j.addr.2007.08.041] [PMID: 18045729]

[24] Yang S, Leong KF, Du Z, Chua CK. The design of scaffolds for use in tissue engineering. Part I. Traditional factors. Tissue Eng 2001; 7(6): 679-89.
[http://dx.doi.org/10.1089/107632701753337645] [PMID: 11749726]

[25] Ma PX, Elisseeff J, Eds. Scaffolding in tissue engineering. Boca Raton, FL, USA: CRC Press 2006.

[26] Alvarez-Urena P, Kim J, Bhattacharyya S, Ducheyne P. Bioactive ceramics and bioactive ceramic composite-based scaffolds. In: Ducheyne P, Ed. Comprehensive Biomaterials II. Oxford: Elsevier 2017; pp. 1-19.
[http://dx.doi.org/10.1016/B978-0-12-803581-8.10136-5]

[27] Baldwin J, Henkel J, Hutmacher DW. Engineering the organ bone. In: Ducheyne P, Ed. Comprehensive Biomaterials II. Oxford: Elsevier 2017; pp. 54-74.
[http://dx.doi.org/10.1016/B978-0-12-803581-8.09342-5]

[28]   Williams DF. The biomaterials conundrum in tissue engineering. Tissue Eng Part A 2014; 20(7-8): 1129-31.
[http://dx.doi.org/10.1089/ten.tea.2013.0769] [PMID: 24417599]

[29]   Freed LE, Guilak F, Guo XE, *et al.* Advanced tools for tissue engineering: Scaffolds, bioreactors, and signaling. Tissue Eng 2006; 12(12): 3285-305.
[http://dx.doi.org/10.1089/ten.2006.12.3285] [PMID: 17518670]

[30]   Gandaglia A, Bagno A, Naso F, Spina M, Gerosa G. Cells, scaffolds and bioreactors for tissue-engineered heart valves: A journey from basic concepts to contemporary developmental innovations. Eur J Cardiothorac Surg 2011; 39(4): 523-31.
[http://dx.doi.org/10.1016/j.ejcts.2010.07.030] [PMID: 21163670]

[31]   Hui JHP, Buhary KS, Chowdhary A. Implantation of orthobiologic, biodegradable scaffolds in osteochondral repair. Orthop Clin North Am 2012; 43(2): 255-61.
[http://dx.doi.org/10.1016/j.ocl.2012.01.002] [PMID: 22480474]

[32]   Vanderleyden E, Mullens S, Luyten J, Dubruel P. Implantable (bio)polymer coated titanium scaffolds: a review. Curr Pharm Des 2012; 18(18): 2576-90.
[http://dx.doi.org/10.2174/138161212800492903] [PMID: 22512448]

[33]   Service RF. Tissue engineers build new bone. Science 2000; 289(5484): 1498-500.
[http://dx.doi.org/10.1126/science.289.5484.1498] [PMID: 10991738]

[34]   Deligianni DD, Katsala ND, Koutsoukos PG, Missirlis YF. Effect of surface roughness of hydroxyapatite on human bone marrow cell adhesion, proliferation, differentiation and detachment strength. Biomaterials 2000; 22(1): 87-96.
[http://dx.doi.org/10.1016/S0142-9612(00)00174-5] [PMID: 11085388]

[35]   Fini M, Giardino R, Borsari V, *et al.* *In vitro* behaviour of osteoblasts cultured on orthopaedic biomaterials with different surface roughness, uncoated and fluorohydroxyapatite-coated, relative to the *in vivo* osteointegration rate. Int J Artif Organs 2003; 26(6): 520-8.
[http://dx.doi.org/10.1177/039139880302600611] [PMID: 12866658]

[36]   Kumar G, Waters MS, Farooque TM, Young MF, Simon CG Jr. Freeform fabricated scaffolds with roughened struts that enhance both stem cell proliferation and differentiation by controlling cell shape. Biomaterials 2012; 33(16): 4022-30.
[http://dx.doi.org/10.1016/j.biomaterials.2012.02.048] [PMID: 22417619]

[37]   Holthaus MG, Treccani L, Rezwan K. Osteoblast viability on hydroxyapatite with well-adjusted submicron and micron surface roughness as monitored by the proliferation reagent WST-1. J Biomater Appl 2013; 27(7): 791-800.
[http://dx.doi.org/10.1177/0885328211426354] [PMID: 22262576]

[38]   Bianchi M, Urquia Edreira ER, Wolke JGC, *et al.* Substrate geometry directs the *in vitro* mineralization of calcium phosphate ceramics. Acta Biomater 2014; 10(2): 661-9.
[http://dx.doi.org/10.1016/j.actbio.2013.10.026] [PMID: 24184857]

[39]   Sadowska JM, Wei F, Guo J, Guillem-Marti J, Ginebra MP, Xiao Y. Effect of nano-structural properties of biomimetic hydroxyapatite on osteoimmunomodulation. Biomaterials 2018; 181: 318-32.
[http://dx.doi.org/10.1016/j.biomaterials.2018.07.058] [PMID: 30098568]

[40]   Huebsch N, Mooney DJ. Inspiration and application in the evolution of biomaterials. Nature 2009; 462(7272): 426-32.
[http://dx.doi.org/10.1038/nature08601] [PMID: 19940912]

[41]   Karageorgiou V, Kaplan D. Porosity of 3D biomaterial scaffolds and osteogenesis. Biomaterials 2005; 26(27): 5474-91.
[http://dx.doi.org/10.1016/j.biomaterials.2005.02.002] [PMID: 15860204]

[42]   Barba A, Maazouz Y, Diez-Escudero A, *et al.* Osteogenesis by foamed and 3D-printed nanostructured calcium phosphate scaffolds: Effect of pore architecture. Acta Biomater 2018; 79: 135-47.

[http://dx.doi.org/10.1016/j.actbio.2018.09.003] [PMID: 30195084]

[43]   Ebaretonbofa E, Evans JRG. High porosity hydroxyapatite foam scaffolds for bone substitute. J Porous Mater 2002; 9(4): 257-63.
[http://dx.doi.org/10.1023/A:1021696711468]

[44]   Hing KA. Bioceramic bone graft substitutes: Influence of porosity and chemistry. Int J Appl Ceram Technol 2005; 2(3): 184-99.
[http://dx.doi.org/10.1111/j.1744-7402.2005.02020.x]

[45]   Malmström J, Adolfsson E, Arvidsson A, Thomsen P. Bone response inside free-form fabricated macroporous hydroxyapatite scaffolds with and without an open microporosity. Clin Implant Dent Relat Res 2007; 9(2): 79-88.
[http://dx.doi.org/10.1111/j.1708-8208.2007.00031.x] [PMID: 17535331]

[46]   Lew KS, Othman R, Ishikawa K, Yeoh FY. Macroporous bioceramics: A remarkable material for bone regeneration. J Biomater Appl 2012; 27(3): 345-58.
[http://dx.doi.org/10.1177/0885328211406459] [PMID: 21862511]

[47]   Ren LM, Todo M, Arahira T, Yoshikawa H, Myoui A. A comparative biomechanical study of bone ingrowth in two porous hydroxyapatite bioceramics. Appl Surf Sci 2012; 262: 81-8.
[http://dx.doi.org/10.1016/j.apsusc.2012.02.060]

[48]   Guda T, Walker JA, Singleton B, *et al.* Hydroxyapatite scaffold pore architecture effects in large bone defects *in vivo.* J Biomater Appl 2014; 28(7): 1016-27.
[http://dx.doi.org/10.1177/0885328213491790] [PMID: 23771772]

[49]   Shao R, Quan R, Zhang L, Wei X, Yang D, Xie S. Porous hydroxyapatite bioceramics in bone tissue engineering: Current uses and perspectives. J Ceram Soc Jpn 2015; 123(1433): 17-20.
[http://dx.doi.org/10.2109/jcersj2.123.17]

[50]   Zhou Y, Chen F, Ho S, Woodruff M, Lim T, Hutmacher D. Combined marrow stromal cell-sheet techniques and high-strength biodegradable composite scaffolds for engineered functional bone grafts. Biomaterials 2007; 28(5): 814-24.
[http://dx.doi.org/10.1016/j.biomaterials.2006.09.032] [PMID: 17045643]

[51]   Vitale-Brovarone C, Baino F, Verné E. High strength bioactive glass-ceramic scaffolds for bone regeneration. J Mater Sci Mater Med 2009; 20(2): 643-53.
[http://dx.doi.org/10.1007/s10856-008-3605-0] [PMID: 18941868]

[52]   Stevens MM. Biomaterials for bone tissue engineering. Mater Today 2008; 11(5): 18-25.
[http://dx.doi.org/10.1016/S1369-7021(08)70086-5]

[53]   Artzi Z, Weinreb M, Givol N, *et al.* Biomaterial resorption rate and healing site morphology of inorganic bovine bone and beta-tricalcium phosphate in the canine: A 24-month longitudinal histologic study and morphometric analysis. Int J Oral Maxillofac Implants 2004; 19(3): 357-68.
[PMID: 15214219]

[54]   Burg KJL, Porter S, Kellam JF. Biomaterial developments for bone tissue engineering. Biomaterials 2000; 21(23): 2347-59.
[http://dx.doi.org/10.1016/S0142-9612(00)00102-2] [PMID: 11055282]

[55]   Ajaal TT, Smith RW. Employing the Taguchi method in optimizing the scaffold production process for artificial bone grafts. J Mater Process Technol 2009; 209(3): 1521-32.
[http://dx.doi.org/10.1016/j.jmatprotec.2008.04.001]

[56]   Daculsi G. Smart scaffolds: The future of bioceramic. J Mater Sci Mater Med 2015; 26(4): 154.
[http://dx.doi.org/10.1007/s10856-015-5482-7] [PMID: 25779511]

[57]   Daculsi G, Miramond T, Borget P, Baroth S. Smart calcium phosphate bioceramic scaffold for bone tissue engineering. Key Eng Mater 2012; 529-530: 19-23.
[http://dx.doi.org/10.4028/www.scientific.net/KEM.529-530.19]

[58]   Vallet-Regí M. Bioceramics: Where do we come from and which are the future expectations. Key Eng Mater 2008; 377: 1-18.
[http://dx.doi.org/10.4028/www.scientific.net/KEM.377.1]

[59]   Sakamoto M, Matsumoto T. Development and evaluation of superporous ceramics bone tissue scaffold materials with triple pore structure a) hydroxyapatite, b) beta-tricalcium phosphate. In: Tal H, Ed. Bone regeneration. Rijeka, Croatia: InTech Europe 2012; pp. 301-20.
[http://dx.doi.org/10.5772/33901]

[60]   Swain SK, Bhattacharyya S, Sarkar D. Fabrication of porous hydroxyapatite scaffold *via* polyethylene glycol-polyvinyl alcohol hydrogel state. Mater Res Bull 2015; 64: 257-61.
[http://dx.doi.org/10.1016/j.materresbull.2014.12.072]

[61]   Balázsi C, Wéber F, Kövér Z, Horváth E, Németh C. Preparation of calcium–phosphate bioceramics from natural resources. J Eur Ceram Soc 2007; 27(2-3): 1601-6.
[http://dx.doi.org/10.1016/j.jeurceramsoc.2006.04.016]

[62]   Gergely G, Wéber F, Lukács I, *et al.* Nano-hydroxyapatite preparation from biogenic raw materials. Cent Eur J Chem 2010; 8: 375-81.

[63]   Mondal S, Mahata S, Kundu S, Mondal B. Processing of natural resourced hydroxyapatite ceramics from fish scale. Adv Appl Ceramics 2010; 109(4): 234-9.
[http://dx.doi.org/10.1179/174367613X13789812714425]

[64]   Lim KT, Suh JD, Kim J, Choung PH, Chung JH. Calcium phosphate bioceramics fabricated from extracted human teeth for tooth tissue engineering. J Biomed Mater Res B Appl Biomater 2011; 99B(2): 399-411.
[http://dx.doi.org/10.1002/jbm.b.31912] [PMID: 21953824]

[65]   Seo DS, Hwang KH, Yoon SY, Lee JK. Fabrication of hydroxyapatite bioceramics from the recycling of pig bone. J Ceram Process Res 2012; 13: 586-9.

[66]   Ho WF, Hsu HC, Hsu SK, Hung CW, Wu SC. Calcium phosphate bioceramics synthesized from eggshell powders through a solid state reaction. Ceram Int 2013; 39(6): 6467-73.
[http://dx.doi.org/10.1016/j.ceramint.2013.01.076]

[67]   González-Rodríguez L, López-Álvarez M, Astray S, Solla EL, Serra J, González P. Hydroxyapatite scaffolds derived from deer antler: Structure dependence on processing temperature. Mater Charact 2019; 155: 109805.
[http://dx.doi.org/10.1016/j.matchar.2019.109805]

[68]   Lim KT, Patel DK, Dutta SD, *et al.* Human teeth-derived bioceramics for improved bone regeneration. Nanomaterials 2020; 10(12): 2396.
[http://dx.doi.org/10.3390/nano10122396] [PMID: 33266215]

[69]   Grigoraviciute-Puroniene I, Zarkov A, Tsuru K, Ishikawa K, Kareiva A. A novel synthetic approach for the calcium hydroxyapatite from the food products. J Sol-Gel Sci Technol 2019; 91(1): 63-71.
[http://dx.doi.org/10.1007/s10971-019-05020-4]

[70]   Tosun GU, Sakhno Y, Jaisi DP. Synthesis of hydroxyapatite nanoparticles from phosphorus recovered from animal wastes. ACS Sustain Chem& Eng 2021; 9: 15117-26.
[http://dx.doi.org/10.1021/acssuschemeng.1c01006]

[71]   Layrolle P, Ito A, Tateishi T. Sol-gel synthesis of amorphous calcium phosphate and sintering into microporous hydroxyapatite bioceramics. J Am Ceram Soc 1998; 81(6): 1421-8.
[http://dx.doi.org/10.1111/j.1151-2916.1998.tb02499.x]

[72]   Ozgür Engin N, Tas AC. Manufacture of macroporous calcium hydroxyapatite bioceramics. J Eur Ceram Soc 1999; 19(13-14): 2569-72.
[http://dx.doi.org/10.1016/S0955-2219(99)00131-4]

[73]   Ahn ES, Gleason NJ, Nakahira A, Ying JY. Nanostructure processing of hydroxyapatite-based

bioceramics. Nano Lett 2001; 1(3): 149-53.
[http://dx.doi.org/10.1021/nl0055299]

[74]   Khalil KA, Kim SW, Dharmaraj N, Kim KW, Kim HY. Novel mechanism to improve toughness of the hydroxyapatite bioceramics using high-frequency induction heat sintering. J Mater Process Technol 2007; 187-188: 417-20.
[http://dx.doi.org/10.1016/j.jmatprotec.2006.11.105]

[75]   Laasri S, Taha M, Laghzizil A, Hlil EK, Chevalier J. The affect of densification and dehydroxylation on the mechanical properties of stoichiometric hydroxyapatite bioceramics. Mater Res Bull 2010; 45(10): 1433-7.
[http://dx.doi.org/10.1016/j.materresbull.2010.06.040]

[76]   Kitamura M, Ohtsuki C, Ogata S, Kamitakahara M, Tanihara M. Microstructure and bioresorbable properties of α-TCP ceramic porous body fabricated by direct casting method. Mater Trans 2004; 45(4): 983-8.
[http://dx.doi.org/10.2320/matertrans.45.983]

[77]   Kawagoe D, Ioku K, Fujimori H, Goto S. Transparent β-tricalcium phosphate ceramics prepared by spark plasma sintering. J Ceram Soc Jpn 2004; 112(1308): 462-3.
[http://dx.doi.org/10.2109/jcersj.112.462]

[78]   Wang CX, Zhou X, Wang M. Influence of sintering temperatures on hardness and Young's modulus of tricalcium phosphate bioceramic by nanoindentation technique. Mater Charact 2004; 52(4-5): 301-7.
[http://dx.doi.org/10.1016/j.matchar.2004.06.007]

[79]   Ioku K, Kawachi G, Nakahara K, *et al.* Porous granules of β-tricalcium phosphate composed of rod-shaped particles. Key Eng Mater 2006; 309-311: 1059-62.
[http://dx.doi.org/10.4028/www.scientific.net/KEM.309-311.1059]

[80]   Kamitakahara M, Ohtsuki C, Miyazaki T. Review paper: Behavior of ceramic biomaterials derived from tricalcium phosphate in physiological condition. J Biomater Appl 2008; 23(3): 197-212.
[http://dx.doi.org/10.1177/0885328208096798] [PMID: 18996965]

[81]   Vorndran E, Klarner M, Klammert U, *et al.* 3D powder printing of β-tricalcium phosphate ceramics using different strategies. Adv Eng Mater 2008; 10(12): B67-71.
[http://dx.doi.org/10.1002/adem.200800179]

[82]   Descamps M, Duhoo T, Monchau F, *et al.* Manufacture of macroporous β-tricalcium phosphate bioceramics. J Eur Ceram Soc 2008; 28(1): 149-57.
[http://dx.doi.org/10.1016/j.jeurceramsoc.2007.05.025]

[83]   Liu Y, Kim JH, Young D, Kim S, Nishimoto SK, Yang Y. Novel template-casting technique for fabricating β-tricalcium phosphate scaffolds with high interconnectivity and mechanical strength and *in vitro* cell responses. J Biomed Mater Res A 2010; 92A(3): 997-1006.
[http://dx.doi.org/10.1002/jbm.a.32443] [PMID: 19296544]

[84]   Carrodeguas RG, De Aza S. α-Tricalcium phosphate: Synthesis, properties and biomedical applications. Acta Biomater 2011; 7(10): 3536-46.
[http://dx.doi.org/10.1016/j.actbio.2011.06.019] [PMID: 21712105]

[85]   Kim IY, Wen J, Ohtsuki C. Fabrication of α-tricalcium phosphate ceramics through two-step sintering. Key Eng Mater 2014; 631: 78-82.
[http://dx.doi.org/10.4028/www.scientific.net/KEM.631.78]

[86]   Bohner M, Santoni BLG, Döbelin N. β-tricalcium phosphate for bone substitution: Synthesis and properties. Acta Biomater 2020; 113: 23-41.
[http://dx.doi.org/10.1016/j.actbio.2020.06.022] [PMID: 32565369]

[87]   Dorozhkin SV. Multiphasic calcium orthophosphate (CaPO$_4$) bioceramics and their biomedical applications. Ceram Int 2016; 42(6): 6529-54.

[http://dx.doi.org/10.1016/j.ceramint.2016.01.062]

[88] LeGeros RZ, Lin S, Rohanizadeh R, Mijares D, LeGeros JP. Biphasic calcium phosphate bioceramics: Preparation, properties and applications. J Mater Sci Mater Med 2003; 14(3): 201-9.
[http://dx.doi.org/10.1023/A:1022872421333] [PMID: 15348465]

[89] Daculsi G, Laboux O, Malard O, Weiss P. Current state of the art of biphasic calcium phosphate bioceramics. J Mater Sci Mater Med 2003; 14(3): 195-200.
[http://dx.doi.org/10.1023/A:1022842404495] [PMID: 15348464]

[90] Dorozhkina EI, Dorozhkin SV. Mechanism of the solid-state transformation of a calcium-deficient hydroxyapatite (CDHA) into biphasic calcium phosphate (BCP) at elevated temperatures. Chem Mater 2002; 14(10): 4267-72.
[http://dx.doi.org/10.1021/cm0203060]

[91] Daculsi G. Biphasic calcium phosphate granules concept for injectable and mouldable bone substitute. Adv Sci Technol 2006; 49: 9-13.
[http://dx.doi.org/10.4028/www.scientific.net/AST.49.9]

[92] Lecomte A, Gautier H, Bouler JM, *et al.* Biphasic calcium phosphate: A comparative study of interconnected porosity in two ceramics. J Biomed Mater Res B Appl Biomater 2008; 84B(1): 1-6.
[http://dx.doi.org/10.1002/jbm.b.30569] [PMID: 17907206]

[93] Daculsi G, Baroth S, LeGeros RZ. 20 years of biphasic calcium phosphate bioceramics development and applications. Ceram Eng Sci Proc 2010; 30: 45-58.

[94] Lukić M, Stojanović Z, Škapin SD, *et al.* Dense fine-grained biphasic calcium phosphate (BCP) bioceramics designed by two-step sintering. J Eur Ceram Soc 2011; 31(1-2): 19-27.
[http://dx.doi.org/10.1016/j.jeurceramsoc.2010.09.006]

[95] Descamps M, Boilet L, Moreau G, *et al.* Processing and properties of biphasic calcium phosphates bioceramics obtained by pressureless sintering and hot isostatic pressing. J Eur Ceram Soc 2013; 33(7): 1263-70.
[http://dx.doi.org/10.1016/j.jeurceramsoc.2012.12.020]

[96] Kermani F, Kargozar S, Tayarani-Najaran Z, Yousefi A, Beidokhti SM, Moayed MH. Synthesis of nano HA/βTCP mesoporous particles using a simple modification in granulation method. Mater Sci Eng C 2019; 96: 859-71.
[http://dx.doi.org/10.1016/j.msec.2018.11.045] [PMID: 30606600]

[97] Li Y, Kong F, Weng W. Preparation and characterization of novel biphasic calcium phosphate powders (α-TCP/HA) derived from carbonated amorphous calcium phosphates. J Biomed Mater Res B Appl Biomater 2009; 89B(2): 508-17.
[http://dx.doi.org/10.1002/jbm.b.31242] [PMID: 18937266]

[98] Sureshbabu S, Komath M, Varma HK. *in situ* formation of hydroxyapatite –alpha tricalcium phosphate biphasic ceramics with higher strength and bioactivity. J Am Ceram Soc 2012; 95(3): 915-24.
[http://dx.doi.org/10.1111/j.1551-2916.2011.04987.x]

[99] Radovanović Ž, Jokić B, Veljović D, *et al.* Antimicrobial activity and biocompatibility of Ag^+- and Cu^{2+}-doped biphasic hydroxyapatite/α-tricalcium phosphate obtained from hydrothermally synthesized Ag^+- and Cu^{2+}-doped hydroxyapatite. Appl Surf Sci 2014; 307: 513-9.
[http://dx.doi.org/10.1016/j.apsusc.2014.04.066]

[100] Oishi M, Ohtsuki C, Kitamura M, *et al.* Fabrication and chemical durability of porous bodies consisting of biphasic tricalcium phosphates. Phosphorus Res Bull 2004; 17(0): 95-100.
[http://dx.doi.org/10.3363/prb1992.17.0_95]

[101] Kamitakahara M, Ohtsuki C, Oishi M, Ogata S, Miyazaki T, Tanihara M. Preparation of porous biphasic tricalcium phosphate and its *in vivo* behavior. Key Eng Mater 2005; 284-286: 281-4.
[http://dx.doi.org/10.4028/www.scientific.net/KEM.284-286.281]

[102] Wang RB, Weng WJ, Deng XL, *et al.* Dissolution behavior of submicron biphasic tricalcium

phosphate powders. Key Eng Mater 2006; 309-311: 223-6.
[http://dx.doi.org/10.4028/www.scientific.net/KEM.309-311.223]

[103] Li Y, Weng W, Tam KC. Novel highly biodegradable biphasic tricalcium phosphates composed of α-tricalcium phosphate and β-tricalcium phosphate. Acta Biomater 2007; 3(2): 251-4.
[http://dx.doi.org/10.1016/j.actbio.2006.07.003] [PMID: 16979393]

[104] Zou C, Cheng K, Weng W, *et al.* Characterization and dissolution–reprecipitation behavior of biphasic tricalcium phosphate powders. J Alloys Compd 2011; 509(24): 6852-8.
[http://dx.doi.org/10.1016/j.jallcom.2011.03.158]

[105] Xie L, Yu H, Deng Y, Yang W, Liao L, Long Q. Preparation and *in vitro* degradation study of the porous dual alpha/beta-tricalcium phosphate bioceramics. Mater Res Innov 2016; 20(7): 530-7.
[http://dx.doi.org/10.1179/1433075X15Y.0000000079]

[106] Albuquerque JSV, Nogueira REFQ, Pinheiro da SIlva TD, Lima DO, Prado da Silva MH. Porous triphasic calcium phosphate bioceramics. Key Eng Mater 2003; 254-256: 1021-4.
[http://dx.doi.org/10.4028/www.scientific.net/KEM.254-256.1021]

[107] Mendonça F, Louro LHL, de Campos JB, Prado da Silva MH. Porous biphasic and triphasic bioceramics scaffolds produced by gelcasting. Key Eng Mater 2007; 361-363: 27-30.
[http://dx.doi.org/10.4028/www.scientific.net/KEM.361-363.27]

[108] Vani R, Girija EK, Elayaraja K, Prakash Parthiban S, Kesavamoorthy R, Narayana Kalkura S. Hydrothermal synthesis of porous triphasic hydroxyapatite/(α and β) tricalcium phosphate. J Mater Sci Mater Med 2009; 20(S1): 43-8.
[http://dx.doi.org/10.1007/s10856-008-3480-8] [PMID: 18560768]

[109] Ahn MK, Moon YW, Koh YH, Kim HE. Production of highly porous triphasic calcium phosphate scaffolds with excellent *in vitro* bioactivity using vacuum-assisted foaming of ceramic suspension (VFC) technique. Ceram Int 2013; 39(5): 5879-85.
[http://dx.doi.org/10.1016/j.ceramint.2013.01.006]

[110] Dorozhkin SV. Self-setting calcium orthophosphate (CaPO$_4$) formulations and their biomedical applications. Adv. Nano-Bio. Mater Dev 2019; 3: 321-421.

[111] Tamimi F, Sheikh Z, Barralet J. Dicalcium phosphate cements: Brushite and monetite. Acta Biomater 2012; 8(2): 474-87.
[http://dx.doi.org/10.1016/j.actbio.2011.08.005] [PMID: 21856456]

[112] Drouet C, Largeot C, Raimbeaux G, *et al.* Bioceramics: Spark plasma sintering (SPS) of calcium phosphates. Adv Sci Technol 2006; 49: 45-50.
[http://dx.doi.org/10.4028/www.scientific.net/AST.49.45]

[113] Ishihara S, Matsumoto T, Onoki T, Sohmura T, Nakahira A. New concept bioceramics composed of octacalcium phosphate (OCP) and dicarboxylic acid-intercalated OCP *via* hydrothermal hot-pressing. Mater Sci Eng C 2009; 29(6): 1885-8.
[http://dx.doi.org/10.1016/j.msec.2009.02.023]

[114] Barinov SM, Komlev VS. Osteoinductive ceramic materials for bone tissue restoration: Octacalcium phosphate (review). Inorg Mater: Appl Res 2010; 1(3): 175-81.
[http://dx.doi.org/10.1134/S2075113310030019]

[115] Moseke C, Gbureck U. Tetracalcium phosphate: Synthesis, properties and biomedical applications. Acta Biomater 2010; 6(10): 3815-23.
[http://dx.doi.org/10.1016/j.actbio.2010.04.020] [PMID: 20438869]

[116] Morimoto S, Anada T, Honda Y, Suzuki O. Comparative study on *in vitro* biocompatibility of synthetic octacalcium phosphate and calcium phosphate ceramics used clinically. Biomed Mater 2012; 7(4): 045020.
[http://dx.doi.org/10.1088/1748-6041/7/4/045020] [PMID: 22740587]

[117] Tamimi F, Nihouannen DL, Eimar H, Sheikh Z, Komarova S, Barralet J. The effect of autoclaving on

the physical and biological properties of dicalcium phosphate dihydrate bioceramics: Brushite vs. monetite. Acta Biomater 2012; 8(8): 3161-9.
[http://dx.doi.org/10.1016/j.actbio.2012.04.025] [PMID: 22522010]

[118] Suzuki O. Octacalcium phosphate (OCP)-based bone substitute materials. Jpn Dent Sci Rev 2013; 49(2): 58-71.
[http://dx.doi.org/10.1016/j.jdsr.2013.01.001]

[119] Komlev VS, Barinov SM, Bozo II, *et al.* Bioceramics composed of octacalcium phosphate demonstrate enhanced biological behavior. ACS Appl Mater Interfaces 2014; 6(19): 16610-20.
[http://dx.doi.org/10.1021/am502583p] [PMID: 25184694]

[120] Zhou H, Yang L, Gbureck U, Bhaduri SB, Sikder P. Monetite, an important calcium phosphate compound–Its synthesis, properties and applications in orthopedics. Acta Biomater 2021; 127: 41-55.
[http://dx.doi.org/10.1016/j.actbio.2021.03.050] [PMID: 33812072]

[121] LeGeros RZ. Calcium phosphates in oral biology and medicine Monographs in oral science. Basel, Switzerland: Karger 1991; Vol. 15.

[122] Narasaraju TSB, Phebe DE. Some physico-chemical aspects of hydroxylapatite. J Mater Sci 1996; 31(1): 1-21.
[http://dx.doi.org/10.1007/BF00355120]

[123] Elliott JC. Structure and chemistry of the apatites and other calcium orthophosphates Studies in inorganic chemistry. Amsterdam, Netherlands: Elsevier 1994; Vol. 18.

[124] Brown PW, Constantz B, Eds. Hydroxyapatite and related materials. Boca Raton, FL, USA: CRC Press 1994.

[125] Amjad Z, Ed. Calcium phosphates in biological and industrial systems. Boston, MA, USA: Kluwer Academic Publishers 1997.

[126] Da Silva RV, Bertran CA, Kawachi EY, Camilli JA. Repair of cranial bone defects with calcium phosphate ceramic implant or autogenous bone graft. J Craniofac Surg 2007; 18(2): 281-6.
[http://dx.doi.org/10.1097/scs.0b013e31802d8ac4] [PMID: 17414276]

[127] Okanoue Y, Ikeuchi M, Takemasa R, *et al.* Comparison of *in vivo* bioactivity and compressive strength of a novel superporous hydroxyapatite with beta-tricalcium phosphates. Arch Orthop Trauma Surg 2012; 132(11): 1603-10.
[http://dx.doi.org/10.1007/s00402-012-1578-4] [PMID: 22760581]

[128] Draenert M, Draenert A, Draenert K. Osseointegration of hydroxyapatite and remodeling-resorption of tricalciumphosphate ceramics. Microsc Res Tech 2013; 76(4): 370-80.
[http://dx.doi.org/10.1002/jemt.22176] [PMID: 23390042]

[129] Okuda T, Ioku K, Yonezawa I, *et al.* The slow resorption with replacement by bone of a hydrothermally synthesized pure calcium-deficient hydroxyapatite. Biomaterials 2008; 29(18): 2719-28.
[http://dx.doi.org/10.1016/j.biomaterials.2008.03.028] [PMID: 18403011]

[130] Daculsi G, Bouler JM, LeGeros RZ. Adaptive crystal formation in normal and pathological calcifications in synthetic calcium phosphate and related biomaterials. Int Rev Cytol 1997; 172: 129-91.
[http://dx.doi.org/10.1016/S0074-7696(08)62360-8] [PMID: 9102393]

[131] Zhu XD, Zhang HJ, Fan HS, Li W, Zhang XD. Effect of phase composition and microstructure of calcium phosphate ceramic particles on protein adsorption. Acta Biomater 2010; 6(4): 1536-41.
[http://dx.doi.org/10.1016/j.actbio.2009.10.032] [PMID: 19857608]

[132] Bohner M. Calcium orthophosphates in medicine: from ceramics to calcium phosphate cements. Injury 2000; 31 (Suppl. 4): D37-47.
[http://dx.doi.org/10.1016/S0020-1383(00)80022-4] [PMID: 11270080]

[133] Bohner M, Loosli Y, Baroud G, Lacroix D. Commentary: Deciphering the link between architecture and biological response of a bone graft substitute. Acta Biomater 2011; 7(2): 478-84.
[http://dx.doi.org/10.1016/j.actbio.2010.08.008] [PMID: 20709195]

[134] Peppas NA, Langer R. New challenges in biomaterials. Science 1994; 263(5154): 1715-20.
[http://dx.doi.org/10.1126/science.8134835] [PMID: 8134835]

[135] Hench LL. Biomaterials: A forecast for the future. Biomaterials 1998; 19(16): 1419-23.
[http://dx.doi.org/10.1016/S0142-9612(98)00133-1] [PMID: 9794512]

[136] Barrère F, Mahmood TA, de Groot K, van Blitterswijk CA. Advanced biomaterials for skeletal tissue regeneration: Instructive and smart functions. Mater Sci Eng Rep 2008; 59(1-6): 38-71.
[http://dx.doi.org/10.1016/j.mser.2007.12.001]

[137] Liu H, Webster TJ. Nanomedicine for implants: A review of studies and necessary experimental tools. Biomaterials 2007; 28(2): 354-69.
[http://dx.doi.org/10.1016/j.biomaterials.2006.08.049] [PMID: 21898921]

[138] Wang C, Duan Y, Markovic B, et al. Proliferation and bone-related gene expression of osteoblasts grown on hydroxyapatite ceramics sintered at different temperature. Biomaterials 2004; 25(15): 2949-56.
[http://dx.doi.org/10.1016/j.biomaterials.2003.09.088] [PMID: 14967527]

[139] Samavedi S, Whittington AR, Goldstein AS. Calcium phosphate ceramics in bone tissue engineering: A review of properties and their influence on cell behavior. Acta Biomater 2013; 9(9): 8037-45.
[http://dx.doi.org/10.1016/j.actbio.2013.06.014] [PMID: 23791671]

[140] Matsumoto T, Okazaki M, Nakahira A, Sasaki J, Egusa H, Sohmura T. Modification of apatite materials for bone tissue engineering and drug delivery carriers. Curr Med Chem 2007; 14(25): 2726-33.
[http://dx.doi.org/10.2174/092986707782023208] [PMID: 17979722]

[141] Chai YC, Carlier A, Bolander J, et al. Current views on calcium phosphate osteogenicity and the translation into effective bone regeneration strategies. Acta Biomater 2012; 8(11): 3876-87.
[http://dx.doi.org/10.1016/j.actbio.2012.07.002] [PMID: 22796326]

[142] Denry I, Kuhn LT. Design and characterization of calcium phosphate ceramic scaffolds for bone tissue engineering. Dent Mater 2016; 32(1): 43-53.
[http://dx.doi.org/10.1016/j.dental.2015.09.008] [PMID: 26423007]

[143] Lodoso-Torrecilla I, Klein Gunnewiek R, Grosfeld EC, et al. Bioinorganic supplementation of calcium phosphate-based bone substitutes to improve in vivo performance: A systematic review and meta-analysis of animal studies. Biomater Sci 2020; 8(17): 4792-809.
[http://dx.doi.org/10.1039/D0BM00599A] [PMID: 32729591]

[144] Ishikawa K, Miyamoto Y, Tsuchiya A, Hayashi K, Tsuru K, Ohe G. Physical and histological comparison of hydroxyapatite, carbonate apatite, and β-tricalcium phosphate bone substitutes. Materials 2018; 11(10): 1993.
[http://dx.doi.org/10.3390/ma11101993] [PMID: 30332751]

[145] Traykova T, Aparicio C, Ginebra MP, Planell JA. Bioceramics as nanomaterials. Nanomedicine 2006; 1(1): 91-106.
[http://dx.doi.org/10.2217/17435889.1.1.91] [PMID: 17716212]

[146] Kalita SJ, Bhardwaj A, Bhatt HA. Nanocrystalline calcium phosphate ceramics in biomedical engineering. Mater Sci Eng C 2007; 27(3): 441-9.
[http://dx.doi.org/10.1016/j.msec.2006.05.018]

[147] Dorozhkin SV. Nanometric calcium orthophosphates (CaPO₄): Preparation, properties and biomedical applications. Adv. Nano-Bio. Mater Dev 2019; 3: 422-513.

[148] Šupová M. Isolation and preparation of nanoscale bioapatites from natural sources: A review. J

Nanosci Nanotechnol 2014; 14(1): 546-63.
[http://dx.doi.org/10.1166/jnn.2014.8895] [PMID: 24730282]

[149]   Zhao J, Liu Y, Sun WB, Zhang H. Amorphous calcium phosphate and its application in dentistry. Chem Cent J 2011; 5: 40.
[http://dx.doi.org/10.1186/1752-153X-5-40]

[150]   Dorozhkin SV. Synthetic amorphous calcium phosphates (ACPs): Preparation, structure, properties, and biomedical applications. Biomater Sci 2021; 9(23): 7748-98.
[http://dx.doi.org/10.1039/D1BM01239H] [PMID: 34755730]

[151]   Venkatesan J, Kim SK. Nano-hydroxyapatite composite biomaterials for bone tissue engineering--a review. J Biomed Nanotechnol 2014; 10(10): 3124-40.
[http://dx.doi.org/10.1166/jbn.2014.1893] [PMID: 25992432]

[152]   Holzmeister I, Schamel M, Groll J, Gbureck U, Vorndran E. Artificial inorganic biohybrids: The functional combination of microorganisms and cells with inorganic materials. Acta Biomater 2018; 74: 17-35.
[http://dx.doi.org/10.1016/j.actbio.2018.04.042] [PMID: 29698705]

[153]   Wu Y, Hench LL, Du J, Choy KL, Guo J. Preparation of hydroxyapatite fibers by electrospinning technique. J Am Ceram Soc 2004; 87(10): 1988-91.
[http://dx.doi.org/10.1111/j.1151-2916.2004.tb06351.x]

[154]   Ramanan SR, Venkatesh R. A study of hydroxyapatite fibers prepared *via* sol–gel route. Mater Lett 2004; 58(26): 3320-3.
[http://dx.doi.org/10.1016/j.matlet.2004.06.030]

[155]   Aizawa M, Porter AE, Best SM, Bonfield W. Ultrastructural observation of single-crystal apatite fibres. Biomaterials 2005; 26(17): 3427-33.
[http://dx.doi.org/10.1016/j.biomaterials.2004.09.044] [PMID: 15621231]

[156]   Park YM, Ryu SC, Yoon SY, Stevens R, Park HC. Preparation of whisker-shaped hydroxyapatite/β-tricalcium phosphate composite. Mater Chem Phys 2008; 109(2-3): 440-7.
[http://dx.doi.org/10.1016/j.matchemphys.2007.12.013]

[157]   Aizawa M, Ueno H, Itatani K, Okada I. Syntheses of calcium-deficient apatite fibres by a homogeneous precipitation method and their characterizations. J Eur Ceram Soc 2006; 26(4-5): 501-7.
[http://dx.doi.org/10.1016/j.jeurceramsoc.2005.07.007]

[158]   Seo DS, Lee JK. Synthesis of hydroxyapatite whiskers through dissolution–reprecipitation process using EDTA. J Cryst Growth 2008; 310(7-9): 2162-7.
[http://dx.doi.org/10.1016/j.jcrysgro.2007.11.028]

[159]   Tas AC. Formation of calcium phosphate whiskers in hydrogen peroxide (H_2O_2) solutions at 90°C. J Am Ceram Soc 2007; 90(8): 2358-62.
[http://dx.doi.org/10.1111/j.1551-2916.2007.01768.x]

[160]   Neira IS, Guitián F, Taniguchi T, Watanabe T, Yoshimura M. Hydrothermal synthesis of hydroxyapatite whiskers with sharp faceted hexagonal morphology. J Mater Sci 2008; 43(7): 2171-8.
[http://dx.doi.org/10.1007/s10853-007-2032-9]

[161]   Yang HY, Yang SF, Chi XP, *et al.* Sintering behaviour of calcium phosphate filaments for use as hard tissue scaffolds. J Eur Ceram Soc 2008; 28(1): 159-67.
[http://dx.doi.org/10.1016/j.jeurceramsoc.2007.04.013]

[162]   Junginger M, Kübel C, Schacher FH, Müller AHE, Taubert A. Crystal structure and chemical composition of biomimetic calcium phosphate nanofibers. RSC Advances 2013; 3(28): 11301-8.
[http://dx.doi.org/10.1039/c3ra23348k]

[163]   Cui YS, Yan TT, Wu XP, Chen QH. Preparation and characterization of hydroxyapatite whiskers. Appl Mech Mater 2013; 389: 21-4.
[http://dx.doi.org/10.4028/www.scientific.net/AMM.389.21]

[164] Lee JH, Kim YJ. Hydroxyapatite nanofibers fabricated through electrospinning and sol–gel process. Ceram Int 2014; 40(2): 3361-9.
[http://dx.doi.org/10.1016/j.ceramint.2013.09.096]

[165] Zhang H, Zhu Q. Synthesis of nanospherical and ultralong fibrous hydroxyapatite and reinforcement of biodegradable chitosan/hydroxyapatite composite. Mod Phys Lett B 2009; 23(31n32): 3967-76.
[http://dx.doi.org/10.1142/S0217984909022071]

[166] Wijesinghe WPSL, Mantilaka MMMGPG, Premalal EVA, *et al.* Facile synthesis of both needle-like and spherical hydroxyapatite nanoparticles: Effect of synthetic temperature and calcination on morphology, crystallite size and crystallinity. Mater Sci Eng C 2014; 42: 83-90.
[http://dx.doi.org/10.1016/j.msec.2014.05.032] [PMID: 25063096]

[167] Zhou WY, Wang M, Cheung WL, Guo BC, Jia DM. Synthesis of carbonated hydroxyapatite nanospheres through nanoemulsion. J Mater Sci Mater Med 2008; 19(1): 103-10.
[http://dx.doi.org/10.1007/s10856-007-3156-9] [PMID: 17577636]

[168] Lim JH, Park JH, Park EK, *et al.* Fully interconnected globular porous biphasic calcium phosphate ceramic scaffold facilitates osteogenic repair. Key Eng Mater 2007; 361-363: 119-22.
[http://dx.doi.org/10.4028/www.scientific.net/KEM.361-363.119]

[169] Kawai T, Sekikawa H, Unuma H. Preparation of hollow hydroxyapatite microspheres utilizing poly(divinylbenzene) as a template. J Ceram Soc Jpn 2009; 117(1363): 340-3.
[http://dx.doi.org/10.2109/jcersj2.117.340]

[170] Cho JS, Ko YN, Koo HY, Kang YC. Synthesis of nano-sized biphasic calcium phosphate ceramics with spherical shape by flame spray pyrolysis. J Mater Sci Mater Med 2010; 21(4): 1143-9.
[http://dx.doi.org/10.1007/s10856-009-3980-1] [PMID: 20052521]

[171] Ye F, Guo H, Zhang H, He X. Polymeric micelle-templated synthesis of hydroxyapatite hollow nanoparticles for a drug delivery system. Acta Biomater 2010; 6(6): 2212-8.
[http://dx.doi.org/10.1016/j.actbio.2009.12.014] [PMID: 20004747]

[172] Itatani K, Tsugawa T, Umeda T, Musha Y, Davies IJ, Koda S. Preparation of submicrometer-sized porous spherical hydroxyapatite agglomerates by ultrasonic spray pyrolysis technique. J Ceram Soc Jpn 2010; 118(1378): 462-6.
[http://dx.doi.org/10.2109/jcersj2.118.462]

[173] Xiao W, Fu H, Rahaman MN, Liu Y, Bal BS. Hollow hydroxyapatite microspheres: A novel bioactive and osteoconductive carrier for controlled release of bone morphogenetic protein-2 in bone regeneration. Acta Biomater 2013; 9(9): 8374-83.
[http://dx.doi.org/10.1016/j.actbio.2013.05.029] [PMID: 23747325]

[174] Rahaman MN, Fu H, Xiao W, Liu Y. Bioactive ceramic implants composed of hollow hydroxyapatite microspheres for bone regeneration. Ceram Eng Sci Proc 2014; 34: 67-76.

[175] Ito N, Kamitakahara M, Ioku K. Preparation and evaluation of spherical porous granules of octacalcium phosphate/hydroxyapatite as drug carriers in bone cancer treatment. Mater Lett 2014; 120: 94-6.
[http://dx.doi.org/10.1016/j.matlet.2014.01.040]

[176] Li Z, Wen T, Su Y, Wei X, He C, Wang D. Hollow hydroxyapatite spheres fabrication with three-dimensional hydrogel template. CrystEngComm 2014; 16(20): 4202-9.
[http://dx.doi.org/10.1039/C3CE42517G]

[177] Kovach I, Kosmella S, Prietzel C, Bagdahn C, Koetz J. Nano-porous calcium phosphate balls. Colloids Surf B Biointerfaces 2015; 132: 246-52.
[http://dx.doi.org/10.1016/j.colsurfb.2015.05.021] [PMID: 26052107]

[178] He F, Tian Y, Fang X, Xu Y, Ye J. Porous calcium phosphate composite bioceramic beads. Ceram Int 2018; 44(11): 13430-3.
[http://dx.doi.org/10.1016/j.ceramint.2018.04.109]

[179] Hettich G, Schierjott RA, Epple M, *et al.* Calcium phosphate bone graft substitutes with high mechanical load capacity and high degree of interconnecting porosity. Materials (Basel) 2019; 12(21): 3471.
[http://dx.doi.org/10.3390/ma12213471] [PMID: 31652704]

[180] Chandanshive B, Dyondi D, Ajgaonkar VR, Banerjee R, Khushalani D. Biocompatible calcium phosphate based tubes. J Mater Chem 2010; 20(33): 6923-8.
[http://dx.doi.org/10.1039/c0jm00145g]

[181] Kamitakahara M, Takahashi H, Ioku K. Tubular hydroxyapatite formation through a hydrothermal process from α-tricalcium phosphate with anatase. J Mater Sci 2012; 47(9): 4194-9.
[http://dx.doi.org/10.1007/s10853-012-6274-9]

[182] Ustundag CB, Kaya F, Kamitakahara M, Kaya C, Ioku K. Production of tubular porous hydroxyapatite using electrophoretic deposition. J Ceram Soc Jpn 2012; 120(1408): 569-73.
[http://dx.doi.org/10.2109/jcersj2.120.569]

[183] Li C, Ge X, Li G, Lu H, Ding R. *In situ* hydrothermal crystallization of hexagonal hydroxyapatite tubes from yttrium ion-doped hydroxyapatite by the Kirkendall effect. Mater Sci Eng C 2014; 45: 191-5.
[http://dx.doi.org/10.1016/j.msec.2014.09.012] [PMID: 25491819]

[184] Zhang YG, Zhu YJ, Chen F, Sun TW. Biocompatible, ultralight, strong hydroxyapatite networks based on hydroxyapatite microtubes with excellent permeability and ultralow thermal conductivity. ACS Appl Mater Interfaces 2017; 9(9): 7918-28.
[http://dx.doi.org/10.1021/acsami.6b13328] [PMID: 28240537]

[185] Zhang YG, Zhu YJ, Xiong ZC, Wu J, Chen F. Bioinspired ultralight inorganic aerogel for highly efficient air filtration and oil-water separation. ACS Appl Mater Interfaces 2018; 10(15): 13019-27.
[http://dx.doi.org/10.1021/acsami.8b02081] [PMID: 29611706]

[186] Dadhich P, Dhara S. Calcium phosphate flowers. Mater Today 2017; 20(10): 657-8.
[http://dx.doi.org/10.1016/j.mattod.2017.10.009]

[187] Nonoyama T, Kinoshita T, Higuchi M, *et al.* Arrangement techniques of proteins and cells using amorphous calcium phosphate nanofiber scaffolds. Appl Surf Sci 2012; 262: 8-12.
[http://dx.doi.org/10.1016/j.apsusc.2011.12.009]

[188] Galea LG, Bohner M, Lemaître J, Kohler T, Müller R. Bone substitute: Transforming β-tricalcium phosphate porous scaffolds into monetite. Biomaterials 2008; 29(24-25): 3400-7.
[http://dx.doi.org/10.1016/j.biomaterials.2008.04.041] [PMID: 18495242]

[189] Tamimi F, Torres J, Gbureck U, *et al.* Craniofacial vertical bone augmentation: A comparison between 3D printed monolithic monetite blocks and autologous onlay grafts in the rabbit. Biomaterials 2009; 30(31): 6318-26.
[http://dx.doi.org/10.1016/j.biomaterials.2009.07.049] [PMID: 19695698]

[190] Sheikh Z, Drager J, Zhang YL, Abdallah MN, Tamimi F, Barralet J. Controlling bone graft substitute microstructure to improve bone augmentation. Adv Healthc Mater 2016; 5(13): 1646-55.
[http://dx.doi.org/10.1002/adhm.201600052] [PMID: 27214877]

[191] Oryan A, Alidadi S, Bigham-Sadegh A. Dicalcium phosphate anhydrous: An appropriate bioceramic in regeneration of critical-sized radial bone defects in rats. Calcif Tissue Int 2017; 101(5): 530-44.
[http://dx.doi.org/10.1007/s00223-017-0309-9] [PMID: 28761974]

[192] Sugiura Y, Ishikawa K. Fabrication of pure octacalcium phosphate blocks from dicalcium hydrogen phosphate dihydrate blocks *via* a dissolution–precipitation reaction in a basic solution. Mater Lett 2019; 239: 143-6.
[http://dx.doi.org/10.1016/j.matlet.2018.12.093]

[193] Kim JS, Jang TS, Kim SY, Lee WP. Octacalcium phosphate bone substitute (Bontree®): from basic research to clinical case study. Appl Sci 2021; 11(17): 7921.

[http://dx.doi.org/10.3390/app11177921]

[194] Sohier J, Daculsi G, Sourice S, de Groot K, Layrolle P. Porous beta tricalcium phosphate scaffolds used as a BMP-2 delivery system for bone tissue engineering. J Biomed Mater Res A 2010; 92A(3): 1105-14.
[http://dx.doi.org/10.1002/jbm.a.32467] [PMID: 19301273]

[195] Stähli C, Bohner M, Bashoor-Zadeh M, Doebelin N, Baroud G. Aqueous impregnation of porous β-tricalcium phosphate scaffolds. Acta Biomater 2010; 6(7): 2760-72.
[http://dx.doi.org/10.1016/j.actbio.2010.01.018] [PMID: 20083239]

[196] Lin K, Chen L, Qu H, Lu J, Chang J. Improvement of mechanical properties of macroporous β-tricalcium phosphate bioceramic scaffolds with uniform and interconnected pore structures. Ceram Int 2011; 37(7): 2397-403.
[http://dx.doi.org/10.1016/j.ceramint.2011.03.079]

[197] Wójtowicz J, Leszczyńska J, Chróścicka A, *et al.* Comparative *in vitro* study of calcium phosphate ceramics for their potency as scaffolds for tissue engineering. Biomed Mater Eng 2014; 24(3): 1609-23.
[http://dx.doi.org/10.3233/BME-140965] [PMID: 24840199]

[198] Simon JL, Michna S, Lewis JA, *et al. In vivo* bone response to 3D periodic hydroxyapatite scaffolds assembled by direct ink writing. J Biomed Mater Res A 2007; 83A(3): 747-58.
[http://dx.doi.org/10.1002/jbm.a.31329] [PMID: 17559109]

[199] Yoshikawa H, Myoui A. Bone tissue engineering with porous hydroxyapatite ceramics. J Artif Organs 2005; 8(3): 131-6.
[http://dx.doi.org/10.1007/s10047-005-0292-1] [PMID: 16235028]

[200] Min SH, Jin HH, Park HY, Park IM, Park HC, Yoon SY. Preparation of porous hydroxyapatite scaffolds for bone tissue engineering. Mater Sci Forum 2006; 510-511: 754-7.
[http://dx.doi.org/10.4028/www.scientific.net/MSF.510-511.754]

[201] Deville S, Saiz E, Nalla RK, Tomsia AP. Strong biomimetic hydroxyapatite scaffolds. Adv Sci Technol 2006; 49: 148-52.
[http://dx.doi.org/10.4028/www.scientific.net/AST.49.148]

[202] Buckley CT, O'Kelly KU. Fabrication and characterization of a porous multidomain hydroxyapatite scaffold for bone tissue engineering investigations. J Biomed Mater Res B Appl Biomater 2010; 93B(2): 459-67.
[http://dx.doi.org/10.1002/jbm.b.31603] [PMID: 20166121]

[203] Qin T, Li X, Long H, Bin S, Xu Y. Bioactive tetracalcium phosphate scaffolds fabricated by selective laser sintering for bone regeneration applications. Materials 2020; 13(10): 2268.
[http://dx.doi.org/10.3390/ma13102268] [PMID: 32423078]

[204] Ramay HRR, Zhang M. Biphasic calcium phosphate nanocomposite porous scaffolds for load-bearing bone tissue engineering. Biomaterials 2004; 25(21): 5171-80.
[http://dx.doi.org/10.1016/j.biomaterials.2003.12.023] [PMID: 15109841]

[205] Chen G, Li W, Zhao B, Sun K. A novel biphasic bone scaffold: β-calcium phosphate and amorphous calcium polyphosphate. J Am Ceram Soc 2009; 92(4): 945-8.
[http://dx.doi.org/10.1111/j.1551-2916.2009.02971.x]

[206] Guo D, Xu K, Han Y. The *in situ* synthesis of biphasic calcium phosphate scaffolds with controllable compositions, structures, and adjustable properties. J Biomed Mater Res A 2009; 88A(1): 43-52.
[http://dx.doi.org/10.1002/jbm.a.31844] [PMID: 18257062]

[207] Sarin P, Lee SJ, Apostolov ZD, Kriven WM. Porous biphasic calcium phosphate scaffolds from cuttlefish bone. J Am Ceram Soc 2011; 94(8): 2362-70.
[http://dx.doi.org/10.1111/j.1551-2916.2011.04404.x]

[208] Kim DH, Kim KL, Chun HH, Kim TW, Park HC, Yoon SY. *In vitro* biodegradable and mechanical

performance of biphasic calcium phosphate porous scaffolds with unidirectional macro-pore structure. Ceram Int 2014; 40(6): 8293-300.
[http://dx.doi.org/10.1016/j.ceramint.2014.01.031]

[209] Marques CF, Perera FH, Marote A, *et al.* Biphasic calcium phosphate scaffolds fabricated by direct write assembly: Mechanical, anti-microbial and osteoblastic properties. J Eur Ceram Soc 2017; 37(1): 359-68.
[http://dx.doi.org/10.1016/j.jeurceramsoc.2016.08.018]

[210] Kon M, Ishikawa K, Miyamoto Y, Asaoka K. Development of calcium phosphate based functional gradient bioceramics. Biomaterials 1995; 16(9): 709-14.
[http://dx.doi.org/10.1016/0142-9612(95)99699-M] [PMID: 7578775]

[211] Wong LH, Tio B, Miao X. Functionally graded tricalcium phosphate/fluoroapatite composites. Mater Sci Eng C 2002; 20(1-2): 111-5.
[http://dx.doi.org/10.1016/S0928-4931(02)00020-6]

[212] Tampieri A, Celotti G, Sprio S, Delcogliano A, Franzese S. Porosity-graded hydroxyapatite ceramics to replace natural bone. Biomaterials 2001; 22(11): 1365-70.
[http://dx.doi.org/10.1016/S0142-9612(00)00290-8] [PMID: 11336309]

[213] Werner J, Linner-Krčmar B, Friess W, Greil P. Mechanical properties and *in vitro* cell compatibility of hydroxyapatite ceramics with graded pore structure. Biomaterials 2002; 23(21): 4285-94.
[http://dx.doi.org/10.1016/S0142-9612(02)00191-6] [PMID: 12194531]

[214] Furuichi K, Oaki Y, Ichimiya H, Komotori J, Imai H. Preparation of hierarchically organized calcium phosphate–organic polymer composites by calcification of hydrogel. Sci Technol Adv Mater 2006; 7(2): 219-25.
[http://dx.doi.org/10.1016/j.stam.2005.10.008]

[215] Wei J, Jia J, Wu F, *et al.* Hierarchically microporous/macroporous scaffold of magnesium–calcium phosphate for bone tissue regeneration. Biomaterials 2010; 31(6): 1260-9.
[http://dx.doi.org/10.1016/j.biomaterials.2009.11.005] [PMID: 19931903]

[216] Ye X, Zhou C, Xiao Z, *et al.* Fabrication and characterization of porous 3D whisker-covered calcium phosphate scaffolds. Mater Lett 2014; 128: 179-82.
[http://dx.doi.org/10.1016/j.matlet.2014.04.142]

[217] Zhao R, Shang T, Yuan B, Zhu X, Zhang X, Yang X. Osteoporotic bone recovery by a bamboo-structured bioceramic with controlled release of hydroxyapatite nanoparticles. Bioact Mater 2022; 17: 379-93.
[http://dx.doi.org/10.1016/j.bioactmat.2022.01.007] [PMID: 35386445]

[218] Zhu YJ, Lu BQ. Deformable biomaterials based on ultralong hydroxyapatite nanowires. ACS Biomater Sci Eng 2019; 5(10): 4951-61.
[http://dx.doi.org/10.1021/acsbiomaterials.9b01183] [PMID: 33455242]

[219] Huang GJ, Yu HP, Wang XL, *et al.* Highly porous and elastic aerogel based on ultralong hydroxyapatite nanowires for high-performance bone regeneration and neovascularization. J. Mater. Chem. B 2021, 9, 1277-1287. Erratum 2021; 9: 7566.
[PMID: 34551056]

[220] Gbureck U, Grolms O, Barralet JE, Grover LM, Thull R. Mechanical activation and cement formation of β-tricalcium phosphate. Biomaterials 2003; 24(23): 4123-31.
[http://dx.doi.org/10.1016/S0142-9612(03)00283-7] [PMID: 12853242]

[221] Gbureck U, Barralet JE, Hofmann M, Thull R. Mechanical activation of tetracalcium phosphate. J Am Ceram Soc 2004; 87(2): 311-3.
[http://dx.doi.org/10.1111/j.1551-2916.2004.00311.x]

[222] Bohner M, Luginbühl R, Reber C, Doebelin N, Baroud G, Conforto E. A physical approach to modify the hydraulic reactivity of α-tricalcium phosphate powder. Acta Biomater 2009; 5(9): 3524-35.

[http://dx.doi.org/10.1016/j.actbio.2009.05.024] [PMID: 19470412]

[223] Hagio T, Tanase T, Akiyama J, Iwai K, Asai S. Formation and biological affinity evaluation of crystallographically aligned hydroxyapatite. J Ceram Soc Jpn 2008; 116(1349): 79-82.
[http://dx.doi.org/10.2109/jcersj2.116.79]

[224] Blawas AS, Reichert WM. Protein patterning. Biomaterials 1998; 19(7-9): 595-609.
[http://dx.doi.org/10.1016/S0142-9612(97)00218-4] [PMID: 9663732]

[225] Kasai T, Sato K, Kanematsu Y, Shikimori M, Kanematsu N, Doi Y. Bone tissue engineering using porous carbonate apatite and bone marrow cells. J Craniofac Surg 2010; 21(2): 473-8.
[http://dx.doi.org/10.1097/SCS.0b013e3181cfea6d] [PMID: 20489453]

[226] Wang L, Fan H, Zhang ZY, *et al.* Osteogenesis and angiogenesis of tissue-engineered bone constructed by prevascularized β-tricalcium phosphate scaffold and mesenchymal stem cells. Biomaterials 2010; 31(36): 9452-61.
[http://dx.doi.org/10.1016/j.biomaterials.2010.08.036] [PMID: 20869769]

[227] Benjumeda Wijnhoven I, Vallejos R, Santibanez JF, Millán C, Vivanco JF. Analysis of cell-biomaterial interaction through cellular bridge formation in the interface between hGMSCs and CaP bioceramics. Sci Rep 2020; 10(1): 16493.
[http://dx.doi.org/10.1038/s41598-020-73428-y] [PMID: 33020540]

[228] Sánchez-Salcedo S, Izquierdo-Barba I, Arcos D, Vallet-Regí M. *In vitro* evaluation of potential calcium phosphate scaffolds for tissue engineering. Tissue Eng 2006; 12(2): 279-90.
[http://dx.doi.org/10.1089/ten.2006.12.279] [PMID: 16548686]

[229] Meganck JA, Baumann MJ, Case ED, McCabe LR, Allar JN. Biaxial flexure testing of calcium phosphate bioceramics for use in tissue engineering. J Biomed Mater Res A 2005; 72A(1): 115-26.
[http://dx.doi.org/10.1002/jbm.a.30213] [PMID: 15558613]

[230] Case ED, Smith IO, Baumann MJ. Microcracking and porosity in calcium phosphates and the implications for bone tissue engineering. Mater Sci Eng A 2005; 390(1-2): 246-54.
[http://dx.doi.org/10.1016/j.msea.2004.08.021]

[231] Tripathi G, Basu B. A porous hydroxyapatite scaffold for bone tissue engineering: Physico-mechanical and biological evaluations. Ceram Int 2012; 38(1): 341-9.
[http://dx.doi.org/10.1016/j.ceramint.2011.07.012]

[232] Sibilla P, Sereni A, Aguiari G, *et al.* Effects of a hydroxyapatite-based biomaterial on gene expression in osteoblast-like cells. J Dent Res 2006; 85(4): 354-8.
[http://dx.doi.org/10.1177/154405910608500414] [PMID: 16567558]

[233] Verron E, Bouler JM. Calcium phosphate ceramics as bone drug-combined devices. Key Eng Mater 2010; 441: 181-201.
[http://dx.doi.org/10.4028/www.scientific.net/KEM.441.181]

[234] Kolmas J, Krukowski S, Laskus A, Jurkitewicz M. Synthetic hydroxyapatite in pharmaceutical applications. Ceram Int 2016; 42(2): 2472-87.
[http://dx.doi.org/10.1016/j.ceramint.2015.10.048]

[235] Parent M, Baradari H, Champion E, Damia C, Viana-Trecant M. Design of calcium phosphate ceramics for drug delivery applications in bone diseases: A review of the parameters affecting the loading and release of the therapeutic substance. J Control Release 2017; 252: 1-17.
[http://dx.doi.org/10.1016/j.jconrel.2017.02.012] [PMID: 28232225]

[236] Mondal S, Pal U. 3D hydroxyapatite scaffold for bone regeneration and local drug delivery applications. J Drug Deliv Sci Technol 2019; 53: 101131.
[http://dx.doi.org/10.1016/j.jddst.2019.101131]

[237] Zhao Q, Zhang D, Sun R, *et al.* Adsorption behavior of drugs on hydroxyapatite with different morphologies: a combined experimental and molecular dynamics simulation study. Ceram. Int. 2019, 45, 19522-19527. Clin Orthop Relat Res 2020; 46: 27909.

[238] Rapoport A, Borovikova D, Kokina A, *et al.* Immobilisation of yeast cells on the surface of hydroxyapatite ceramics. Process Biochem 2011; 46(3): 665-70.
[http://dx.doi.org/10.1016/j.procbio.2010.11.009]

[239] Ghiasi B, Sefidbakht Y, Mozaffari-Jovin S, *et al.* Hydroxyapatite as a biomaterial – a gift that keeps on giving. Drug Dev Ind Pharm 2020; 46(7): 1035-62.
[http://dx.doi.org/10.1080/03639045.2020.1776321] [PMID: 32476496]

[240] Mastrogiacomo M, Muraglia A, Komlev V, *et al.* Tissue engineering of bone: Search for a better scaffold. Orthod Craniofac Res 2005; 8(4): 277-84.
[http://dx.doi.org/10.1111/j.1601-6343.2005.00350.x] [PMID: 16238608]

[241] Quarto R, Mastrogiacomo M, Cancedda R, *et al.* Repair of large bone defects with the use of autologous bone marrow stromal cells. N Engl J Med 2001; 344(5): 385-6.
[http://dx.doi.org/10.1056/NEJM200102013440516] [PMID: 11195802]

[242] Vacanti CA, Bonassar LJ, Vacanti MP, Shufflebarger J. Replacement of an avulsed phalanx with tissue-engineered bone. N Engl J Med 2001; 344(20): 1511-4.
[http://dx.doi.org/10.1056/NEJM200105173442004] [PMID: 11357154]

[243] Morishita T, Honoki K, Ohgushi H, Kotobuki N, Matsushima A, Takakura Y. Tissue engineering approach to the treatment of bone tumors: three cases of cultured bone grafts derived from patients' mesenchymal stem cells. Artif Organs 2006; 30(2): 115-8.
[http://dx.doi.org/10.1111/j.1525-1594.2006.00190.x] [PMID: 16433845]

[244] Eniwumide JO, Yuan H, Cartmell SH, Meijer GJ, de Bruijn JD. Ectopic bone formation in bone marrow stem cell seeded calcium phosphate scaffolds as compared to autograft and (cell seeded) allograft. Eur Cell Mater 2007; 14: 30-9.
[http://dx.doi.org/10.22203/eCM.v014a03] [PMID: 17674330]

[245] Zuolin J, Hong Q, Jiali T. Dental follicle cells combined with beta-tricalcium phosphate ceramic: A novel available therapeutic strategy to restore periodontal defects. Med Hypotheses 2010; 75(6): 669-70.
[http://dx.doi.org/10.1016/j.mehy.2010.08.015] [PMID: 20800363]

[246] Ge S, Zhao N, Wang L, *et al.* Bone repair by periodontal ligament stem cell-seeded nanohydroxyapatite-chitosan scaffold. Int J Nanomedicine 2012; 7: 5405-14.
[http://dx.doi.org/10.2147/IJN.S36714] [PMID: 23091383]

[247] Farré-Guasch E, Bravenboer N, Helder M, Schulten E, ten Bruggenkate C, Klein-Nulend J. Blood vessel formation and bone regeneration potential of the stromal vascular fraction seeded on a calcium phosphate scaffold in the human maxillary sinus floor elevation model. Materials 2018; 11(1): 161.
[http://dx.doi.org/10.3390/ma11010161] [PMID: 29361686]

[248] Tanaka T, Komaki H, Chazono M, *et al.* Basic research and clinical application of beta-tricalcium phosphate (β-TCP). Morphologie 2017; 101(334): 164-72.
[http://dx.doi.org/10.1016/j.morpho.2017.03.002] [PMID: 28462796]

[249] Díaz-Bertrana C, Lafuente P, Fontecha P, Durall I, Franch J. Beta-tricalcium phosphate as a synthetic cancellous bone graft in veterinary orthopaedics. Vet Comp Orthop Traumatol 2006; 19(4): 196-204.
[http://dx.doi.org/10.1055/s-0038-1633001] [PMID: 17143391]

[250] Gasthuys F, Cornelissen M, Schacht E, Vlaminck L, Vertenten G. Enhancing bone healing and regeneration: present and future perspectives in veterinary orthopaedics. Vet Comp Orthop Traumatol 2010; 23(3): 153-62.
[http://dx.doi.org/10.3415/VCOT-09-03-0038] [PMID: 20422117]

[251] Kinoshita Y, Maeda H. Recent developments of functional scaffolds for craniomaxillofacial bone tissue engineering applications. Scientific World J. 2013; 2013: p. 863157.
[http://dx.doi.org/10.1155/2013/863157]

CHAPTER 9

Carbon-Nanostructures for Tissue Engineering and Cancer Therapy

Seyede Atefe Hosseini[1], Saeid Kargozar[2], Anuj Kumar[3] and Hae-Won Kim[4,5,6,*]

[1] Department of Medical Biotechnology and Nanotechnology, Faculty of Medicine, Mashhad University of Medical Sciences, Mashhad, Iran

[2] Department of Radiation Oncology, Simmons Comprehensive Cancer Center, UT Southwestern Medical Center, 5323 Harry Hines Blvd, Dallas, TX 75390, USA

[3] School of Materials Science and Technology, Indian Institute of Technology (BHU), Varanasi 221005, India

[4] Institute of Tissue Regeneration Engineering (ITREN), Dankook University, Cheonan 330-714, Republic of Korea

[5] Department of Nanobiomedical Science & BK21 PLUS NBM Global Research Center for Regenerative Medicine, Dankook University, Cheonan 330-714, Republic of Korea

[6] Department of Biomaterials Science, School of Dentistry, Dankook University, Cheonan 330-714, Republic of Korea

Abstract: Carbon nanostructures have enticed significant attention in biomedical areas over the past few decades owing to their unique electrical, physical, and optical features, biocompatibility, and versatile functionalization chemistry. These nanostructures can be categorized into diverse groups based on their morphology, including fullerenes, nanotubes (*e.g.*, single-walled carbon nanotube (SWCNT) and multi-walled carbon nanotube (MWCNT)), nanodiamonds, nanodots, graphite, and graphene derivatives. Emerging biomedical trends indicate the usefulness of carbon nanostructures in gene/drug delivery, cancer theranostics, and tissue engineering and regenerative medicine, either alone or in combination with other biocompatible materials. This chapter presents a comprehensive overview of various types of carbon family nanostructures and their characteristics. We further highlight how these properties are being utilized for various medical applications.

Keywords: Biocompatibility, Biomedical imaging, Cancer therapy, Carbon nanostructure, Drug delivery, Gene delivery, Graphene, Graphite, Fullerenes, Multi-walled carbon nanotube (MWCNT), Nanodiamonds, Nanodots, Scaffold, Single-walled carbon nanotubes (SWCNT), Tissue engineering.

* **Corresponding author Hae-Won Kim:** Institute of Tissue Regeneration Engineering (ITREN), Dankook University, Cheonan 330-714, Republic of Korea; Tel: +82 41 550 3081; E-mail: kimhw@dku.edu

INTRODUCTION

In recent years, the rapid advances in the nanobiomedical research, mainly the utilization of engineered nanomaterials have fetched many fascinating ideas and opportunities to diagnose and treat diseases and utilization in tissue engineering applications [1]. Nanostructures are materials with different structures that have at least one dimension in the range of nanometers (1-100 nm). Self-assembly of nanomaterials is a common phenomenon in nanotechnology and refers to a spontaneous assembly of components to form an intricate nanostructure without significant external intervention [2]. Since it provides the direction for the aggregation of very small structures, this phenomenon is extremely useful to reform individually into organized patterns that often give different roles to the materials. Several factors can affect self-assembly such as particle size, particle shape, and their interactions [3]. On this point, self-assembled nanomaterials demonstrate a typical mechanism of induced noncovalent interactions [4]. These nano-assembled structures were applied to create nanostructures of hierarchical protein, one-dimensional (1D) structures (nanowires/strings/tubules), two-dimensional (2D) structures (networks/nanorings), and three-dimensional (3D) structures (crystalline frames and hydrogels) [5]. The most fundamental self-assembled 3D nanostructures are primary forms of carbon materials that play a crucial role in the development of the latest nanotechnologies with suitable mechanical and multifunctional surface properties, outstanding optical activity, and high aspect ratio [6, 7].

Elemental carbon continues to astonish with the bonding diversity that leads to its different forms with distinct physico-chemical, mechanical, and biological characteristics. Nano-carbons are regarded as artificially composed structures with a modifiable construction since the 1990s following their discovery [8 - 10]. Carbon is a very adaptable material with a wide range of arrangements and allotropes (clusters, crystallites, or molecules) (Table. **1**). The hybridization of carbon (sp3, sp2, and sp1) and its bonding around the atoms determine the kind of allotrope. All allotropic modifications of carbon are formed on a nanometer-scale and independent of their synthesis methods. Carbon nanostructures (CN) or nano-carbons consist of sp^2 carbon atoms with different spatial arrangements and mainly include fullerenes (F, 0D), carbon nanotubes (CNT, 1D), graphene (G, 2D), and graphite/diamond/Mackay crystals (3D) (Fig. **1**) [11]. In addition, carbon nanocones, carbon nanohorns, carbon nanofibers, carbon nano-onion, carbon nanodot, nanocraters, and nanoscale carbon toroidal structures are other known structures of nano-carbons [12].

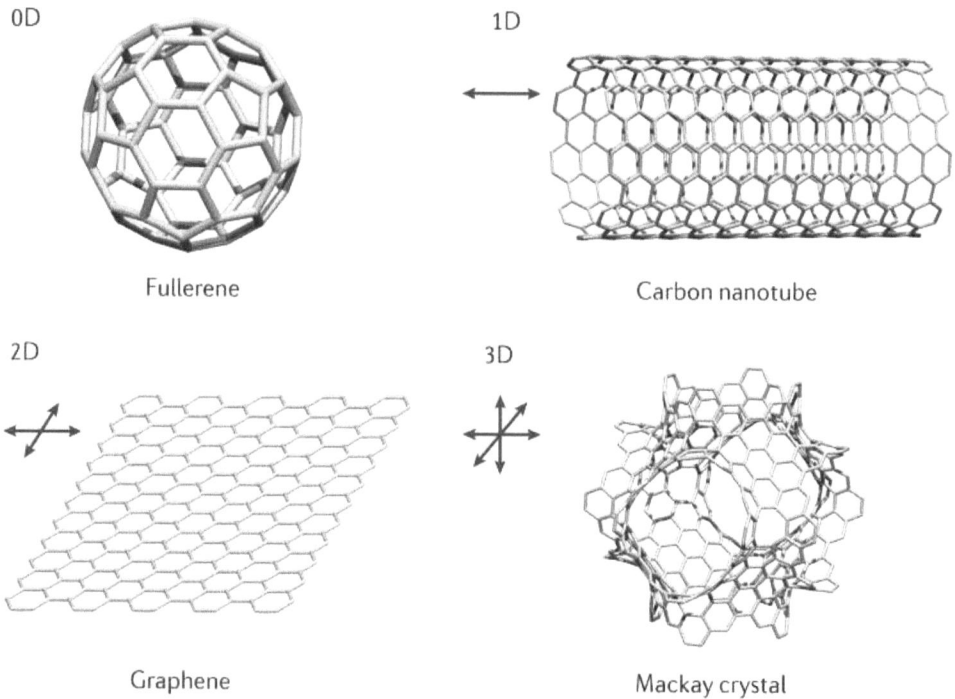

Fig. (1). Classification of typical nano-carbon structures based on their dimensions. Fullerenes (**0D**), carbon nanotubes (**1D**), graphene (**2D**) and Mackay crystals (**3D**) represent. Reproduced from [16].

There are several approaches for the synthesis of nano-carbons which are divided into two main classes the bottom-up and top-down approaches [13 - 15]; however, currently applied methods result in mixtures of particles with a range of structures and characteristics. Thus, a big challenge in the field of nano-carbon science is to obtain pure nano-carbons and the ability to synthesize structurally uniformed nano-carbons, ideally as single molecules, which is critical for the progression of functional materials. In this regard, organic synthesis (bottom-up construction) is a promising approach to attaining precise nano-carbons with atomic design [16]. These carbon-based nanostructures, which display unique forms and features, are practical in numerous biosystems like nanocarriers, diagnostic probes, and biomarkers for adjusting and controlling biological processes at the cellular and subcellular level [17 - 19]. Moreover, manufactured carbon nanomaterials, including fullerenes, carbon nanotubes (CNTs), and graphene, are very useful for several medical applications such as nanomedicine and drug/gene delivery [20]. Interestingly, several studies have also shown the feasibility of particles prepared using this method in tissue regeneration (*e.g.*, the skin, cartilage, bone, heart, nerve, lung, and vascular tissues) due to improved biological properties (*e.g.*, cell adhesion, proliferation, cell differentiation, *etc.*) [21]. In this chapter, we will introduce various forms of nanostructured carbons and focus on their applications in tissue engineering and cancer theragnostics.

Nanocarbon Family

Graphite

Graphite is a mineral discovered in igneous and metamorphic rocks that is entirely made up of sp^2p^z hybridized carbon atoms with π-electrons. It has a very low specific gravity and is a very soft polymorphic material showing hexagonal, turbostratic, and rhombohedral structures [22].

Graphite is also anisotropic that illustrates a good thermal and electrical conductor within the layers (due to the in-plane metallic bonding) [23] and a poor thermal and electrical conductor perpendicular to the layers (due to the weak Van Der Waals interactions between the layers). For instance, graphite (archaically referred to as plumbago) can maintain its strength and firmness up to a temperature of more than 3600°C [24]. Because of its electrical conductivity, graphite may be employed as electrochemical electrodes and electric brushes (owing to the in-plane covalent bonding). However, the poor Van Der Waals interactions among its layers leads to the weak thermal and electrical conductivity perpendicular to them [25]. As a result of this anisotropy, the carbon layers can slide against one another quite easily, thus making it an excellent lubricant and pencil material. Furthermore, graphite can undergo chemical reactions by allowing the reactant (known as intercalating) to reside between the graphene layers and forming compounds (known as intercalation compounds) [26]. In this regard, various research activities have demonstrated that these special features offer a diversity of applications.

Graphite can be obtained naturally or created in a laboratory. The three primary forms of natural graphite are [27] (I) crystalline small flake graphite (or flake graphite), (II) crystalline vein or lump graphite, and (III) amorphous graphite (very fine flake graphite). These subgroups exhibit various physical characteristics, chemical compositions, appearances, and impurities that result from the kind of precursor materials used as well as the manufacturing technique of graphite [24]. Naturally occurring graphite is commonly found in nature as a result of the metamorphism of sedimentary carbon compounds [28, 29] and has a close relationship with quartz and other silicate minerals. On this point, typical metamorphic settings provide the proper environment for graphite formation including hydrothermal vein-type graphite, regional metamorphism, coal seam metamorphism, and tiny-particle graphite in igneous settings [30].

Amorphous carbon is a kind of carbon that has a structure comparable to graphite, except that there is no lack of long-range order. The AB stacking order is typically missing, and the layers aren't always flat. In this class of graphite family, heat treatment enhances the degree of crystallinity (known as degree of graphitization). Many carbons employed in practice, such as carbon fibers, are not entirely graphitic but have a range of graphitization degrees depending on the heat treatment temperature [26]. Another class is flake graphite which is the most prevalent kind of graphite with the top market share in the world. It has also future needs in technologies like lithium-ion batteries, fuel cells, graphene, electronics, and other structural materials are also raising interest in graphite materials.

As mentioned above, graphite can also be prepared *via* synthetic processes through the heat treatment of coal tar pitch or petroleum coke [31]. This graphite is termed highly oriented or ordered pyrolytic graphite (HOPG) that has been treated in the range of 2500–3000 °C [32, 33]. At this processing condition, any impurities (*e.g.*, pyrite, vanadium) can be reduced to generate a high-quality graphite material [34]. In fact, the heat treatment procedure can be particularly effective in purifying a graphite substance. Today, synthetic graphite is a prominent material for the anode in batteries. However the high cost of their production serves as a major economic factor for the creation of new natural graphite sources for energy storage and conversion devices [22].

Diamonds

Diamond, unlike graphite, is an extremely hard wide-bandgap insulator that is generally considered to be one of the most perfect crystals found in nature [35]. Its name comes from the Greek ἀδάμας word (the indomitable one), indicating its extraordinary resistance to mechanical, thermal, and chemical stress. Also, it was formerly valued as a rare gem mineral with near-mythical qualities [36]. Like cyclohexane molecules, diamond comprises tetrahedrally bonded carbon atoms that are covalently connected to form a six-membered ring in a chair conformation. The corrugated layers generated by these rings can be used to describe the structure of a diamond. These layers are entirely saturated "diaphane" units, and the phrase "graphene" is used to describe a single plane of sp^2 bonded carbon atoms [37]. In a diamond, the six-membered rings shift halfway across their diagonal as the orientated diaphane units stack normal to the cubic axis. For the cubic diamond polytype, this stacking results in a cubic packing arrangement of the carbon atoms.

Diamonds are transparent through most of the electromagnetic spectra and have strangely high thermal conductivity. These features are a result of tetrahedrally linked sp^3 bonded networks of carbon atoms, and the crystallite size also has a big impact on them [38]. High surface area to volume fractions result in enhanced irregularity, hydrogen content, sp^2 bonding, and scattering of phonons and electrons. Most of these characteristics are shared by all low-dimensional materials, but the inclusion of carbon allotropes generates sp^2 bonding, which is a substantial drawback compared to amorphous silicon. Increased sp^2 bonding leads to increased irregularity, a considerably more complicated density of states within the bandgap, a decrease of Young's modulus, and increased optical absorption, among other things [39].

Despite these disadvantages, the properties of nanodiamond (ND) particles and films (sizes below 10 nm) significantly differ from the bulk diamond, mostly due not only to the very low dimensions owing to size impacts but also due to the contribution of sp^2 bonding. They consist of tetrahedrally bonded carbon atoms in the form of a 3D cubic lattice that have identified powerful systems for different applications and basic science research. Understanding the fundamental properties of these materials enables researchers for considerably more effective use of their properties in specific applications [40, 41]. NDs can be manufactured by diverse methods including detonation, laser ablation, chemical vapor deposition (CVD), ultrasound cavitation, and high-energy ball milling of micro diamonds created using high pressure and high temperature (HPHT). Two important properties of them are thermal conductivity and electron/hole mobility, both obviously being restricted by grain boundary scattering of phonons or electrons/holes, respectively. Recently, NDs have been employed as abrasive materials for time-being and have been found in more complicated applications such as biomarkers and single-photon sources [42].

Moreover, NDs are considered an excellent candidate for various biomedical applications due to exceptional optical and mechanical properties including low coefficient of friction, high hardness, wear resistance, as well as high thermal and chemical resistance. For example, nitrogen-vacancy (NV) cores of NDs emit fluorescence at 550-800 nm. This trait was effectively used to create fluorescent probes for single-particle tracking in a heterogeneous environment. Owing to these special optical and spectroscopic properties, NDs have been approved in bio-imaging approaches. In addition, NDs can attach easily to different chemical compounds, ligands, and drug molecules. This nanostructure is developed as a suitable pH-sensitive drug delivery for intravenous drug administration in which release drug in pH-response [43]. Hence, NDs are being effectively utilized in gene and drug delivery (also for water-insoluble therapeutics). In the fields of theragnostics and selective targeting, interest in NDs is steadily increasing. They are also employed in the production of biodegradable bone scaffolds for tissue engineering strategies [44 - 47]. Furthermore, different studies have revealed that surface modification (*e.g.*, either with carboxyl, hydroxyl, amine, or ester terminations, as well as with biomolecules) and doping with transition metals (*e.g.*, Si and N) can pursue to improve and control the NDs properties for biomedical utilizations [48]. After functionalization, this class surpasses CNTs in terms of biocompatibility.

Detonation nanodiamonds (DNDs) and fluorescent nanodiamonds are two types of nanodiamonds that have been categorized based on their size and production method (FNDs). In particular, detonation is a common and low-cost technology that can manufacture single-crystalline NDs with an average size of 4-5 nm, following a post-processing procedure, such as purification, de-aggregation, and fractionation. DNDs are usually fabricated by shocks with hexogen and trinitrotoluene-like explosives. On the other hand, FNDs possess a broader size distribution and flexibility in the sp^2/sp^3 bonds that are formed by HPHT [49]. However, the core of particles intended to aggregate into bigger agglomerates that are hard to disperse is one of the most serious problems with such materials. This is a particular issue for material growth that is synthesized by the detonation method, in which the particles are strongly joined by sp^2 bonding created during the detonation shockwave. Therefore, the purification of ND particles is a complex and active research area that will define the utilizations of these particles [50].

Another group in this family is carbon nano-onions (CNOs) that are derived from the ND particles under a vacuum and can serve as potential carriers for drugs. They are polyhedral and quasi-spherical shaped graphitic layers close together. These materials have unique chemical and physical characteristics, that are not found in any other type of carbon-based nanostructures [51, 52]. Several pre-clinical research activities have confirmed the biocompatibility of CNOs. They exhibit some further advantages over other types of carbon-based nanoparticles, such as the ease of manufacturing, excellent purity with a limited poly-dispersibility index, and better stability, and therefore they do not require any specific storage conditions. In addition, CNOs allow both covalent and non-covalent functionalization that enables the adhesion of more functional moieties [53].

Fullerene

Fullerenes are a kind of hollow-cage carbon cluster composed of even numbers of triply coordinated carbons organized in 12 pentagonal rings and several hexagonal (sometimes also heptagonal) rings [54]. Each carbon in their structures is sp^2 hybridized and is bonded to other carbon atoms through one double bond and two single bonds. Different shapes of these molecules have been created including, (I) spherical (buckyballs), (II) ellipsoid, (III) tubular (nanotubes), or (IV) combination shapes (nanobuds). Among them, C_{60} is the most common and the smallest form of fullerene in which pentagons (five-membered rings) do not overlap in the edge [55]. In their structure, each vertex of a truncated icosahedron is replaced by a pentagon. With a carbon atom at each hexagon's corners and a bond along each edge, this process also changes each of the 20 former triangular faces into a hexagon (a six-membered ring). Structures that prevent edge-sharing pentagons are extremely stable because the five-membered rings that are responsible for the closure tend to be the focal point of molecular strain. C_{60}s act physically and chemically as electron-deficient alkenes rather than electron-rich systems. They can easily accept electrons and are soluble in organic solvents [56]. Moreover, chemical modification of fullerenes by the addition of hydroxyl group can enhance their water solubility. This method can be used to create novel materials or a bioactive redox medication [57].

Fullerenes can be manufactured by various strategies, which mainly comprise the vaporization of graphite or similar carbon the vaporization of graphite or similar carbon sources, and include pyrolysis, arc-evaporation, laser ablation, or radio-frequency plasma. In addition, the purification of fullerenes needs large volumes of organic solvents owing to their low solubility [58]. Hence, recent research has developed efficient methods such as green procedures for their preparation. On the other hand, the utilization of microwaves [59] can be useful in decreasing reaction temperatures and time duration; however, even this convenient technology has not solved many problems faced by the industry to generate fullerenes at a low cost. It's worth mentioning that fullerenes must be derivatized to be water-soluble in biomedical utilizations; hence, prospects for green synthesis of fullerene may lie in the production of such derivatives. For instance, glutathione and hydrophilic polydopamine were applied to the solubilization of fullerenes *via* simple mixing in water, followed by dialysis and freeze-drying [60]. Moreover, Sono-chemical treatment in the water of a blend of gallium oxide and fullerene generated nanostructured hybrids with the potential to be used in the field of sensing [61]. Whereas these strategies simply emphasize derivation without addressing the synthesis of the core structure, a new advancement involved the catalytic conversion of plastic waste into a magnetic fullerene-based composite related to the key role of ferrocene, which acted both as a magnetic and catalyst nanoparticle precursor [62].

Chemical reactivity is another characteristic of fullerenes that can add to polymer structures to construct new co-polymers with specific mechanical and physical characteristics. Their application has also been growing as photodetectors and polymer transistors (Organic Field-Effect Transistors (OFETS)). Fullerenes can be used to build sensors that detect chemical vapors like biological molecules or carbon monoxide since their electrical resistance changes considerably when other molecules are bound to the carbon atoms [63]. In the field of medical science, they can easily enter across cell membranes and act as gene and drug vehicles as well as tissue engineering. The antioxidant activity and sensing ability of fullerenes are the most fascinating properties for medical applications, and the creation of targeted drug delivery systems is the most exciting prospective use of fullerenes in nano-pharmacology. In this regard, these potent antioxidants react quickly and efficiently with free radicals, which are frequently the reason for cell damage or death. They act like "radical sponges", as they can absorb and neutralize up to 20 free radicals per each fullerene [64, 65]. Furthermore, targeted imaging, therapy, and diagnostics are all continuously expanding the use of functionalized fullerenes.

Carbon Nanotubes

Besides graphite, fullerene, and diamond, the quasi-one-dimensional nanotube is another type of carbon nanostructure that was first reported by Ijima in 1991 [8]. CNTs are carbon allotropes created from graphite with cylindrical or tubular structures. These tubes include at least two layers, frequently many more, and have an outside diameter ranging from 3 to 30 nm, so they can be considered nearly 1D structures [66]. Each nanotube is a single molecule made up of millions of atoms that may be tens of micrometers long with diameters as thin as 0.7 nanometers [67]. They can differ in length, diameter, layer count, and chirality. Also, they show many attractive properties including special chemical, mechanical, and electrical properties owing to the unique combination of strength, rigidity, and elasticity [68, 69]. For instance, CNTs have an extremely high ratio of surface area over volume owing to their diameter in the nanoscale. This feature is appealing for biomedical applications, where a high aspect ratio is widely desired as more molecules can be loaded onto the nanotubes and interact with cells and tissues. Moreover, CNTs possess a high optical absorbance in the near-infrared (NIR) regime [70] that allows their detection using infrared fluorescence microscopy. They can act as microwave absorbers [71] and can provide microwave-based thermal drug release. Furthermore, CNTs are one of the most powerful materials discovered [72] with a very high thermal conductivity [73].

CNTs are mainly synthetized by three techniques; (I) arc-discharge [74, 75], (II) laser ablation [76], and (III) CVD [77, 78]. The fundamental factors for the production of nanotubes are a source of carbon, a catalyst, and adequate energy. The common principle of these methods is to add energy to a carbon source to create fragments (single C atoms or groups) that can recombine to produce CNT [79]. The first two techniques employ graphite rods as a carbon source, while CVD uses hydrocarbons like ethylene and methane. One of the major challenges in incorporating CNTs into biological systems is the lack of solubility in physiological fluids due to their high hydrophobicity. Hence, several methods have been developed for the solubilization of them *via* surface modification [80 - 82]. Furthermore, owing to their electrical conductivity and capability for electrical stimulation of cells, they can improve cellular behaviors (*e.g.*, adhesion, growth, and differentiation). Nonetheless, free CNTs may be cytotoxic as a result of their ability to cause oxidative damage, as well as possible contamination with catalysts (transition metals like Ni, Fe, and Y), used during their preparation. On this point, approaches such as arc-discharge evaporation of graphite rods have been developed for the synthesis of metal-free CNTs [83]. Thus, various kinds of CNTs have been generated based on different methods for their preparation of them that are classified into the following two types, *i.e.*, (I) a single cylindrical graphene sheet (also known as a single-walled CNT or SWNT) and (II) several concentric graphene sheets (multiwalled CNTs, referred to as MWCNTs or MWNTs). Few-walled CNTs (FWCNTs) and double-walled CNTs are further types of MWNTs (DWCNTs) [83]. The structure of pure SWCNT can be apparent as a rolled-up tubular shell of graphene with sp^2 hybridization. The SWCNTs typically have just 10 atoms around the circumference with only one atom thick for each tube [84]. They are frequently used in numerous applications of nanotechnology, but the main difficulty is the high production cost [85]. Also, SWCNTs have low solubility and are difficult to disperse in the aqueous phase due to strong π-π interactions. Therefore, more research is required to develop a cost-effective synthesis technique. Chemical modifications have been proposed to solubilize and disperse SWCNTs in water to enhance their pharmaceutical and biomedical applications [86]. Nanohorns are a modified form of SWCNTs that are made of a single hexagonal ring of carbon. Due to their unique property and increased diameter and length, they are utilized in drug delivery and targeting medications for malignant tumors [87]. Moreover, SWCNTs are widely used in the biomedical area as diagnostic tools and scaffolds for cell growth [88].

DWCNTs are identical to SWNTs in terms of shape and properties. Coupling these features into the nanotube is critical for supplementing new functionality [89]. DWCNTs are an emerging group of carbon nanostructures consisting of a double-layer structure and two carbon nanotubes comprising two graphene sheets folded on top of each other. Compared with SWCNTs, DWCNTs have higher chemical resistance, mechanical strength, and thermal stability. This class also reveals interesting electronic and optical properties [90], hence is used in the development of biosensors [91]. For tissue engineering applications, DWCNTs provide a suitable surface texture for growing neuron cells rather than a SiO_2 surface. Therefore, experimental studies have indicated that cell differentiation improves after culturing neurons on DWCNT [92].

Another group in the CNTs family is MWCNT, which is composed of a stack of graphene sheets rolled up into concentric cylinders with diameters between 5 and 50 nm. They are composed of many SWNTs stacked one inside the other and possess more intricate electronic properties [93, 94]. For instance, they are considered superior substances in the case of signal/noise ratio as compared to SWCNT and DWCNT. The structural intricacy and diversity in this family are created by wrapping one outside of another with an interlayer dispersion of 3.4Å. The structure of MWCNTs can be well clarified using the Parchment and Russian-Doll model. Many researchers have also reported that a slight modification in structure can reduce their favorite material properties [68].

In addition to the two different main structures, there are three possible types of CNTs including (I) armchair CNTs, (II) zigzag CNTs, and (III) chiral CNTs. The difference in these kinds of CNTs is formed depending on how the graphite is "rolled up" during their formation process. The choice of the rolling axis relative to the radius of the closing cylinder and the hexagonal network of the graphene she*et al*lows for diverse types of SWCNTs [79]. The goal of recent research has been established to improve the quality of catalytically-created nanotubes [95, 96]. On this point, significant factors were considered for the more massive production of nanotubes, which are summarized as increased purity and quality, surface chemistry, dimension chirality control, and dispersibility in aqueous media. Furthermore, CNTs have some special properties including a special length/diameter aspect ratio (up to 10^7), mechanical strength, chemical and thermal stability [97], gas adsorption, semiconducting behavior, and heat conduction that inspire their use in a wide range of applications. In medical research, the peculiarities of CNTs have been employed to develop pH- or temperature-triggered delivery systems for DNA, proteins, anti-tumor medications [98], or vaccines [99], as well as to improve the electrical and structural characteristics of scaffolds (*e.g.*, myocardium, bone, and neural scaffolds) [100]. In addition, these carbon nanostructures have been used to manufacture the sensors for nucleic acids, glucose, and cancer biomarkers [101], and also patterned surfaces for promoting cell growth and differentiation.

Despite many successful biomedical applications, there is a growing concern about their safety. Some experimental studies have documented increased cytotoxicity of CNTs due to their agglomeration, cellular uptake, and induced oxidative stress [102, 103]. These contradictory results regarding the CNTs' biocompatibility mostly stem from their variability including surface properties, size, and functionalization, and also testing protocols including *in vitro* vs *in vivo*, types of cells, tissues, and animals tested. Furthermore, insufficient removal of metal catalysts employed to manufacture CNTs has been associated with higher cytotoxicity [103]. On the other hand, most animal studies on CNTs have demonstrated no toxicity and stated the renal clearance from the body. While, small portions of CNTs may accumulate in some organs (*e.g.*, spleen, liver, and lungs) and induce inflammation [104]. However, based on various cell culture studies, the cytotoxicity looks to be more pronounced and extremely variable at the cellular level [105]. Thus, more systematic biological evaluations are needed regarding the continued growth of different CNTs to determine their precise pharmacokinetics, cytotoxicity, and optimal dosages. It should be mentioned that cytotoxicity can efficiently be diminished by surface functionalization of CNTs with biocompatible polymers or surfactants to avoid aggregation [106].

Graphene

Graphene is the thinnest material in nature with a one-atom-thick layer of graphite, but it is considerably strong (about 100 times stronger than steel). Its structure is composed of sp^2 hybridized carbon atoms arranged as a 2D material involving six side rings (hexagons); on each corner of a ring, there is a single carbon atom. These honeycomb lattice and their electrons are presented in aromatic conjugated domains [107]. Therefore, it can be considered as an atomic-scale chicken wire created from carbon atoms and their bonds (The carbon-carbon bond length is about 1.42 Å) [108]. Graphene is a basic building block for other carbon allotropes, such as graphite, fullerenes, and CNTs. It shows unique chemical, physical, and electronic properties including high thermal and electrical conductivity, large surface area, high elastic strength, high charge carrier mobility, optical transparency, and excellent biological/mechanical attributes. These characteristics of the graphene family can have various applications, ranging from batteries, light-emitting diodes, and supercapacitors to biomedical uses [109].

Graphene family are categorized based on their chemical modification during preparations or the number of layers in the sheet [9]. In the top-down approaches, graphite nanomaterials are prepared *via* treatment by electrochemical or mechanical exfoliation, sonication or intercalation, and also a slicing nanotube. However, the bottom-up methods for their preparation include the epitaxial growth of graphene on silicon carbide, the dry ice process, the growth of graphene from carbon-metal melts, and deposition approaches such as CVD [21]. In this regard, several types of GFNs were formed with diverse purity, lateral dimensions, surface chemistry, number of layers, defect density, and composition. Some of the common graphene families include single layer graphene (SLG), bi-layer graphene, multilayer graphene, graphene oxide (GO), and reduced graphene oxide (rGO). Graphene can also be produced in the form of nanoplatelets, nanoflakes, nanoscrolls, nanoribbons, and graphene quantum dots (GQDs) [110]. SLG as the name implies is an isolated single layer of carbon atoms that are joined together in a 2D structure. It is produced through either tightly controlled growth on surfaces or repetitive mechanical exfoliation (like silicon carbide) [111] through CVD. However, several research works have shown that the synthesis of single layer defect-free graphene is problematic in bulk. Furthermore, owing to the highly reactive surface, it is also difficult to isolate it in the gas phase and suspend it in solutions [21]. Hence, several alterations have been made to develop chemically and physically modified graphene, including the creation of multilayer graphene, GO, rGO, and layered GO, especially for biomedical applications. GO is a highly oxidative and water-soluble form of graphene that can be attained *via* the exfoliation of GO. This graphene and its derivatives are widely utilized in sol-gel chemistry and are found to be appropriate in the formation of biocompatible nanostructures [68]. On the other hand, reduced GO has been prepared for the improvement of its electrical properties. Its preparation methods include thermal, chemical, or pressure reduction, and even *via* bacteria-mediated reduction of GO. As mentioned above, graphene nanoribbons (GNRs) and quantum dots are other members of the graphene family. They are formed due to the employment of graphene as a semiconductor and the interaction of graphene and electrons. These new quasi-particles exhibit ballistic transport where electrons can travel up to a micrometer without scattering [112].

GNRs are finite, planer, and quasi-one-dimensional graphene structures with usually less than 1000 nm in length and about 10 nm in width that are prepared by either physical or chemical approaches. In the physical methods, the sheets of graphene are divided into fine tiles with a higher aspect ratio, subsequently, a fixed bandgap is viewed with less than 10 nm wide. With the growth in width, their behavior shifts from semiconductors to semimetals [113]. Owing to higher thermal and electrical conductivity, they can be used as a substitute for copper in integrated circuits. GNRs can also be fabricated by chemical synthesis either by a liquid phase of exfoliation or by etching graphene with high resolution electron beam lithography [114]. The oxide-functionalized GNRs possess a high potential in therapeutic applications like cancer treatment, drug delivery, and DNA-based systems. Importantly, they are practically non-toxic in human and environmental health [115].

GQDs are zero-dimensional nanomaterials that can be produced by cutting the graphene into a fragment in the range of 2-20 nm. They have ultra-small size, appropriate photostability, and high water solubility, and are eco-friendly and less toxic compared to typical QDs [113]. Due to these unique properties, GQDs are utilized for numerous biomedical applications. Moreover, they overcome the challenges related to other kinds of fluorophores thanks to features like photostability, photoluminescence, and better renal clearance. Therefore, these nanomaterials show fluorescent activity and are effectively applied in bioimaging. Consequently, they were used to develop a fluorescent probe for observing the cellular dynamics *in vitro* and *in vivo* and tumor imaging [116]. For instance, redox fluorescent probes were synthetized by using GQDs that are useful to monitor the real-time oxidative stress-induced dynamic [117]. In conclusion, graphene-based nanostructures show great promise in a wide range of industrial and biomedical applications including gene, drug, and protein delivery, bioimaging, biosensor construction, photothermal therapy, antimicrobial treatment, and also a scaffold for tissue engineering and regenerative medicine [118, 119].

Table 1. Summarizing nano-carbon family members with their structure and preparation methods.

Schematic View	Entity	Subgroup Family	Structure	Preparation Methods
	Graphite	Natural	Hexagonal	Metamorphism of sedimentary carbon compounds.
		Synthetic	Rhombohedral	Heat-treatment of petroleum coke or coal tar pitch.
			Turbostratic	
	Diamonds	-	Crystal (diamond cubic)	Detonation
				Laser ablation
				CVD
				Ultrasound cavitation
				Microdiamonds
				High energy ball milling
	Fullerene	-	Spherical (buckyballs)	Vaporization
			Ellipsoid	Pyrolysis
			(nanotubes)	Arc-evaporation
			Combination shape (nanobuds)	Laser ablation
			Tubular	Radio-frequency plasma

(Table 1) cont.....

	Carbon nanotube	SWCNTS	Single graphite sheet is rolled round itself	Arc discharge
		MWCNTS	Graphite sheets are arranged in concentric layers	Laser ablation
				Chemical vapor deposition
				Flame synthesis
				Silane solution
				Nebulized spray pyrolysis
	Graphene	SLG	Honeycomb lattice	Mechanical exfoliation
		Bi-layer graphene		Chemical exfoliation
		Multilayer graphene		Chemical synthesis
		GO		Pyrolysis
		rGO		Epitaxial growth
				CVD

Chemical vapor deposition, CVD; Single-walled carbon nanotube, SWCNTs; Multi-walled carbon nanotube, MWCNTs; Single layer graphene, SLG; Graphene oxide, GO; reduced graphene oxide, rGO.

Application in Drug Delivery

The design of carbon-based nanostructures with enhanced safety profiles for encapsulation of drugs proposes alternative pharmacotherapy options for current drugs (*e.g.*, anti-cancer drugs). The ideal nanosystem for drug delivery must intrinsically contain specific targeting functional groups, trigger a precise biological response and be detectable. Nanomaterials can trap cytotoxic chemo-drugs to prevent their metabolism, degradation, and interaction with normal cells on the way to the tumor site. Either passive or active targeting techniques can accumulate these encapsulated drug carriers at the tumor site. In the passive targeting technique, the nanosystems (*e.g.*, nanocarriers) are generated to passively accumulate at the tumor location through the increased permeability and retention (EPR) effect. Nanocarriers penetrate tumor *via* the leaky and fenestrated vasculature made by tumor angiogenesis, and the lack of lymphatic outflow of tumor tissue leads to the nanocarriers remaining inside the tumor mass for longer [120]. The stimuli applied in targeted drug delivery are classified as endogenous (internal stimuli or biological molecules) or exogenous (physical external stimuli). Light, temperature, magnetic fields, electrical fields, and ultrasound are some exogenous stimuli employed to initiate the drug release. However, pH-, enzyme-, hypoxia-, and redox-responsive are prevalent endogenous stimuli utilized for cancer targeting [121, 122].

All of these efficacies are possible with carbon nanoparticles that make them promising nanovectors for directed drug delivery and imaging contrast agent [123]. After loading the genes/drugs, the carbon nanostructures can accumulate into tumors *via* active targeting (possessing a targeting ligand for overexpressed receptors in the vasculature/tumor) or improved EPR effect. The pharmacokinetic behavior of gene/drug cargos on these nanomaterials differs from free molecules, which may improve tumor killing effectiveness, while reducing toxicity in healthy tissue. Moreover, the addition of stimuli responsive polymers to carbon nanostructures can be greatly helpful for controllable and selective drug release [124]. Thus, they can be used to overcome the limitations of free therapeutics and negative biological barriers including systemic, cellular, and microenvironmental barriers that are heterogeneous across patients and diseases [125, 126]. For instance, a "small nanoneedle" method allows CNTs to readily overcome a variety of biological barriers, penetrating through the plasma membrane and entering the cytoplasm, which makes it easier to carry and deliver medications or other cargo into the intended tissues.

Drugs are often enclosed in nanostructures by immersing carbon nanomaterials in a drug-containing solution or attaching the drug to them by physisorption or chemical conjugation [127]. As compared to conventional drug carrier systems, nanostructures provide a large specific surface area and delocalize surface p-conjugated electrons. These properties can support high drug loading of less soluble drugs through p–p stacking and hydrophobic interactions [128]. When the surface of nanostructures has been chemically modified with functional agents, ionic interactions lead to the adsorption of the therapeutic molecules. Furthermore, drug molecules can be carried out by nanoparticles *via* chemical bonds such as double sulfide and ester bonds, which present ideal stimuli-responsive release patterns for the treatment [129, 130]. One of the major goals of carbon-nanostructure research is to provide prolonged or regulated medication release with minimal dosage frequency. On this point, drug release from nanoparticles is affected by different factors including drug solubility, temperature, pH, desorption of the surface-bound or adsorbed drug, degradation of the matrix, and drug diffusion through the matrix. Nanoparticles generally interact with plasma membranes and enter the cells through endocytosis or passive diffusion across the phospholipid bilayer. In addition, targeted delivery to the nucleus has been carried out *via* the functionalization of carbon nanocomposites to produce stimuli sensitive carriers that release drugs in the cytosol [21]. For this purpose, a variety of these structures have been utilized including graphene, fullerenes, CNTs, CNOs, carbon quantum dots, and magnetic carbon nanostructures (MCNs) (Table. **2**).

Graphene

Graphene is a carbon-based honeycomb lattice at the atomic scale that has been investigated as one of the most promising carbon nanocomposites for drug delivery because of its favorable characteristics, including high biocompatibility, minimal cytotoxicity, and unique physicochemical properties in optics, electrics, chemistry, and mechanics. Furthermore, hydrophobic and π–π stacking interactions and high-specific surface area, for encapsulation of small molecules on the 2D planar surface are other ideal properties in the gene/drug delivery area. Also, negative charges of free π electrons in the surface have been used to condense proteins and genes. The presence of free electrons and numerous functional groups on the surface of graphene and GO provides several opportunities for covalent bonds of chemically varied small molecules and proteins. The use of functionalized graphene in targeted delivery and release using external stimuli like pH, magnetic fields, or near infrared radiations is a new drug/gene delivery platform. For instance, functionalized GO with a cancer-targeting molecule folic acid (FA) loaded with camptothecin and doxorubicin was approved to improve anticancer activity and cancer-targeting ability as compared with drugs used with unmodified GO or drugs delivered alone [131]. Similarly, the capacity for the endosomal escape of loaded cargo into cytosol provides the potential for efficient gene and drug delivery to the nucleus [21]. They also possess several advantages over CNTs, including a lower cost, ease of manufacturing and customization, and a greater drug loading ratio with two exterior surfaces [132]. Hence, graphene and its derivatives (*e.g.*, GO) have been extensively studied for drug delivery during the last decade.

Fullerenes

Fullerenes as nanostructured carbon cages can serve as vectors or drug delivery scaffolds either with covalent or noncovalent linkages between the fullerene and a bioactive material [133]. The main promises of these nanocarriers are controlling behavior, the drug release rate, immuno-neutrality, higher drug loading, and capability to bypass mononuclear phagocytic system, long circulating nature, substantial biocompatibility, and tissue extraction by virtue of improved EPR effect. In this regard, different studies have indicated that they have turned out to be able to act as medication vehicles after suitable functionalization (*e.g.*, attaching hydrophilic moieties) [134]. Two types of fullerenes including exohedral and endohedral fullerene have been applied in nanomedicine area as a delivery system. Exohedral fullerenes have additional atoms, ions, or clusters attached their outer sphere's structure; however, endohedral fullerenes have additional ions, atoms, or clusters enclosed within their inner spheres structure. Endohedral fullerene and its derivatives can be used to transport ions or atoms in biological systems. For instance, metallofullerene is utilized as a medication delivery system depending on the composition and characteristics of the trapped metal within its structure. The derivative surface of exohedral fullerene is directly attached to the activated drugs *via* covalent bonds. These fullerenes offer unique bio-function with biological body and drug release with selective targeting at the same time, especially to cancerous cells (*e.g.*, C_{60}-paclitaxel fullerene) [135]. C_{60} fullerenes (CFs) not only are utilized as a drug delivery system but are also produced to possess neuroprotective, antiviral, anti-inflammatory, antioxidant, and magnetic resonance imaging (MRI) contrast properties. Although, concerns like anticipated tissue toxicity, reduction from the biological system, sterility issues, stability of the final product, and commercial viability pose challenges in the appropriate application of CFs as ideal drug transporters [136]. In addition, they have high tendency to bond covalently with various biological substances such as cholesterol, sugar, and carbohydrates [137].

Carbon Nanotubes

It is well documented that CNTs are the most significant and commonly utilized class of carbon nanostructures for drug and gene delivery. They are not only used in the cellular imaging with diagnostic effects in nanomedicine but also show promise as one of the potential cargos for targeted delivery. They improve the pharmacological profiles, while reducing toxicological impacts of delivered drugs because of their favorable properties. More importantly, CNTs provide some interesting benefits as compared to spherical nanoparticle. In fact, their large inner volume allows the loading of small molecules while their outer surface can be chemically modified and functionalized to load proteins and genes for effective delivery. Nowadays, both SWCNTs and MWCNTs are being modified and turned out to be useful in the delivery of nuclear acids, peptides, proteins, and drugs [138, 139]. For example, oxidized SWNTs can be functionalized at their carboxylic groups with peptide, proteins, peptide nucleic acid, sugar moieties, oligonucleotide, and poly oxide derivatives. The functionalized SWNTs can transport these molecular cargoes into mammalian cells, indicating their potential application as carriers [140]. When biomolecules attach to SWNTs, they are protected from degradation in mammalian systems, demonstrating superior biostability compared with free molecules. On the other hand, the bonding of drugs to these appropriate carriers substantially improves their bioavailability due to their enhanced resistance time in blood circulation and increased solubility. Furthermore, drugs accumulated in the selective site of pathological zone, also known as therapeutic-effects-related sites, can increase the drug's therapeutic efficacy. The unique ability of CNTs to pass through cell membranes is another characteristic that opens the door for using them as carriers to transport therapeutic chemicals into the cytoplasm and, in many cases, into the nucleus. The inherent spectroscopic properties of CNTs like Raman and photoluminescence offer additional benefits for tracking and real-time monitoring of drug delivery efficacy *in vivo* [141].

Aforementioned carbon-based nanostructures have also been utilized for the delivery of nucleic acid-based materials such as gene and siRNA across biological barriers. Since genetic materials are unable to pass biological membranes, viral or non-viral vectors must be used to transport the gene and internalize it into the cell. Nonviral vectors are less effective [142] and have a shorter lifespan than viral vectors [143]; but they are considerably safer [144]. On this point, many researchers have suggested functionalized SWNTs as appropriate nonviral carriers of macromolecules that exhibit high DNA expression compared to naked DNA [145]. These complexes like other CNTs internalize macromolecules into living cells *via* energy-dependent endocytosis [146]. Due to their high surface area, they can conjugate or adsorb with a wide variety of therapeutic materials. Therefore, CNTs can be surface engineered (*e.g.*, functionalized) in order to improve their dispersibility in the aqueous phase or to provide the proper functional groups that can attach to the desired therapeutic molecules or the target tissue to elicit a therapeutic impact [147] (Fig. **2**).

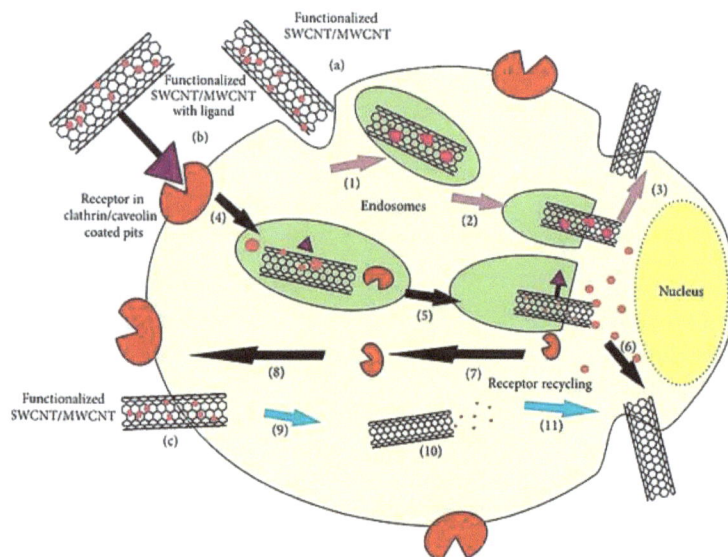

Fig. (2).   Internalization mechanism of various types of CNTs. **a)** endocytosis of functionalized SWCNT/MWCNT loaded with drug and the formation endosome for drug release, and release of free CNT outside the cell; **b)** Receptor-mediated internalization of functionalized SWCNT/MWCNT with ligand loaded with drug and then the formation of early endosome followed by late endosome which is responsible for drug release, and recycling of the receptor; **c)** internalization of functionalized SWCNT/MWCNT and drug release in the cytoplasm, and then the release of free CNT outside the cell. Reproduced from [148].

In addition to DNA delivery into mammalian cells, they have demonstrated efficient siRNA [149] or plasmid DNA delivery into several species of plants (*e.g.*, wheat, arugula, and cotton) and result in high protein expression levels without transgene integration [150]. Moreover, literature has interestingly demonstrated that CNTs are useful materials for transdermal drug delivery. The fundamental goal of a transdermal drug delivery system is to transfer medicines into systemic circulation at a predetermined rate *via* the skin, with minimum inter- and interpatient variation [151]. CNTs are not directly adsorbed inside the organism; instead they are administered outside the stratum corneum of the skin, and only the active therapeutic material is intended to cross the body barriers [152]. For instance, researchers have prepared CNT-framed membranes by self-assembly of highly thermo-conductive CNTs that hybridize with chitosan. These constructs indicate highly effective drug loading and drug-releasing properties and can have the potential for use as a skin heat signal responsive patch type transdermal drug delivery system in the pharmaceutical field [153].

Carbon Quantum Dots (CQDs)

CQDs have recently achieved much attention for bioimaging and drug delivery applications owing to their excellent biocompatibility, flexible surface chemistry, tiny size, amazing electromagnetic, optical properties, drug loading capacity, and low toxicity [154]. Their luminescent properties enable both cellular and systemic drug release and QD as a vehicle transport to be monitored in real-time. Nowadays, there are highly valuable drug delivery applications for CQDs, including anticancer, antimicrobial and neurodegenerative agent [155]. For instance, research studies have shown that luminescent CQDs can enter into the nuclei of both cancer cells and cancer stem cells. Additionally, the anticancer drug-loaded CQDs (*e.g.*, doxorubicin-loaded CQDs) exhibit a good therapeutic impact by reducing these cells [156].

Magnetic Carbon Nanostructures (MCNs)

MCNs are fascinating building blocks for designing bottom-up medication delivery systems. Their magnetic behavior can be used to concentrate on drug-loaded constructions within the target tissue under the control of an external magnetic field and hold the drug as long as the drug is released. Due to the type of magnetic source, drug delivery can occur to tissues near the body surface or internal target tissues (*e.g.*, magnets that are located in the target tissue). Since drugs can be bound to vast carbonic surfaces or carried in the internal cavities of hollow nanocontainers, there are several approaches to MCN-mediated drug delivery. Another possibility for magnetic behavior in the drug delivery field is magnetically induced heating which can accelerate both the cellular uptake and drug diffusion at the target site [157].

Application in Tissue Engineering

Carbon-based nanostructures are important candidates for the manufacture of synthetic scaffolds in tissue engineering and regenerative medicine [158]. In terms of chemical composition and physical structure, artificial scaffolds for tissue regeneration must be comparable to a native extracellular matrix (ECM). In fact, the majority of cells reside in 3D microenvironments *in vivo*; as a result, due to the physiological importance of the structure, cells cultured on 3D constructions may behave identical to cells on natural ECM. Carbon nanostructures are considered a physical analogue of ECM constituents such as collagen fibers due to their similar dimensions. In this regard, several technologies including the solvent casting method, hydrothermal technology, and electrospinning process were used to create 3D scaffolds incorporating carbon nanomaterials [159]. On the other hand, because of their superior mechanical capabilities, carbon nanostructures can play a key role in the reinforcement of organic/inorganic artificial scaffolds by generating π–π stacking and electrostatic interactions between carbon nanostructures and these components [160]. Moreover, their high conductivity maybe useful for providing thermal and electrical stimulation in the artificial scaffolds [161].

An ideal construct for tissue engineering should be biocompatible and encourage the attachment, growth, and differentiation of mammalian cells. While non-biodegradability and long-term toxicity are two main barriers to the biomedical applications of carbon nanostructures. On this matter, researchers have developed various surface functionalization strategies for carbon nanostructures that not only can attenuate their toxicity and long-term safety but also provide biocompatibility and cell adhesion moieties [162, 163]. Nowadays, carbon-based nanomaterials are being generated with suitable biocompatibility and capability of supporting growth and differentiation of a variety of cells (*e.g.*, osteoblasts, neurons, and stem cells). It is of great interest that reduction state and surface roughness of carbon nanostructures can modulate the differential behavior of many cells [164]. Carbon-based nanostructures including SWCNTs, MWCNTs, fullerene, nanodiamond, and graphene are the most important structures that are utilized in tissue engineering due to their unique chemical, mechanical and electronic properties (Table **2**) [165]. This section highlights two common types of these carbon nanomaterials, *i.e.*, CNTs and graphene.

Carbon Nanotubes (CNTs)

CNTs are regarded as potent materials in tissue engineering with a wide range applications of cell tracking, gene delivery, and microenvironment sensing, and fabricating composite scaffolds. Furthermore, the utilization of CNTs for magnetic, optical resonance, radiotracer contrast agents were mentioned for providing superior means of evaluating tissue formation. The monitoring and alteration of intra/intercellular processes would be helpful for the design of appropriate engineered tissue and organs [166]. CNTs can be coupled with scaffolds for supporting structural reinforcement and subsequently improved mechanical properties. Some commonly used constructs, such as fibrous scaffolds and hydrogels, are intrinsically soft to mimic the natural tissues stiffness and often lack structural support and strength. The incorporation of CNTs into the mentioned scaffolds was demonstrated to increase their mechanical characteristics [167]. SWCNTs and MNCNTs have been claimed to have a strength almost 100 times that of steel while having a specific weight roughly 6 times lower than steel. Thus, they are being utilized for reinforcing several natural and synthetic polymers for potential applications in hard tissue engineering. When CNTs are incorporated into a polymer matrix, it can resemble the inorganic phase of the bone and generates nanoscale irregularities on the surface of 2D materials as well as in the pores of 3D materials, leading to facilitated cell growth and adhesion (Fig. **3**) [168]. In this sense, fluorinated or carboxylated SWNTs promoted both biomineral formation and the expression of key molecules in the mineralization process including bone osteocalcin and sialoprotein [169]. Furthermore, SWNT monolayers could up-regulate the expression of osteogenic proteins and genes and also increase adhesion, proliferation, and osteogenic differentiation of cells (*e.g.*, human mesenchymal stem cells (hMSCs) [170]. Furthermore, CNTs can be used for imparting unique properties such as electrical conductivity into the scaffolds that aid in directing cell growth. Since most utilized biomaterials in tissue engineering are comprised of non-conductive polymers, the prepared constructs are electrically insulating [171 - 173]. However, conductive scaffolds that can successfully transport electrical impulses across tissue constructs for optimal electrophysiological function would be extremely beneficial in certain applications such as cardiac and brain tissue repair and regeneration. For instance, electrical stimulation of neural stem cells (NSCs) cultured on CNT-laminin composite films resulted in enhanced action potentials and differentiation of NSCs into functional neural networks [172]. In addition, it has been shown that culturing cardiomyocytes on gelatin hydrogels reinforced with CNT enhances their electrophysiological activities and eventually develops functional cardiac tissue [174]. Overall, CNTs can play an important role as a unique biomaterial for monitoring and creating engineered tissue.

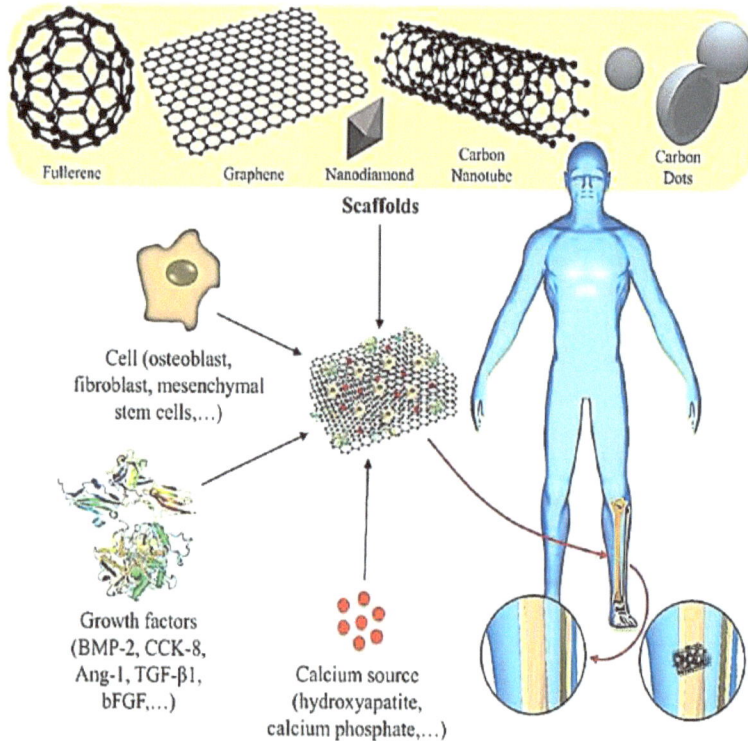

Fig. (3). Different types of carbon-based nanostructures as scaffolds in bone tissue regeneration application. Application of CNTs, Graphene, fullerene, CDs, and NDs as matrices for various bone forming cells, growth factors, and calcium sources. Reproduced from [175].

Graphene

Graphene-based composites have been applied for stem cell engineering, wound healing, tissue engineering, and regenerative medicine [21]. According to recently reported studies, graphene-based nanostructures combined with nano/microfabrication methods may lead to the production of scaffolds with characteristics tailored to the target tissue/organ. The electrical and mechanical properties of graphene nanomaterials can be used to reinforce tissue engineered (TE) constructs. Despite the fact that hydrogels have viscoelastic and transport characteristics that are similar to natural tissues, their weak mechanical properties limit their usage in several tissue engineering applications [176]. Graphene has exceptional mechanical qualities including high strength, elasticity, and flexibility, and may be tailored to perform a variety of functions on flat surfaces. Therefore, it can be potentially employed as a reinforcement material in biodegradable films, hydrogels, electrospun fibers, and other tissue engineering scaffolds [177]. 3D scaffolds made of graphene, or its derivatives have been effectively manufactured by hydrothermal reduction, filtration of graphene-based nanomaterials, or supramolecular interaction control. For instance, a one-step hydrothermal technique was used to create self-assembled hydrogels of graphene with an interconnected 3D porous network [178]. The successful fabrication of self-assembled graphene hydrogels was made possible through the interactions of graphene sheets *via* stacking and water encapsulation by residual oxygenation functional groups of hydrothermally rGO [179]. As an illustration, the incorporation of GO into gelatin methacrylate (GelMa) hydrogels could enhance their electrical and mechanical characteristics with no side effects on encapsulated fibroblasts [177].

To obtain suitable scaffolds for tissue regeneration, the production of microenvironments comparable to ECM is critical to improve attachment, proliferation, and differentiation cells. On this point, it is believed that surface oxygen content and roughness, which affect serum protein adsorption, lead to improved cellular behavior on graphene derivatives. Plentiful studies have reported that graphene composites such as graphene, GO, or rGO, can support the adhesion and proliferation of mammalian cells (*e.g.*, fibroblasts, human osteoblasts, adenocalcinoma cells, as well as stem cells) [180]. For example, cultured neural stem cells (NSC) on graphene foams were demonstrated to support NSC proliferation and differentiation towards neuronal lineages, beside the capability for electrical stimulation of differentiated NSCs [181]. Moreover, cultured myoblasts on GO show high myotube fusion/maturation index and remarkably enhance multinucleate myotube formation, myogenic differentiation, myogenic protein expression, and expression of differentiation-specific genes (Troponin T, myogenin, MyoD, and MHC). Thus, they can induce myogenic differentiation and present a potential for skeletal tissue engineering applications [182].

In the field of bone tissue engineering, composite materials made of graphene and its derivatives and calcium phosphates can enhance the regeneration of tissue due to excellent mechanical properties and bone compatibility of graphene. For instance, composites of GO/graphene–$CaCO_3$ induce the formation of vaterite crystals *via* desired interactions with Ca^{2+} ions. These hybrid materials significantly increased the synthesis of hydroxyapatite when incubated in a simulated body fluid (SBF). So, the GO/graphene-HAp composites formed support high viability of osteoblasts with elongated morphology [183].

In addition to these properties, functionalization of graphene with peptides or proteins can be useful for tissue regeneration applications [184]. Functionalized graphene with pyrenebutanoic acid–succinimidyl ester (PYR–NHS) was previously approved for micropatterning of laminin that is an ECM protein utilized for neuronal guidance (Fig. **4**), [185].

Application in Cancer Theragnostics

Nanoscale structures have revolutionized biomedicine, particularly in the diagnosis, detection, and treatment of cancer. In fact, they provide the possibility of detecting different cancers at an early stage with even very low levels of cancer biomarkers, when treatment is far more effective. Moreover, nanostructures can improve the specificity and precision of traditional diagnostic methods such as MRI, single-photon-emission computed topography (SPECT), positron emission topography (PET), as well as electrochemical and optical biosensors [187 - 189]. Nanomaterials have been described as one of the most promising platforms for drug delivery and cancer imaging because of its potential to integrate diagnostic and therapy in a single material. One material that combines diagnosis and treatment is called a "theragnostic". Furthermore, they have great potential in improving the efficacy of cancer treatment options. Nanostructures are also able to mediate cancer treatment using external energy sources, such as light for photodynamic therapy (PDT) and photothermal therapy (PTT), magnetic hypothermia, or radiation for radio-sensitizing therapy [190].

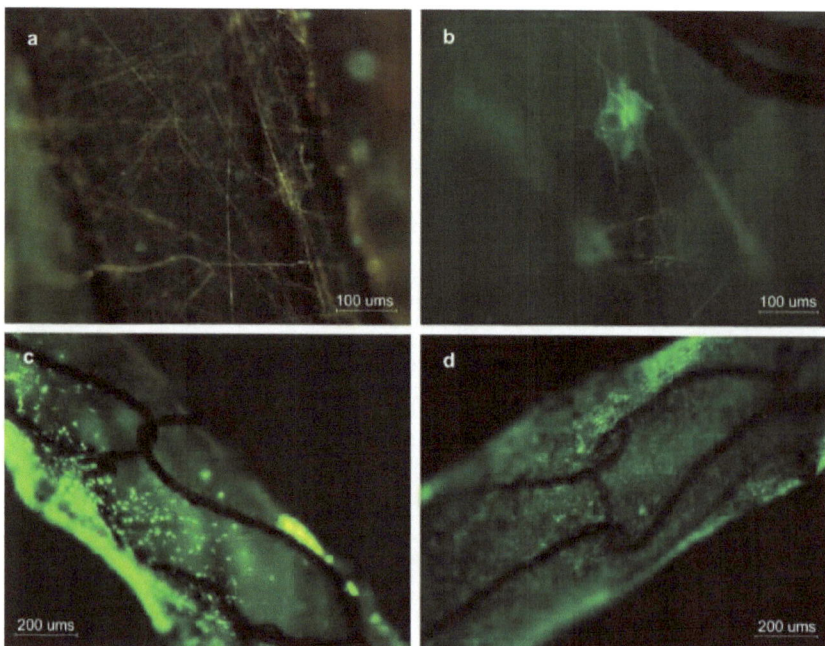

Fig. (4). Adhesion and proliferation of viable cells on multi-walled carbon nanotube (MWCNT) scaffold. Fluorescent optical micrographs exhibiting viable cell adhesion and proliferation on MWCNT/PLGA composites after **(a)** 3 days, **(b)** 6 days (phalloidin stained), **(c)** 6 days, and **(d)** 17 days. Reproduced from [186].

Carbon-based nanostructures, including fullerenes, CNTs, graphene, and nanodiamonds, have been intensively explored for cancer theranostics due to their unique physicochemical features (Fig. **5**). They provide a novel technique for developing the next-generation cancer theranostic agents that can increase the detection limits/quality of multi-function cancer diagnostics and give improved treatment approaches [191 - 193]. Their unique characteristics include several advantages as follows. (I) They can be coupled to several imaging agents and highly concentrated specific moieties due to their large specific surface areas, and consequently the limit of detection/sensitivity towards cancerous cells can be considerably enhanced that is crucial to cancer diagnosis. (II) Versatile surface functionalities provide huge opportunities for constructing a high-specificity anti-cancer drug delivery system [194, 195]. (III) They transform incoming radiation such as near-infrared spectroscopy (NIR), microwave, and radio wave into heat, which can be used to kill malignant cells by thermal ablation [196]. (IV) They have the capability of releasing therapeutic molecules to tumor microenvironment (TME) by physical and chemical stimuli. Owing to their varied sizes, they can deeply penetrate into acidic TME of tumors which is facilitated *via* the EPR effect, the unique phenomenon for accumulating drugs in tumors and increase vascular permeability in cancerous sites [197]. (V) These carbon nanostructures are capable for multimodality for cancer theranostics, which means they can carry out both diagnostic and therapeutic (drug delivery/hyperthermia) tasks at the same time [198].

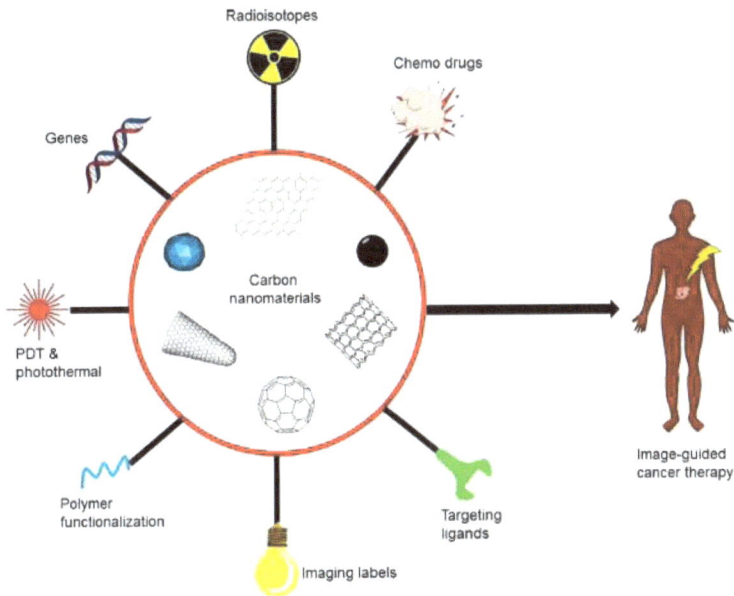

Fig. (5). Various application of carbon nanomaterials in cancer theranostic. Reproduced from [197].

Owing to their many aforementioned advantages, much research efforts have been dedicated to carbonic nanostructures as platforms for diagnosis, drug delivery, and treatment leading to the aim of personalized cancer therapy. The combination of imaging labels with treatment in the same platform led to precise delineation of tumor location, and the optimal drug dosage as well as therapeutic time frame can be identified by obtaining the real-time drug distribution information *in vivo*. By screenings the variation in target proteins and receptors, image-guided carbon nanoparticles can assess the therapeutic effectiveness of a particular therapy [199]. Thus, much research has been employed for taking benefit from carbon nanostructures in different approaches for cancer therapy including anti-angiogenesis, immune-stimulation, cancer stem cells inhibition, anti-hypoxia, microenvironment-based stimuli-responsive drug delivery, and photodynamic and photothermal therapy. Furthermore, it is reported that they are also solving the challenges plagued with common cancer therapy such as chemoresistance [200]. For instance, fullerene is the first class of carbon nanostructures frequently utilized for cancer theranostic applications from diagnostic to treatment. In this regard, the studies have shown IONP-decorated fullerene (C_{60}) composite that is linked to folic acid (FA), a commonly utilized tumor-targeting agent, has been shown actively target breast cancer cells and malignant tumors in mouse models. This hybrid nanoplatform with multifunctional properties was produced for cancer diagnostics, PDT, radiofrequency (RF) thermal treatment (RTT), and magnetic targeting applications. Thus, they not only acted as a strong tumor diagnostic MRI contrast agent but also as a powerful photosensitizer and potent agent for photothermal tumor ablation. Furthermore, these multi-functional nanoplatforms by excellent active tumor targeting and magnetic targeting properties are able to selectively kill cancerous cells in highly localized areas [201]. In addition, CNTs have been investigated in almost every single cancer treatment modality, including drug delivery with small nanomolecules, gene therapy, photodynamic therapy, thermal therapy, photodynamic therapy, and lymphatic targeted chemotherapy and reveal a great promise in targeted drug delivery systems, diagnostic methods and in bioanalytical applications [202, 203].

In the following section, we summarize the state-of-the art application of various carbon nanostructures in cancer imaging and therapy (Table. **2**).

Cancer Imaging

Most frequently utilized imaging agents are unable to penetrate the cell membranes. Carbon-based nanomaterials can be effective to transport such contrast agents intracellularly for cell monitoring with high selectivity, and thus they have great capacity in cancer diagnosis applications [204, 205]. On this point, it is well documented that carbon nanostructures have the potential to serve as nanocarriers for a variety of contrast-enhancing substances for different cancer imaging modalities such as computed tomography (CT), SPECT, PET, MRI, photoacoustic imaging (PAI), ultrasound imaging, photoacoustic tomography (PAT), thermoacoustic tomography (TAT), fluorescence/photoacoustic imaging, and Raman imaging [197]. For instance, carbon nanostructures are desirable PAI agents because they can transform laser energy into acoustic signals. The distribution and metabolism of carbonaceous compounds can also be monitored *in vivo* using intrinsic Raman vibration signals from those materials. The following are some carbon nanostructures that have high potential in cancer imaging. CNTs can be employed in some imaging methods such as MRI, PET, NIR, or multimodality imaging, because of the structural uniqueness and chemical versatility. The intrinsic emission of CNTs in the second NIR region (1100-1400 nm, NIRII) can be readily utilized for *in vivo* tumor detection. Several studies have indicated that CNTs generate high signal-to-noise ratios, consequently produced high-resolution tumor visualization [206]. For instance, M13 phage-conjugated SWNTs with a stable display of tumor-targeting sequence were successfully applied for NIR fluorescence imaging of prostate tumors in mice [207].

About the graphene family, the researchers have reported the use of GO as a platform to produce SPECT imaging agents that can be conjugated with trastuzumab, anti-HER2 (human epidermal growth factor receptor, and 2) antibody to increase uptake in HER2-positive tumors. The findings showed that GO-trastuzumab can detect primary tumor site as well as metastasis location in lymph [208].

Another carbon nanostructure is NDs that are effective imaging agents owing to their bright inherent fluorescence after adding nitrogen and easy modification with MRI detectable isotopes (*e.g.*, ^{15}N or ^{13}C) or an organic fluorophore for MRI or fluorescence imaging. The quantity of nitrogen incorporated into the ND enables fluorescence intensity and emission control, which can be useful for single molecule monitoring inside a cell [209]. The intrinsic fluorescence intensity from a ND is considerably higher compared with an Alex Fluor dye (bright, fluorescent dyes) at the same wavelength, thus the ND does not require further functionalization to provide powerful imaging capabilities [210]. However, to prepare NDs for cancer detection/imaging applications, different imaging labels have simultaneously been combined into NDs such as iron nanoparticles and positron-emitting isotope ^{18}F [211]. Like NDs, carbon-dots have inherent photoluminescence that permits them to be utilized as imaging agents that possess colorful and bright luminescence with a significant excitation wavelength dependency (tunable) [212]. Aside from fluorescence imaging, carbon-dots can be readily employed for multimodal imaging. For instance, after doping with gadolinium, carbon-dots have been employed for both florescence imaging and MRI [213].

Cancer Therapy

Carbon nanostructures can be used in cancer treatment as agents that directly mediate tumor cell killing, when combined with external energy sources including PDT, PTT, or magnetic hypothermia. These surface-engineered carbon nanomaterials can be used for targeted cancer therapy using hyperthermia, which is one of the other potential methods to cure cancer through thermal ablation, due to their inherent optical (absorption in the NIR or UV areas) and thermal properties [214].

In hyperthermia technique, tumors are selectively destroyed in a short time by applying heat (temperature range 41-47°C), while maintaining the surrounding healthy tissue undamaged [215]. In comparison to traditional therapies such as radiation and chemotherapy, hyperthermia therapy is known to have minor negative impacts on healthy cells owing to their excellent heat dissipation ability [216]. Various types of energy sources including laser and radio frequency energy can be applied to attain the hyperthermia effect for cancer treatment. Furthermore, several studies have indicated that combining hyperthermia with radiation (radio frequency, photothermal) or/and chemotherapy can improve cancer treatment effectiveness. It can also be utilized to make tumor tissues more responsive to radiation or to destroy them that is immune to radiation. Hyperthermia has also been shown to boost the effects of some anticancer medications by increasing perfusion in the cancer tissue, resulting in an elevated local oxygen concentration that may create perfect circumstances for radiation-induced tumor cell destruction [217].

On the other hand, other promising approaches that can be combined with carbon nanomaterials, such as PDT (laser source) and thermoacoustic (TA)/photoacoustic (PA) therapy (ultrasound sources, microwave, pulsed laser) have been investigated for cancer therapy. PDT uses reactive oxygen species (ROS) like singlet oxygen generated from a photosensitizer (PS) under laser irradiation to kill tumor cells. For example, surface- derivative fullerene is able to generate a ROS by light irradiation. In addition, several fullerene derivatives can act as excellent photosensitizers for PDT under laser irradiation and serve as a "radical sponge", displaying anticancer capabilities by themselves [218]. Moreover, it has been documented that the combination of CNTs and NIR radiation can be utilized for the hyperthermia therapy of primary brain cancers (glioblastoma multiforme). In these studies, CNTs internalize in glioma cells and generate heat when exposed to NIR radiation, resulting in necrotic cell death and tumor shrinkage only after one treatment. Nevertheless, minimal effects on healthy cells have been documented owing to their minimal uptake of CNTs [219].

Moreover, carbon nanomaterials can also be used as nanocarriers for radiopharmaceuticals or radioisotopes to increase their function. External radiation therapy is a common treatment that is combined with chemotherapy and/or surgery. In spite of its efficacy in killing cancerous cells, it also suffers from non-specific side effects on normal tissues. An alternative approach to target cancer cells is using radionuclides as an internal radiation source; therefore, it is essential to accumulate as many as possible radiation-emitting atoms around and within cancer cells and tumor [220]. Nanomaterials can play a key role in accomplishingthe goal oftransporting a diversity of radionuclides to the target site. Compared to other nanomaterials, carbon nanostructures provide some benefits for radiopharmaceuticals' stability during radiolabeling process and the versatile surface modification chemistry, harsh chemical modification, and the possibility to tailor the size, shape, and surface properties [221]. For this goal, the graphene family has enticed significant attention among different carbon nanomaterials because of their unique inherent properties. Although, other symmetric allotropes of carbon nanomaterials, such as NDs, fullerenes, and CQDs may circulate simply in the blood circulation [220].

Table. 2. Medical application of carbon nanostructures.

Carbon based Nanostructure	*In Vitro /In Vivo* Model	Highlights of the Study	Refs.
Gene and Drug Delivery			
PEI-g-GNR	HeLa cells	Effectively transferring LNA-m-MB into the cells to detect the target miRNA.	[222]
PCD	293T cells	Good biocompatibility and the bright photoluminescence property for bioimaging and gene delivery.	[223]
CQD	3T6 cells	Simultaneously play dual functions as non-viral gene vectors and bioimaging probes.	[224]
SWNT	PC-3 cells/mice	SWNT-PEI/siRNA/NGR can efficiently penetrate cell membrane, induce apoptosis, suppress cells proliferation, and improve the efficacy of photothermal therapy.	[225]
O-GNR	HeLa and HUVEC	Effective and versatile non-viral vectors for a wide range size of genetic materials in primary and secondary cell types for gene therapy.	[226]

Carbon based Nanostructure	In Vitro /In Vivo Model	Highlights of the Study	Refs.
Diamond	HeLa cell	Maintaining the biological activity of the accompanying cargo and high loading of bioactive moiety.	[227]
Fullerene (C60)	NIH 3T3 and HEK 293 cells	Successful intracellular DNA uptake, intracellular transport, and gene expression.	[228]
SWNT- Qdot -cisplatin-EGF	HNSCC cells/mice	Targeted killing of cancer cells.	[229]
MWNT−HCPT	MKN-28 cells/mice	Enhance cell uptake, high drug accumulation in the tumor site, and long blood circulation.	[230]
MWCNT- drug loaded liposomes conjugate	CCL-186 and HEK 294	Efficient cell uptake, high drug loading capacity of nano-liposomes and enhancement of drug delivery to diseased cells and reduction in drug interaction with normal tissues.	[231]
DOX-O-MWNTs-PEG-ANG	C6 cells/mice	High loading anticancer drug of doxorubicin and accumulation in tumors.	[232]
CDDP-SWNHox	NCI-H460/BALB/c nu/nu mice	Increase anti-cancer efficacy and suppress the tumor growth.	[233]
NGO-PEG-DOX	EMT6 cells/ Balb/c mice	Improve the therapeutic efficacy of cancer treatment *via* combination of the local specific chemotherapy with external NIR photothermal therapy	[234]
PEG–BPEI–rGO–DOX	PC3 cells	Photothermally stimulated cytosolic DOX delivery through endosome disruption.	[235]
Tissue Engineering			
SWNT	Swiss mice	Form fiber like structures and induce granuloma formation.	[236]
MWNT	hASC/mice	Promote concentration of proteins to form inductive bone.	[237]
CNT-GelMA	Cardiac cells	Improve cardiac cell adhesion, organization, electroactivity, mechanical integrity and cell-cell coupling.	[238]
CNT	Neonatal hippocampal neuron	The growth of neuronal circuits, rise in network activity and the improvement in the efficacy of neural signal transmission due to high electrical conductivity.	[239]
Graphene	Human osteoblast cells and mesenchymal stromal cells	Better cell adhesion and proliferation.	[240]

(Table 2) cont.....

Carbon based Nanostructure	In Vitro /In Vivo Model	Highlights of the Study	Refs.
GO-PLGA/Col	Human dermal fibroblasts	Effective scaffold supporting tissue regeneration.	[241]
CNT	hNSCs	Synergistic cues for the differentiation of hNSCs in physiological solution and an optimal nanotopography at the same time with good biocompatibility.	[242]
Fullerene	hMSC	Multi-potency retention and regenerative capacity of cultured cells.	[243]
MWCNT/Chitosan	C2C12 cells/mice	Promote the ectopic bone formation at muscle tissue.	[244]
ultra-short SWCNT	MSC	Scaffold with high strength, tunable porosity and unique mechanical properties for bone tissue engineering applications.	[245]
PLA/MWCNT	Osteoblasts	Provide electrical stimulation with an appropriate DC value imparted on conductive substrate for bone tissue engineering.	[246]
Cancer Theranostic			
SWNTs-PEG-RGD peptide	Mice	Long blood circulation times, low uptake by the RES and efficient targeting of integrin positive tumor.	[247]
Nanodiamond	A549 and 3T3-L1 cells	Labeling and tracking of cancer and stem cells.	[248]
SWCNT	live 3T3 cells	Modulate to unique fingerprint agents by the degree to the emission band intensity or wavelength, multiplexed optical detection, and the first label-free tool to optically discriminate between genotoxins.	[249]
Nano-sized GO	B16F0 cells/Mice	Simultaneously *in vivo* fluorescent imaging as well as combined NmPDT and NmPTT effects.	[250]
ICG/rGO or ICG/HArGO	KB cells/ nude mice	Image-guided and synergistic photothermal antitumor therapy.	[251]
RGO	Cancerous cells/mice	Enhance cellular uptake, improve radio-therapeutic efficacy against cancer cells and induce effective photothermal heating of tumor under NIR light irradiation.	[252]
Fullerene C70/photosensitizer (Chlorin e6, Ce6) nanovesicles	A549 cells/ mice	Highly efficient imaging and photodynamic therapy of tumor.	[253]

(Table 2) cont.....

Carbon based Nanostructure	*In Vitro /In Vivo* Model	Highlights of the Study	Refs.
Cu-NOTA-RGO-TRC105	4T1 cells/mice	Specific targeting of tumor, and integrate photothermal therapy and imaging.	[254]
Gd@C82 (OH)22	MDA-MB-231 cells/BALB/c nude mice	Reversal of phenotype of EMT in cancer cells.	[255]
Gd@C82 (OH)22	HMEC cells/athymic BALB/c mice	Downregulation of 10 proangiogenic factors in mice model.	[256]
C60 (OH)20	MCF-7 cells/BALB/c mice	Downregulation of TNF-α, PDGF, and VEGF by 20–40% in the EMT-6 tumor metastasis model	[257]
β-Alanine–Gd@C82 (OH)22	BALB/c mice	Radiofrequency-mediated destruction of tumor vasculature.	[258]
Chitosan and alinomycin loaded HA/SWNTs	AGS cells	Decreased population of gastric cancer stem cells and their ability to form mammosphere.	[259]
PEG–DOX–SWCNT Thermo/pH-sensitive drug release	HeLa cells	Efficient release of drug with NIR light and at pH of 5.0.	[260]
GO–BSA	HUVECs/Chick chorioallantoic membrane	Successful targeting of VEGF, resulting in reduced angiogenesis.	[261]

Polyethylenimine-grafted graphene nanoribbon, PEI-g-GNR; locked nucleic acid modified molecular beacon, LNA-m-MB; microRNA, miRNA; polyethyleneimine-based carbon dots, PCD; Carbon quantum dots, CQDs; Single-walled carbon nanotubes, SWNTs; Oxidized graphene nanoribbons, O-GNRs; Henrietta Lacks, HeLa; Human umbilical vein endothelial cells, HUVEC; HEK 293, human embryonic kidney 293; Reticuloendothelial system, RES; Arginine–glycine–aspartic acid, RGD; Epidermal growth factor, EGF; 10-hydroxycamptothecin, HCPT; doxorubicin oxidized multi-walled carbon nanotubes modified with angiopep-2, DOX-O-MWNTs-PEG-ANG; Cisplatin, CDDP; single-wall carbon nanohorns with holes opened (SWNHox); doxorubicin-loaded PEGylated nanographene oxide, NGO-PEG-DOX; Near-infrared, NIR; Reduced graphene oxide, RGO; Nanomaterial-mediated photodynamic therapeutic, NmPDT; Photothermal therapy, NmPTT; Gelatin methacrylate, GelMA; Graphene oxide, GO; Poly lactic-co-glycolic acid, PLGA; Collagen, Col; Human neural stem cells, hNSCs; Human mesenchymal stem cell, hMSC; Poly-DL-lactide, PLA;

CONCLUDING REMARKS

Enormous developments have been made in nano-carbon science because of advances in synthetic strategies and methodologies as well as measurement devices and techniques. In addition, self-assembly of nanostructural materials is theoretically valuable and has formed new resources to revolutionize biological and biomedical sciences. Extensive research efforts have promoted the status of carbon-based nanostructures as one of the most widely used group of nanomaterials. Owing to their outstanding chemical, physical, mechanical, electrical, thermal, and optical properties, they have been investigated in various industrial areas. Different areas of biomedical engineering have also been benefited extremely from carbon nanoparticles in recent years because incorporating them is effective not only as injectable nanoscale instrument but also as components to considerably improve the function of existing biomaterials significantly.

Besides, it is well documented that carbon nanostructures such as graphite, graphene, carbon nanotubes (CNTs), diamond, and fullerene are used successfully in tissue engineering, gene/drug delivery, and cancer theranostic. Some nano-carbon allotropes including graphene, fullerenes, CNOs, carbon quantum dots, and magnetic carbon nanostructures especially CNTs, have been utilized for controllable drug release and gene delivery. Moreover, carbon nanostructures such as graphene and CNTs have high potential as scaffolds for tissue engineering and regenerative medicine. More studies demonstrated that they can be applied to regulate cellular behavior. Also, in the field of cancer theranostics, they have shown promising approaches in cancer imaging and therapy due to unique characteristics and numerous advantages.

Although significant progress has been achieved in this field, it is still in its early stages and a number of challenges related to the usage of these carbon-based nanostructures must be addressed before they can be used as active therapeutic treatments. For example: (1) it is commonly noticed that if a tumor is too small, it cannot be easily diagnosed. As a result, it is vital to develop unique targeted nanomaterials with high-efficacy imaging moieties that can overcome diagnostic difficulties in case of extremely small tumors. (2) An ideal medication delivery system should be able to penetrate and destroy malignant cells while leaving healthy cells undamaged [1]. Thus, a medication delivery system with active targeting and high loading effectiveness should be created. Most importantly, an understanding of the targeting mechanism should be well established [262]. (3) The delivery of local heat to cancer cells in order to kill them by hyperthermia is difficult, hence efforts should be made to build nanosystems that can efficiently and selectively target cancer cells, which are efficient enough to provide the necessary heat for cancer cell killing [263]. Most *in vivo* hyperthermia treatments are performed using NIR radiation which has a limited penetration ability towards tumors [264, 265]. On the other hand, longer wavelength, electromagnetic or RF radiation may be necessary for the treatment of deep-seated tumor cells, hence other strategies should be developed [266, 267]. (4) One of the other major limitations is the long-term toxicity of these carbon nanostructures which need to be extensively studied *in vivo* using more relevant animal models [268, 269]. (5) The toxicity and pharmacokinetics of carbon nanomaterials are influenced by a number of factors, including physiochemical and structural properties, exposure dosage and duration, cell type, mechanism, residual catalyst, and production technique.

REFERENCES

[1]   Liu Z, Robinson JT, Tabakman SM, Yang K, Dai H. Carbon materials for drug delivery & cancer therapy. Mater Today 2011; 14(7-8): 316-23.
[http://dx.doi.org/10.1016/S1369-7021(11)70161-4]

[2]   Whitesides GM, Grzybowski B. Self-assembly at all scales. Science 2002; 295(5564): 2418-21.
[http://dx.doi.org/10.1126/science.1070821] [PMID: 11923529]

[3]   Whitesides GM. Self-assembling materials. Sci Am 1995; 273(3): 146-9.

[4]   Fendler JH. Self-assembled nanostructured materials. Chem Mater 1996; 8(8): 1616-24.
[http://dx.doi.org/10.1021/cm960116n]

[5]   Mason TO, Shimanovich U. Fibrous protein self-assembly in biomimetic materials. Adv Mater 2018; 30(41): 1706462.
[http://dx.doi.org/10.1002/adma.201706462] [PMID: 29883013]

[6]   Liao G, He F, Li Q, *et al.* Emerging graphitic carbon nitride-based materials for biomedical applications. Prog Mater Sci 2020; 112: 100666.
[http://dx.doi.org/10.1016/j.pmatsci.2020.100666]

[7]   Thiruvengadathan RAR, Somnath C. A sundriyal,poonam %a bhattacharya,shantanu, carbon nanostructures.

[8]   Iijima S. Helical microtubules of graphitic carbon. Nature 1991; 354(6348): 56-8.
[http://dx.doi.org/10.1038/354056a0]

[9]   Novoselov KS, *et al.* Electric field effect in atomically thin carbon films. science 2004; 306(5969): 666-9.

[10]   Kroto HW. C 60: buckminsterfullerene. nature 1985; 318(6042): 162-3.

[11]   Kumar N, Kumbhat S. Carbon-based nanomaterials. Essentials in nanoscience and nanotechnology 2016; 189-236.
[http://dx.doi.org/10.1002/9781119096122.ch5]

[12]   Dunlap RA. Other crystalline allotropes of carbon. Novel Microstructures for Solids. Morgan & Claypool Publishers 2018; pp. 8-1-9.
[http://dx.doi.org/10.1088/2053-2571/aae653ch8]

[13]   Sinnott SB, Andrews R. Carbon nanotubes: Synthesis, properties, and applications. Crit Rev Solid State Mater Sci 2001; 26(3): 145-249.
[http://dx.doi.org/10.1080/20014091104189]

[14]   Kato T, Hatakeyama R. Direct growth of short single-walled carbon nanotubes with narrow-chirality distribution by time-programmed plasma chemical vapor deposition. ACS Nano 2010; 4(12): 7395-400.
[http://dx.doi.org/10.1021/nn102379p] [PMID: 21082841]

[15]   Han MY, Özyilmaz B, Zhang Y, Kim P. Energy band-gap engineering of graphene nanoribbons. Phys Rev Lett 2007; 98(20): 206805.
[http://dx.doi.org/10.1103/PhysRevLett.98.206805] [PMID: 17677729]

[16]   Segawa Y, Ito H, Itami K. Structurally uniform and atomically precise carbon nanostructures. Nat Rev Mater 2016; 1(1): 15002.
[http://dx.doi.org/10.1038/natrevmats.2015.2]

[17]   Krauss TD. Nanotubes light up cells. Nat Nanotechnol 2009; 4(2): 85-6.
[http://dx.doi.org/10.1038/nnano.2008.425] [PMID: 19197307]

[18]   Jiang H. Chemical preparation of graphene-based nanomaterials and their applications in chemical and biological sensors. Small 2011; 7(17): 2413-27.
[http://dx.doi.org/10.1002/smll.201002352] [PMID: 21638780]

[19]   Swierczewska M, Choi KY, Mertz EL, *et al.* A facile, one-step nanocarbon functionalization for biomedical applications. Nano Lett 2012; 12(7): 3613-20.
[http://dx.doi.org/10.1021/nl301309g] [PMID: 22694219]

[20]   Wang J, Hu Z, Xu J, Zhao Y. Therapeutic applications of low-toxicity spherical nanocarbon materials. NPG Asia Mater 2014; 6(2): e84-4.
[http://dx.doi.org/10.1038/am.2013.79]

[21]   Goenka S, Sant V, Sant S. Graphene-based nanomaterials for drug delivery and tissue engineering. J Control Release 2014; 173: 75-88.
[http://dx.doi.org/10.1016/j.jconrel.2013.10.017] [PMID: 24161530]

[22]   Jara AD, Betemariam A, Woldetinsae G, Kim JY. Purification, application and current market trend of natural graphite: A review. Int J Min Sci Technol 2019; 29(5): 671-89.
[http://dx.doi.org/10.1016/j.ijmst.2019.04.003]

[23]   Huai W, Zhang C, Wen S. Graphite-based solid lubricant for high-temperature lubrication. Friction 2021; 9(6): 1660-72.
[http://dx.doi.org/10.1007/s40544-020-0456-2]

[24]   Pierson HO. Handbook of carbon, graphite, diamonds and fullerenes: processing, properties and applications. William Andrew 2012.

[25]   Luo T, Wang Q. Effects of graphite on electrically conductive cementitious composite properties: A review. Materials 2021; 14(17): 4798.
[http://dx.doi.org/10.3390/ma14174798] [PMID: 34500888]

[26]   Chung DDL. Review graphite. J Mater Sci 2002; 37(8): 1475-89.
[http://dx.doi.org/10.1023/A:1014915307738]

[27]   Li H, Feng Q, Ou L, Long S, Cui M, Weng X. Study on washability of microcrystal graphite using float–sink tests. Int J Min Sci Technol 2013; 23(6): 855-61.
[http://dx.doi.org/10.1016/j.ijmst.2013.10.012]

[28]   Lipson HS, Stokes A. The structure of graphite. Proc R Soc Lond A Math Phys Sci 1942; 181(984): 101-5.
[http://dx.doi.org/10.1098/rspa.1942.0063]

[29]   Weis PL, Friedman I, Gleason JP. The origin of epigenetic graphite: evidence from isotopes. Geochim Cosmochim Acta 1981; 45(12): 2325-32.
[http://dx.doi.org/10.1016/0016-7037(81)90086-7]

[30]   Miyashiro A. Metamorphism and metamorphic belts. Springer Science & Business Media 2012.

[31]   Zhao H, Ren J, He X, Li J, Jiang C, Wan C. Purification and carbon-film-coating of natural graphite as anode materials for Li-ion batteries. Electrochim Acta 2007; 52(19): 6006-11.
[http://dx.doi.org/10.1016/j.electacta.2007.03.050]

[32]   Wissler M. Graphite and carbon powders for electrochemical applications. J Power Sources 2006; 156(2): 142-50.
[http://dx.doi.org/10.1016/j.jpowsour.2006.02.064]

[33]   Kwiecińska B, Petersen HI. Graphite, semi-graphite, natural coke, and natural char classification—ICCP system. Int J Coal Geol 2004; 57(2): 99-116.
[http://dx.doi.org/10.1016/j.coal.2003.09.003]

[34]   Zaghib K, Song X, Guerfi A, Rioux R, Kinoshita K. Purification process of natural graphite as anode for Li-ion batteries: chemical *versus* thermal. J Power Sources 2003; 119-121: 8-15.
[http://dx.doi.org/10.1016/S0378-7753(03)00116-2]

[35] Preston GD. Structure of diamond. Nature 1945; 155(3925): 69-70.
[http://dx.doi.org/10.1038/155069a0]

[36] Németh P, McColl K, Garvie LAJ, Salzmann CG, Murri M, McMillan PF. Complex nanostructures in diamond. Nat Mater 2020; 19(11): 1126-31.
[http://dx.doi.org/10.1038/s41563-020-0759-8] [PMID: 32778814]

[37] Geim AK, Novoselov KS. The rise of graphene. Nat Mater 2007; 6(3): 183-91.
[http://dx.doi.org/10.1038/nmat1849] [PMID: 17330084]

[38] Hazen RM, Downs RT, Jones AP, Kah L. Carbon mineralogy and crystal chemistry. Rev Mineral Geochem 2013; 75(1): 7-46.
[http://dx.doi.org/10.2138/rmg.2013.75.2]

[39] Williams OA. Nanocrystalline diamond. Diamond Related Materials 2011; 20(5-6): 621-40.
[http://dx.doi.org/10.1016/j.diamond.2011.02.015]

[40] Mochalin VN, Shenderova O, Ho D, Gogotsi Y. The properties and applications of nanodiamonds. Nat Nanotechnol 2012; 7(1): 11-23.
[http://dx.doi.org/10.1038/nnano.2011.209] [PMID: 22179567]

[41] Chandran M. Chapter six - synthesis, characterization, and applications of diamond films. In: Yaragalla S, Ed. Carbon-Based Nanofillers and Their Rubber Nanocomposites. Elsevier 2019; pp. 183-224.
[http://dx.doi.org/10.1016/B978-0-12-813248-7.00006-7]

[42] Williams OA. Nanocrystalline diamond. Diamond Related Materials 2011; 20(5-6): 621-40.
[http://dx.doi.org/10.1016/j.diamond.2011.02.015]

[43] Tsai LW, Lin YC, Perevedentseva E, Lugovtsov A, Priezzhev A, Cheng CL. Nanodiamonds for medical applications: Interaction with blood *in vitro* and *in vivo*. Int J Mol Sci 2016; 17(7): 1111.
[http://dx.doi.org/10.3390/ijms17071111] [PMID: 27420044]

[44] Narayan RJ, Boehm RD, Sumant AV. Medical applications of diamond particles & surfaces. Mater Today 2011; 14(4): 154-63.
[http://dx.doi.org/10.1016/S1369-7021(11)70087-6]

[45] Perevedentseva E, Lin YC, Jani M, Cheng CL. Biomedical applications of nanodiamonds in imaging and therapy. Nanomedicine 2013; 8(12): 2041-60.
[http://dx.doi.org/10.2217/nnm.13.183] [PMID: 24279492]

[46] Chauhan S, Jain N, Nagaich U. Nanodiamonds with powerful ability for drug delivery and biomedical applications: Recent updates on *in vivo* study and patents. J Pharm Anal 2020; 10(1): 1-12.
[http://dx.doi.org/10.1016/j.jpha.2019.09.003] [PMID: 32123595]

[47] Turcheniuk K, Mochalin VN. Biomedical applications of nanodiamond (Review). Nanotechnology 2017; 28(25): 252001.
[http://dx.doi.org/10.1088/1361-6528/aa6ae4] [PMID: 28368852]

[48] Hanada K. Detonation nanodiamond: Perspective and applications. Surf Eng 2009; 25(7): 487-9.
[http://dx.doi.org/10.1179/174329409X433939]

[49] Bondon N, Raehm L, Charnay C, Boukherroub R, Durand JO. Nanodiamonds for bioapplications, recent developments. J Mater Chem B Mater Biol Med 2020; 8(48): 10878-96.
[http://dx.doi.org/10.1039/D0TB02221G] [PMID: 33156316]

[50] Danilenko VV. On the history of the discovery of nanodiamond synthesis. Springer 2004.
[http://dx.doi.org/10.1134/1.1711431]

[51] Plonska-Brzezinska ME. Carbon nano-onions: A review of recent progress in synthesis and applications. ChemNanoMat 2019; 5(5): 568-80.
[http://dx.doi.org/10.1002/cnma.201800583]

[52]   Bhinge S. Carbon nano-onions–an overview. J Pharm Chem Chem Sci 2017; 1: 1-2.

[53]   Bartkowski M, Giordani S. Carbon nano-onions as potential nanocarriers for drug delivery. Dalton Trans 2021; 50(7): 2300-9.
[http://dx.doi.org/10.1039/D0DT04093B] [PMID: 33471000]

[54]   Wudl F. Fullerene materials. J Mater Chem 2002; 12(7): 1959-63.
[http://dx.doi.org/10.1039/b201196d]

[55]   Hirsch A. Fullerenes and related structures. Springer 2003; Vol. 199.

[56]   Kumar CS, Hormes J, Leuschner C. Nanofabrication towards biomedical applications: techniques, tools, applications, and impact. John Wiley & Sons 2006.

[57]   Wang Z, Lu Z, Zhao Y, Gao X. Oxidation-induced water-solubilization and chemical functionalization of fullerenes C_{60}, Gd@C_{60} and Gd@C_{82} : atomistic insights into the formation mechanisms and structures of fullerenols synthesized by different methods. Nanoscale 2015; 7(7): 2914-25.
[http://dx.doi.org/10.1039/C4NR06633B] [PMID: 25565281]

[58]   Goodarzi S, Da Ros T, Conde J, Sefat F, Mozafari M. Fullerene: Biomedical engineers get to revisit an old friend. Mater Today 2017; 20(8): 460-80.
[http://dx.doi.org/10.1016/j.mattod.2017.03.017]

[59]   Hetzel R, Manning T, Lovingood D, Strouse G, Phillips D. Production of fullerenes by microwave synthesis. Fuller Nanotub Carbon Nanostruct 2012; 20(2): 99-108.
[http://dx.doi.org/10.1080/1536383X.2010.533300]

[60]   Zhang X, Ma Y, Fu S, Zhang A. Facile synthesis of water-soluble fullerene (c_{60}) nanoparticles *via* mussel-inspired chemistry as efficient antioxidants. Nanomaterials 2019; 9(12): 1647.
[http://dx.doi.org/10.3390/nano9121647] [PMID: 31756936]

[61]   Afreen S, Zhu JJ. Effect of switching ultrasonic amplitude in preparing a hybrid of fullerene (C60) and gallium oxide (Ga2O3). Ultrason Sonochem 2020; 67: 105178.
[http://dx.doi.org/10.1016/j.ultsonch.2020.105178] [PMID: 32464503]

[62]   Elessawy NA, El-Sayed EM, Ali S, Elkady MF, Elnouby M, Hamad HA. One-pot green synthesis of magnetic fullerene nanocomposite for adsorption characteristics. J Water Process Eng 2020; 34: 101047.
[http://dx.doi.org/10.1016/j.jwpe.2019.101047]

[63]   Harris PJF. Fullerene Polymers: A Brief Review 2020; 6(4): 71.
[http://dx.doi.org/10.3390/c6040071]

[64]   Bakry R, Vallant RM, Najam-ul-Haq M, *et al.* Medicinal applications of fullerenes. Int J Nanomedicine 2007; 2(4): 639-49.
[PMID: 18203430]

[65]   Nel A, Xia T, Mädler L, Li N. Toxic potential of materials at the nanolevel. Science 2006; 311(5761): 622-7.
[http://dx.doi.org/10.1126/science.1114397] [PMID: 16456071]

[66]   Aqel A, El-Nour KMMA, Ammar RAA, Al-Warthan A. Carbon nanotubes, science and technology part (I) structure, synthesis and characterisation. Arab J Chem 2012; 5(1): 1-23.
[http://dx.doi.org/10.1016/j.arabjc.2010.08.022]

[67]   Meyyappan M, Delzeit L, Cassell A, Hash D. Carbon nanotube growth by PECVD: a review. Plasma Sources Sci Technol 2003; 12(2): 205-16.
[http://dx.doi.org/10.1088/0963-0252/12/2/312]

[68]   Debnath SK, Srivastava R. Drug delivery with carbon-based nanomaterials as versatile nanocarriers: Progress and prospects. Frontiers in Nanotechnology 2021; 3: 644564.
[http://dx.doi.org/10.3389/fnano.2021.644564]

[69]   Li C, Chou T-W. Modeling of Carbon Nanotubes and Their Composites. 2006; pp. 55-65.
[http://dx.doi.org/10.1007/1-4020-3951-4_6]

[70]   O'Connell MJ, Bachilo SM, Huffman CB, *et al.* Band gap fluorescence from individual single-walled carbon nanotubes. Science 2002; 297(5581): 593-6.
[http://dx.doi.org/10.1126/science.1072631] [PMID: 12142535]

[71]   Sharon M, Pradhan D, Zacharia R, Puri V. Application of carbon nanomaterial as a microwave absorber. J Nanosci Nanotechnol 2005; 5(12): 2117-20.
[http://dx.doi.org/10.1166/jnn.2005.186] [PMID: 16430149]

[72]   Yu MF, Files BS, Arepalli S, Ruoff RS. Tensile loading of ropes of single wall carbon nanotubes and their mechanical properties. Phys Rev Lett 2000; 84(24): 5552-5.
[http://dx.doi.org/10.1103/PhysRevLett.84.5552] [PMID: 10990992]

[73]   Fujii M, Zhang X, Xie H, *et al.* Measuring the thermal conductivity of a single carbon nanotube. Phys Rev Lett 2005; 95(6): 065502.
[http://dx.doi.org/10.1103/PhysRevLett.95.065502] [PMID: 16090962]

[74]   Ebbesen TW, Ajayan PM. Large-scale synthesis of carbon nanotubes. Nature 1992; 358(6383): 220-2.
[http://dx.doi.org/10.1038/358220a0]

[75]   Journet C. Maser KW, Bernier P. Large-scale production of single-walled carbon nanotubes by the electric-arc technique. nature 1997; 388(6644): 756-8.

[76]   Thess A, Lee R, Nikolaev P. Crystalline ropes of metallic carbon nanotubes. science 1996; 273(5274): 483-7.

[77]   Kong J, Cassell AM, Dai H. Chemical vapor deposition of methane for single-walled carbon nanotubes. Chem Phys Lett 1998; 292(4-6): 567-74.
[http://dx.doi.org/10.1016/S0009-2614(98)00745-3]

[78]   Su M, Zheng B, Liu J. A scalable CVD method for the synthesis of single-walled carbon nanotubes with high catalyst productivity. Chem Phys Lett 2000; 322(5): 321-6.
[http://dx.doi.org/10.1016/S0009-2614(00)00422-X]

[79]   Saifuddin N, Raziah AZ, Junizah AR. Carbon nanotubes: A review on structure and their interaction with proteins. J Chem 2013; 2013: 1-18.
[http://dx.doi.org/10.1155/2013/676815]

[80]   Banerjee S, Kahn MGC, Wong SS. Rational chemical strategies for carbon nanotube functionalization. Chemistry 2003; 9(9): 1898-908.
[http://dx.doi.org/10.1002/chem.200204618] [PMID: 12740836]

[81]   Wang J, Musameh M, Lin Y. Solubilization of carbon nanotubes by Nafion toward the preparation of amperometric biosensors. J Am Chem Soc 2003; 125(9): 2408-9.
[http://dx.doi.org/10.1021/ja028951v] [PMID: 12603125]

[82]   Wang Y, Iqbal Z, Malhotra SV. Functionalization of carbon nanotubes with amines and enzymes. Chem Phys Lett 2005; 402(1-3): 96-101.
[http://dx.doi.org/10.1016/j.cplett.2004.11.099]

[83]   Trache D, Thakur VK. Nanocellulose and nanocarbons based hybrid materials: Synthesis, characterization and applications. Multidisciplinary Digital Publishing Institute 2020.

[84]   Dresselhaus MS, Dresselhaus G, Jorio A. Unusual properties and structure of carbon nanotubes. Annu Rev Mater Res 2004; 34(1): 247-78.
[http://dx.doi.org/10.1146/annurev.matsci.34.040203.114607]

[85]   Raval JP, Joshi P, Chejara DR. Carbon nanotube for targeted drug delivery.Applications of nanocomposite Materials in drug delivery. Elsevier 2018; pp. 203-16.
[http://dx.doi.org/10.1016/B978-0-12-813741-3.00009-1]

[86]   Liang F, Chen B. A review on biomedical applications of single-walled carbon nanotubes. Curr Med Chem 2010; 17(1): 10-24.
[http://dx.doi.org/10.2174/092986710789957742] [PMID: 19941481]

[87]   Beg S. Rahman M, Jain A. Emergence in the functionalized carbon nanotubes as smart nanocarriers for drug delivery applications 2018.
[http://dx.doi.org/10.1016/B978-0-12-813691-1.00004-X]

[88]   Lalwani G, Gopalan A, D'Agati M, *et al.* Porous three-dimensional carbon nanotube scaffolds for tissue engineering. J Biomed Mater Res A 2015; 103(10): 3212-25.
[http://dx.doi.org/10.1002/jbm.a.35449] [PMID: 25788440]

[89]   Bhatt A, Jain A, Jain R. Carbon Nanotubes: A Promising carrier for drug delivery and targeting. 2016; pp. 465-501.

[90]   Shen C, Brozena AH, Wang Y. Double-walled carbon nanotubes: Challenges and opportunities. Nanoscale 2011; 3(2): 503-18.
[http://dx.doi.org/10.1039/C0NR00620C] [PMID: 21042608]

[91]   Punbusayakul N, Talapatra S, Ajayan PM, Surareungchai W. Label-free as-grown double wall carbon nanotubes bundles for Salmonella typhimuriumimmunoassay. Chem Cent J 2013; 7(1): 102.
[http://dx.doi.org/10.1186/1752-153X-7-102] [PMID: 23764320]

[92]   Béduer A, Seichepine F, Flahaut E, Loubinoux I, Vaysse L, Vieu C. Elucidation of the role of carbon nanotube patterns on the development of cultured neuronal cells. Langmuir 2012; 28(50): 17363-71.
[http://dx.doi.org/10.1021/la304278n] [PMID: 23190396]

[93]   Lin T, Bajpai V, Ji T, Dai L. Chemistry of carbon nanotubes. Aust J Chem 2003; 56(7): 635-51.
[http://dx.doi.org/10.1071/CH02254]

[94]   Dresselhaus G, Dresselhaus MS, Saito R. Physical properties of carbon nanotubes. World scientific 1998.

[95]   Ajayan PM. Nanotubes from Carbon. Chem Rev 1999; 99(7): 1787-800.
[http://dx.doi.org/10.1021/cr970102g] [PMID: 11849010]

[96]   Harris PJ, Harris PJF. Carbon nanotube science: synthesis, properties and applications. Cambridge university press 2009.
[http://dx.doi.org/10.1017/CBO9780511609701]

[97]   Iijima S. Carbon nanotubes: Past, present, and future. Physica B 2002; 323(1-4): 1-5.
[http://dx.doi.org/10.1016/S0921-4526(02)00869-4]

[98]   Iannazzo D, Piperno A, Pistone A, Grassi G, Galvagno S. Recent advances in carbon nanotubes as delivery systems for anticancer drugs. Curr Med Chem 2013; 20(11): 1333-54.
[http://dx.doi.org/10.2174/0929867311320110001] [PMID: 23432581]

[99]   Hassan HAFM, Diebold SS, Smyth LA, Walters AA, Lombardi G, Al-Jamal KT. Application of carbon nanotubes in cancer vaccines: Achievements, challenges and chances. J Control Release 2019; 297: 79-90.
[http://dx.doi.org/10.1016/j.jconrel.2019.01.017] [PMID: 30659906]

[100]   Veetil JV, Ye K. Tailored carbon nanotubes for tissue engineering applications. Biotechnol Prog 2009; 25(3): 709-21.
[http://dx.doi.org/10.1002/btpr.165] [PMID: 19496152]

[101]   Sireesha M, Jagadeesh Babu V, Kranthi Kiran AS, Ramakrishna S. A review on carbon nanotubes in biosensor devices and their applications in medicine. Nanocomposites 2018; 4(2): 36-57.
[http://dx.doi.org/10.1080/20550324.2018.1478765]

[102]   Yang ST, Luo J, Zhou Q, Wang H. Pharmacokinetics, metabolism and toxicity of carbon nanotubes for biomedical purposes. Theranostics 2012; 2(3): 271-82.
[http://dx.doi.org/10.7150/thno.3618] [PMID: 22509195]

[103] Lam C, James JT, McCluskey R, Arepalli S, Hunter RL. A review of carbon nanotube toxicity and assessment of potential occupational and environmental health risks. Crit Rev Toxicol 2006; 36(3): 189-217.
[http://dx.doi.org/10.1080/10408440600570233] [PMID: 16686422]

[104] Firme CP III, Bandaru PR. Toxicity issues in the application of carbon nanotubes to biological systems. Nanomedicine 2010; 6(2): 245-56.
[http://dx.doi.org/10.1016/j.nano.2009.07.003] [PMID: 19699321]

[105] Sato Y, Yokoyama A, Shibata K, *et al.* Influence of length on cytotoxicity of multi-walled carbon nanotubes against human acute monocytic leukemia cell line THP-1 *in vitro* and subcutaneous tissue of rats *in vivo*. Mol Biosyst 2005; 1(2): 176-82.
[http://dx.doi.org/10.1039/b502429c] [PMID: 16880981]

[106] Wick P, Manser P, Limbach L, *et al.* The degree and kind of agglomeration affect carbon nanotube cytotoxicity. Toxicol Lett 2007; 168(2): 121-31.
[http://dx.doi.org/10.1016/j.toxlet.2006.08.019] [PMID: 17169512]

[107] Allen MJ, Tung VC, Kaner RB. Honeycomb carbon: A review of graphene. Chem Rev 2010; 110(1): 132-45.
[http://dx.doi.org/10.1021/cr900070d] [PMID: 19610631]

[108] Tserpes KI, Silvestre N. Modeling of carbon nanotubes, graphene and their composites. Springer 2014.
[http://dx.doi.org/10.1007/978-3-319-01201-8]

[109] Abergel DSL, Apalkov V, Berashevich J, Ziegler K, Chakraborty T. Properties of graphene: A theoretical perspective. Adv Phys 2010; 59(4): 261-482.
[http://dx.doi.org/10.1080/00018732.2010.487978]

[110] Yousefi N, Lu X, Elimelech M, Tufenkji N. Environmental performance of graphene-based 3D macrostructures. Nat Nanotechnol 2019; 14(2): 107-19.
[http://dx.doi.org/10.1038/s41565-018-0325-6] [PMID: 30617310]

[111] Sprinkle M, Ruan M, Hu Y, *et al.* Scalable templated growth of graphene nanoribbons on SiC. Nat Nanotechnol 2010; 5(10): 727-31.
[http://dx.doi.org/10.1038/nnano.2010.192] [PMID: 20890273]

[112] Grebinyk A, Prylutska S, Chepurna O, *et al.* Synergy of chemo-and photodynamic therapies with C60 fullerene-doxorubicin nanocomplex. Nanomaterials 2019; 9(11): 1540.
[http://dx.doi.org/10.3390/nano9111540] [PMID: 31671590]

[113] Crevillen AG, Escarpa A, García CD. Carbon-based nanomaterials in analytical chemistry 2018.
[http://dx.doi.org/10.1039/9781788012751-00001]

[114] Abetz V, Boschetti-de-Fierro A. Polymer Science: A Comprehensive Reference, 10 Volume Set. The Netherlands: Elsevier Amsterdam 2012.

[115] Mousavi SM, Soroshnia S, Hashemi SA, *et al.* Graphene nano-ribbon based high potential and efficiency for DNA, cancer therapy and drug delivery applications. Drug Metab Rev 2019; 51(1): 91-104.
[http://dx.doi.org/10.1080/03602532.2019.1582661] [PMID: 30784324]

[116] Pan D, Zhang J, Li Z, Wu M. Hydrothermal route for cutting graphene sheets into blue-luminescent graphene quantum dots. Adv Mater 2010; 22(6): 734-8.
[http://dx.doi.org/10.1002/adma.200902825] [PMID: 20217780]

[117] Li N, Than A, Sun C, *et al.* Monitoring dynamic cellular redox homeostasis using fluorescence-switchable graphene quantum dots. ACS Nano 2016; 10(12): 11475-82.
[http://dx.doi.org/10.1021/acsnano.6b07237] [PMID: 28024361]

[118] Tadyszak K, Wychowaniec J, Litowczenko J. Biomedical applications of graphene-based structures. Nanomaterials 2018; 8(11): 944.

[http://dx.doi.org/10.3390/nano8110944] [PMID: 30453490]

[119] Bellet P, Gasparotto M, Pressi S, *et al.* Graphene-based scaffolds for regenerative medicine. Nanomaterials 2021; 11(2): 404.
[http://dx.doi.org/10.3390/nano11020404] [PMID: 33562559]

[120] Liu Y, Sun D, Fan Q, *et al.* The enhanced permeability and retention effect based nanomedicine at the site of injury. Nano Res 2020; 13(2): 564-9.
[http://dx.doi.org/10.1007/s12274-020-2655-6] [PMID: 33154805]

[121] Li L, Yang Z, Chen X. Recent advances in stimuli-responsive platforms for cancer immunotherapy. Acc Chem Res 2020; 53(10): 2044-54.
[http://dx.doi.org/10.1021/acs.accounts.0c00334] [PMID: 32877161]

[122] Alsehli M. Polymeric nanocarriers as stimuli-responsive systems for targeted tumor (cancer) therapy: Recent advances in drug delivery. Saudi Pharm J 2020; 28(3): 255-65.
[http://dx.doi.org/10.1016/j.jsps.2020.01.004] [PMID: 32194326]

[123] Mendes RG, Bachmatiuk A, Büchner B, Cuniberti G, Rümmeli MH. Carbon nanostructures as multi-functional drug delivery platforms. J Mater Chem B Mater Biol Med 2013; 1(4): 401-28.
[http://dx.doi.org/10.1039/C2TB00085G] [PMID: 32260810]

[124] Liu Y, Jiao F, Qiu Y, *et al.* The effect of Gd@C82(OH)22 nanoparticles on the release of Th1/Th2 cytokines and induction of TNF-α mediated cellular immunity. Biomaterials 2009; 30(23-24): 3934-45.
[http://dx.doi.org/10.1016/j.biomaterials.2009.04.001] [PMID: 19403166]

[125] Mitchell MJ, Billingsley MM, Haley RM, Wechsler ME, Peppas NA, Langer R. Engineering precision nanoparticles for drug delivery. Nat Rev Drug Discov 2021; 20(2): 101-24.
[http://dx.doi.org/10.1038/s41573-020-0090-8] [PMID: 33277608]

[126] Poon W, Kingston BR, Ouyang B, Ngo W, Chan WCW. A framework for designing delivery systems. Nat Nanotechnol 2020; 15(10): 819-29.
[http://dx.doi.org/10.1038/s41565-020-0759-5] [PMID: 32895522]

[127] Aram E, Mehdipour-Ataei S. Carbon-based nanostructured composites for tissue engineering and drug delivery. Int J Polym Mater 2021; 70(16): 1167-88.
[http://dx.doi.org/10.1080/00914037.2020.1785456]

[128] Ezzati Nazhad Dolatabadi J, Omidi Y, Losic D. Carbon nanotubes as an advanced drug and gene delivery nanosystem. Curr Nanosci 2011; 7(3): 297-314.
[http://dx.doi.org/10.2174/157341311795542444]

[129] Pattnaik S, Swain K, Lin Z. Graphene and graphene-based nanocomposites: Biomedical applications and biosafety. J Mater Chem B Mater Biol Med 2016; 4(48): 7813-31.
[http://dx.doi.org/10.1039/C6TB02086K] [PMID: 32263772]

[130] Mohajeri M, Behnam B, Sahebkar A. Biomedical applications of carbon nanomaterials: Drug and gene delivery potentials. J Cell Physiol 2019; 234(1): 298-319.
[http://dx.doi.org/10.1002/jcp.26899] [PMID: 30078182]

[131] Zhang L, Xia J, Zhao Q, Liu L, Zhang Z. Functional graphene oxide as a nanocarrier for controlled loading and targeted delivery of mixed anticancer drugs. Small 2010; 6(4): 537-44.
[http://dx.doi.org/10.1002/smll.200901680] [PMID: 20033930]

[132] Liu J, Cui L, Losic D. Graphene and graphene oxide as new nanocarriers for drug delivery applications. Acta Biomater 2013; 9(12): 9243-57.
[http://dx.doi.org/10.1016/j.actbio.2013.08.016] [PMID: 23958782]

[133] Bolskar R. Fullerenes for drug delivery. 2012.

[134] Xu ZP, Zeng QH, Lu GQ, Yu AB. Inorganic nanoparticles as carriers for efficient cellular delivery. Chem Eng Sci 2006; 61(3): 1027-40.

[http://dx.doi.org/10.1016/j.ces.2005.06.019]

[135] Yaroslavtsev AB. Nanomaterials for electrical energy storage. In Comprehensive nanoscience and nanotechnology. 2019; pp. 165-206.
[http://dx.doi.org/10.1016/B978-0-12-803581-8.10426-6]

[136] Kumar M, Raza K. C60-fullerenes as drug delivery carriers for anticancer agents: Promises and hurdles. Pharm Nanotechnol 2017; 5(3): 169-79.
[PMID: 29361902]

[137] Zakharian TY, Seryshev A, Sitharaman B, Gilbert BE, Knight V, Wilson LJ. A fullerene-paclitaxel chemotherapeutic: synthesis, characterization, and study of biological activity in tissue culture. J Am Chem Soc 2005; 127(36): 12508-9.
[http://dx.doi.org/10.1021/ja0546525] [PMID: 16144396]

[138] Pantarotto D, Briand JP, Prato M, Bianco A. Translocation of bioactive peptides across cell membranes by carbon nanotubesElectronic supplementary information (ESI) available: details of the synthesis and characterization, cell culture, TEM, epifluorescence and confocal microscopy images of CNTs 1, 2 and fluorescein. See http://www.rsc.org/suppdata/cc/b3/b311254c/. Chem Commun 2004; (1): 16-7.
[http://dx.doi.org/10.1039/b311254c] [PMID: 14737310]

[139] Lu H, Wang J, Wang T, Zhong J, Bao Y, Hao H. Recent progress on nanostructures for drug delivery applications. J Nanomater 2016; 2016: 1-12.
[http://dx.doi.org/10.1155/2016/5762431]

[140] Yang R, Yang X, Zhang Z, *et al.* Single-walled carbon nanotubes-mediated *in vivo* and *in vitro* delivery of siRNA into antigen-presenting cells. Gene Ther 2006; 13(24): 1714-23.
[http://dx.doi.org/10.1038/sj.gt.3302808] [PMID: 16838032]

[141] Zhang W, Zhang Z, Zhang Y. The application of carbon nanotubes in target drug delivery systems for cancer therapies. Nanoscale Res Lett 2011; 6(1): 555.
[http://dx.doi.org/10.1186/1556-276X-6-555] [PMID: 21995320]

[142] Fortunati E, Bout A, Antonia Zanta M, Valerio D, Scarpa M. *In vitro* and *in vivo* gene transfer to pulmonary cells mediated by cationic liposomes. Biochim Biophys Acta Gene Struct Expr 1996; 1306(1): 55-62.
[http://dx.doi.org/10.1016/0167-4781(95)00217-0] [PMID: 8611625]

[143] Rochat T, Morris MA. Gene therapy for cystic fibrosis by means of aerosol. J Aerosol Med 2002; 15(2): 229-35.
[http://dx.doi.org/10.1089/089426802320282356] [PMID: 12184873]

[144] Ledley FD. Non-viral gene therapy. Curr Opin Biotechnol 1994; 5(6): 626-36.
[http://dx.doi.org/10.1016/0958-1669(94)90085-X] [PMID: 7765746]

[145] Pantarotto D, Singh R, McCarthy D, *et al.* Functionalized carbon nanotubes for plasmid DNA gene delivery. Angew Chem Int Ed 2004; 43(39): 5242-6.
[http://dx.doi.org/10.1002/anie.200460437] [PMID: 15455428]

[146] Kam NWS, O'Connell M, Wisdom JA, Dai H. Carbon nanotubes as multifunctional biological transporters and near-infrared agents for selective cancer cell destruction. Proc Natl Acad Sci USA 2005; 102(33): 11600-5.
[http://dx.doi.org/10.1073/pnas.0502680102] [PMID: 16087878]

[147] Elhissi AMA, Ahmed W, Hassan IU, Dhanak VR, D'Emanuele A. Carbon nanotubes in cancer therapy and drug delivery. J Drug Deliv 2012; 2012: 1-10.
[http://dx.doi.org/10.1155/2012/837327] [PMID: 22028974]

[148] Karimi M. Carbon Nanotubes in Drug and Gene Delivery. Morgan & Claypool Publishers 2017.
[http://dx.doi.org/10.1088/978-1-6817-4261-8]

[149] Demirer GS, Zhang H, Goh NS, Pinals RL, Chang R, Landry MP. Carbon nanocarriers deliver siRNA

to intact plant cells for efficient gene knockdown. Sci Adv 2020; 6(26): eaaz0495.
[http://dx.doi.org/10.1126/sciadv.aaz0495] [PMID: 32637592]

[150] Demirer GS, Zhang H, Goh NS, González-Grandío E, Landry MP. Carbon nanotube–mediated DNA delivery without transgene integration in intact plants. Nat Protoc 2019; 14(10): 2954-71.
[http://dx.doi.org/10.1038/s41596-019-0208-9] [PMID: 31534231]

[151] D, N.V., N. Shrestha, and J. Sharma, Transdermal drug delivery system: An overview. International Journal of Research in Pharmaceutical Sciences, 2012. 3.

[152] Bhunia T, Giri A, Nasim T, Chattopadhyay D, Bandyopadhyay A. Uniquely different PVA-xanthan gum irradiated membranes as transdermal diltiazem delivery device. Carbohydr Polym 2013; 95(1): 252-61.
[http://dx.doi.org/10.1016/j.carbpol.2013.02.043] [PMID: 23618267]

[153] Bianco A, Kostarelos K, Prato M. Applications of carbon nanotubes in drug delivery. Curr Opin Chem Biol 2005; 9(6): 674-9.
[http://dx.doi.org/10.1016/j.cbpa.2005.10.005] [PMID: 16233988]

[154] Yuan F, Li S, Fan Z, Meng X, Fan L, Yang S. Shining carbon dots: Synthesis and biomedical and optoelectronic applications. Nano Today 2016; 11(5): 565-86.
[http://dx.doi.org/10.1016/j.nantod.2016.08.006]

[155] Nair A, Haponiuk JT, Thomas S, Gopi S. Natural carbon-based quantum dots and their applications in drug delivery: A review. Biomed Pharmacother 2020; 132: 110834.
[http://dx.doi.org/10.1016/j.biopha.2020.110834] [PMID: 33035830]

[156] Su W, Guo R, Yuan F, *et al.* Red-emissive carbon quantum dots for nuclear drug delivery in cancer stem cells. J Phys Chem Lett 2020; 11(4): 1357-63.
[http://dx.doi.org/10.1021/acs.jpclett.9b03891] [PMID: 32017568]

[157] Boncel S, Herman AP, Walczak KZ. Magnetic carbon nanostructures in medicine. J Mater Chem 2012; 22(1): 31-7.
[http://dx.doi.org/10.1039/C1JM13734D]

[158] Langer R. Tissue engineering. Science 1993; 260: 920-6.

[159] Ku SH, Lee M, Park CB. Carbon-based nanomaterials for tissue engineering. Adv Healthc Mater 2013; 2(2): 244-60.
[http://dx.doi.org/10.1002/adhm.201200307] [PMID: 23184559]

[160] Loos MR, Manas-Zloczower I. Reinforcement efficiency of carbon nanotubes—myth and reality. In: Loos M, Ed. Carbon Nanotube Reinforced Composites. Oxford: William Andrew Publishing 2015; pp. 233-46.
[http://dx.doi.org/10.1016/B978-1-4557-3195-4.00009-6]

[161] Islam M. Lantada AD, Mager D, Korvink JG. Carbon-based materials for articular tissue engineering: from innovative scaffolding materials toward engineered living carbon. Advanced Healthcare Materials 2022; 11(1): 2101834.

[162] Lin Y, Taylor S, Li H, *et al.* Advances toward bioapplications of carbon nanotubes. J Mater Chem 2004; 14(4): 527-41.
[http://dx.doi.org/10.1039/b314481j]

[163] Kuila T, Bose S, Mishra AK, Khanra P, Kim NH, Lee JH. Chemical functionalization of graphene and its applications. Prog Mater Sci 2012; 57(7): 1061-105.
[http://dx.doi.org/10.1016/j.pmatsci.2012.03.002]

[164] Wang Y, Lee WC, Manga KK, *et al.* Fluorinated graphene for promoting neuro-induction of stem cells. Adv Mater 2012; 24(31): 4285-90.
[http://dx.doi.org/10.1002/adma.201200846] [PMID: 22689093]

[165] Cha C, Shin SR, Annabi N, Dokmeci MR, Khademhosseini A. Carbon-based nanomaterials:

Multifunctional materials for biomedical engineering. ACS Nano 2013; 7(4): 2891-7.
[http://dx.doi.org/10.1021/nn401196a] [PMID: 23560817]

[166]   Harrison BS, Atala A. Carbon nanotube applications for tissue engineering. Biomaterials 2007; 28(2): 344-53.
[http://dx.doi.org/10.1016/j.biomaterials.2006.07.044] [PMID: 16934866]

[167]   Sahithi K, Swetha M, Ramasamy K, Srinivasan N, Selvamurugan N. Polymeric composites containing carbon nanotubes for bone tissue engineering. Int J Biol Macromol 2010; 46(3): 281-3.
[http://dx.doi.org/10.1016/j.ijbiomac.2010.01.006] [PMID: 20093139]

[168]   Shin SR, Bae H, Cha JM, *et al.* Carbon nanotube reinforced hybrid microgels as scaffold materials for cell encapsulation. ACS Nano 2012; 6(1): 362-72.
[http://dx.doi.org/10.1021/nn203711s] [PMID: 22117858]

[169]   Armentano I, Álvarez-Pérez MA, Carmona-Rodríguez B, Gutiérrez-Ospina I, Kenny JM, Arzate H. Analysis of the biomineralization process on SWNT-COOH and F-SWNT films. Mater Sci Eng C 2008; 28(8): 1522-9.
[http://dx.doi.org/10.1016/j.msec.2008.04.012]

[170]   Baik KY. Park SY, Heo K, Lee KB, Hong S. Carbon nanotube monolayer cues for osteogenesis of mesenchymal stem cells. small 2011; 7(6): 741-5.

[171]   Lau C, Cooney MJ, Atanassov P. Conductive macroporous composite chitosan-carbon nanotube scaffolds. Langmuir 2008; 24(13): 7004-10.
[http://dx.doi.org/10.1021/la8005597] [PMID: 18517231]

[172]   Kam NWS, Jan E, Kotov NA. Electrical stimulation of neural stem cells mediated by humanized carbon nanotube composite made with extracellular matrix protein. Nano Lett 2009; 9(1): 273-8.
[http://dx.doi.org/10.1021/nl802859a] [PMID: 19105649]

[173]   Worsley MA, Kucheyev SO, Kuntz JD, Hamza AV, Satcher JH Jr, Baumann TF. Stiff and electrically conductive composites of carbon nanotube aerogels and polymers. J Mater Chem 2009; 19(21): 3370-2.
[http://dx.doi.org/10.1039/b905735h]

[174]   Shin SR, Jung SM, Zalabany M, *et al.* Carbon-nanotube-embedded hydrogel sheets for engineering cardiac constructs and bioactuators. ACS Nano 2013; 7(3): 2369-80.
[http://dx.doi.org/10.1021/nn305559j] [PMID: 23363247]

[175]   Eivazzadeh-Keihan R, Maleki A, de la Guardia M, *et al.* Carbon based nanomaterials for tissue engineering of bone: Building new bone on small black scaffolds: A review. J Adv Res 2019; 18: 185-201.
[http://dx.doi.org/10.1016/j.jare.2019.03.011] [PMID: 31032119]

[176]   Sant S, Hancock MJ, Donnelly JP, Iyer D, Khademhosseini A. Biomimetic gradient hydrogels for tissue engineering. Can J Chem Eng 2010; 88(6): 899-911.
[http://dx.doi.org/10.1002/cjce.20411] [PMID: 21874065]

[177]   Zhang L, Wang Z, Xu C, *et al.* High strength graphene oxide/polyvinyl alcohol composite hydrogels. J Mater Chem 2011; 21(28): 10399-406.
[http://dx.doi.org/10.1039/c0jm04043f]

[178]   Xu Y, Sheng K, Li C, Shi G. Self-assembled graphene hydrogel *via* a one-step hydrothermal process. ACS Nano 2010; 4(7): 4324-30.
[http://dx.doi.org/10.1021/nn101187z] [PMID: 20590149]

[179]   Geng J, Jung HT. Porphyrin functionalized graphene sheets in aqueous suspensions: From the preparation of graphene sheets to highly conductive graphene films. J Phys Chem C 2010; 114(18): 8227-34.
[http://dx.doi.org/10.1021/jp1008779]

[180]   Ku SH, Lee M, Park CB. Carbon-based nanomaterials for tissue engineering. Adv Healthc Mater

2013; 2(2): 244-60.
[http://dx.doi.org/10.1002/adhm.201200307] [PMID: 23184559]

[181]   Li N, Zhang Q, Gao S, *et al.* Three-dimensional graphene foam as a biocompatible and conductive scaffold for neural stem cells. Sci Rep 2013; 3(1): 1604.
[http://dx.doi.org/10.1038/srep01604] [PMID: 23549373]

[182]   Ku SH, Park CB. Myoblast differentiation on graphene oxide. Biomaterials 2013; 34(8): 2017-23.
[http://dx.doi.org/10.1016/j.biomaterials.2012.11.052] [PMID: 23261212]

[183]   Kim S, Ku SH, Lim SY, Kim JH, Park CB. Graphene-biomineral hybrid materials. Adv Mater 2011; 23(17): 2009-14.
[http://dx.doi.org/10.1002/adma.201100010] [PMID: 21413084]

[184]   Bae H, Chu H, Edalat F, *et al.* Development of functional biomaterials with micro- and nanoscale technologies for tissue engineering and drug delivery applications. J Tissue Eng Regen Med 2014; 8(1): 1-14.
[http://dx.doi.org/10.1002/term.1494] [PMID: 22711442]

[185]   Kodali VK, Scrimgeour J, Kim S, *et al.* Nonperturbative chemical modification of graphene for protein micropatterning. Langmuir 2011; 27(3): 863-5.
[http://dx.doi.org/10.1021/la1033178] [PMID: 21182241]

[186]   Edwards SL, Church JS, Werkmeister JA, Ramshaw JAM. Tubular micro-scale multiwalled carbon nanotube-based scaffolds for tissue engineering. Biomaterials 2009; 30(9): 1725-31.
[http://dx.doi.org/10.1016/j.biomaterials.2008.12.031] [PMID: 19124155]

[187]   Cuenca AG, Jiang H, Hochwald SN, Delano M, Cance WG, Grobmyer SR. Emerging implications of nanotechnology on cancer diagnostics and therapeutics. Cancer 2006; 107(3): 459-66.
[http://dx.doi.org/10.1002/cncr.22035] [PMID: 16795065]

[188]   Ferrari M. Cancer nanotechnology: Opportunities and challenges. Nat Rev Cancer 2005; 5(3): 161-71.
[http://dx.doi.org/10.1038/nrc1566] [PMID: 15738981]

[189]   Grobmyer SR. Iwakuma N, Sharma P, Moudgil BM. What is cancer nanotechnology? In: Grobmyer SR, Moudgil BM, Eds. Cancer Nanotechnology: Methods and Protocols. Totowa, NJ: Humana Press 2010; pp. 1-9.
[http://dx.doi.org/10.1007/978-1-60761-609-2_1]

[190]   Grodzinski P, Silver M, Molnar LK. Nanotechnology for cancer diagnostics: Promises and challenges. Expert Rev Mol Diagn 2006; 6(3): 307-18.
[http://dx.doi.org/10.1586/14737159.6.3.307] [PMID: 16706735]

[191]   Hassan HAFM, Smyth L, Wang JTW, *et al.* Dual stimulation of antigen presenting cells using carbon nanotube-based vaccine delivery system for cancer immunotherapy. Biomaterials 2016; 104: 310-22.
[http://dx.doi.org/10.1016/j.biomaterials.2016.07.005] [PMID: 27475727]

[192]   Sharma A, Hong S, Singh R, Jang J. Single-walled carbon nanotube based transparent immunosensor for detection of a prostate cancer biomarker osteopontin. Anal Chim Acta 2015; 869: 68-73.
[http://dx.doi.org/10.1016/j.aca.2015.02.010] [PMID: 25818141]

[193]   Battigelli A, Ménard-Moyon C, Bianco A. Carbon nanomaterials as new tools for immunotherapeutic applications. J Mater Chem B Mater Biol Med 2014; 2(37): 6144-56.
[http://dx.doi.org/10.1039/C4TB00563E] [PMID: 32262133]

[194]   Augustine S, Singh J, Srivastava M, Sharma M, Das A, Malhotra BD. Recent advances in carbon based nanosystems for cancer theranostics. Biomater Sci 2017; 5(5): 901-52.
[http://dx.doi.org/10.1039/C7BM00008A] [PMID: 28401206]

[195]   Feng H, Qian Z. Functional carbon quantum dots: A versatile platform for chemosensing and biosensing. Chem Rec 2017; 18.
[PMID: 29171708]

[196] Song J, Qu J, Swihart MT, Prasad PN. Near-IR responsive nanostructures for nanobiophotonics: Emerging impacts on nanomedicine. Nanomedicine 2016; 12(3): 771-88.
[http://dx.doi.org/10.1016/j.nano.2015.11.009] [PMID: 26656629]

[197] Chen D, Dougherty CA, Zhu K, Hong H. Theranostic applications of carbon nanomaterials in cancer: Focus on imaging and cargo delivery. J Control Release 2015; 210: 230-45.
[http://dx.doi.org/10.1016/j.jconrel.2015.04.021] [PMID: 25910580]

[198] Zhang R, Cheng K, Antaris AL, *et al.* Hybrid anisotropic nanostructures for dual-modal cancer imaging and image-guided chemo-thermo therapies. Biomaterials 2016; 103: 265-77.
[http://dx.doi.org/10.1016/j.biomaterials.2016.06.063] [PMID: 27394161]

[199] Janib SM, Moses AS, MacKay JA. Imaging and drug delivery using theranostic nanoparticles. Adv Drug Deliv Rev 2010; 62(11): 1052-63.
[http://dx.doi.org/10.1016/j.addr.2010.08.004] [PMID: 20709124]

[200] Wang L, Sun Q, Wang X, *et al.* Using hollow carbon nanospheres as a light-induced free radical generator to overcome chemotherapy resistance. J Am Chem Soc 2015; 137(5): 1947-55.
[http://dx.doi.org/10.1021/ja511560b] [PMID: 25597855]

[201] Shi J, Wang L, Gao J, *et al.* A fullerene-based multi-functional nanoplatform for cancer theranostic applications. Biomaterials 2014; 35(22): 5771-84.
[http://dx.doi.org/10.1016/j.biomaterials.2014.03.071] [PMID: 24746227]

[202] Ji SR, Liu C, Zhang B, *et al.* Carbon nanotubes in cancer diagnosis and therapy. Biochim Biophys Acta 2010; 1806(1): 29-35.
[PMID: 20193746]

[203] Curcio M. Cirillo G, Saletta F, *et al.* Carbon nanohorns as effective nanotherapeutics in cancer therapy. C 2021; 7(1): 3.

[204] Kim JW, Galanzha EI, Shashkov EV, Moon HM, Zharov VP. Golden carbon nanotubes as multimodal photoacoustic and photothermal high-contrast molecular agents. Nat Nanotechnol 2009; 4(10): 688-94.
[http://dx.doi.org/10.1038/nnano.2009.231] [PMID: 19809462]

[205] De La Zerda A, Zavaleta C, Keren S, *et al.* Carbon nanotubes as photoacoustic molecular imaging agents in living mice. Nat Nanotechnol 2008; 3(9): 557-62.
[http://dx.doi.org/10.1038/nnano.2008.231] [PMID: 18772918]

[206] Welsher K, Liu Z, Sherlock SP, *et al.* A route to brightly fluorescent carbon nanotubes for near-infrared imaging in mice. Nat Nanotechnol 2009; 4(11): 773-80.
[http://dx.doi.org/10.1038/nnano.2009.294] [PMID: 19893526]

[207] Yi H, Ghosh D, Ham MH, *et al.* M13 phage-functionalized single-walled carbon nanotubes as nanoprobes for second near-infrared window fluorescence imaging of targeted tumors. Nano Lett 2012; 12(3): 1176-83.
[http://dx.doi.org/10.1021/nl2031663] [PMID: 22268625]

[208] Cornelissen B, Able S, Kersemans V, *et al.* Nanographene oxide-based radioimmunoconstructs for *in vivo* targeting and SPECT imaging of HER2-positive tumors. Biomaterials 2013; 34(4): 1146-54.
[http://dx.doi.org/10.1016/j.biomaterials.2012.10.054] [PMID: 23171545]

[209] Bradac C, Gaebel T, Naidoo N, *et al.* Observation and control of blinking nitrogen-vacancy centres in discrete nanodiamonds. Nat Nanotechnol 2010; 5(5): 345-9.
[http://dx.doi.org/10.1038/nnano.2010.56] [PMID: 20383128]

[210] Fu CC, Lee HY, Chen K, *et al.* Characterization and application of single fluorescent nanodiamonds as cellular biomarkers. Proc Natl Acad Sci 2007; 104(3): 727-32.
[http://dx.doi.org/10.1073/pnas.0605409104] [PMID: 17213326]

[211] Chang YR, Lee HY, Chen K, *et al.* Mass production and dynamic imaging of fluorescent nanodiamonds. Nat Nanotechnol 2008; 3(5): 284-8.

[http://dx.doi.org/10.1038/nnano.2008.99] [PMID: 18654525]

[212] Sun YP, Zhou B, Lin Y, *et al.* Quantum-sized carbon dots for bright and colorful photoluminescence. J Am Chem Soc 2006; 128(24): 7756-7.
[http://dx.doi.org/10.1021/ja062677d] [PMID: 16771487]

[213] Xu Y, Jia XH, Yin XB, He XW, Zhang YK. Carbon quantum dot stabilized gadolinium nanoprobe prepared *via* a one-pot hydrothermal approach for magnetic resonance and fluorescence dual-modality bioimaging. Anal Chem 2014; 86(24): 12122-9.
[http://dx.doi.org/10.1021/ac503002c] [PMID: 25383762]

[214] Chen X-Q, Yu J-G, Jiao F-P, *et al.* Irradiation-mediated carbon nanotubes' use in cancer therapy. J Cancer Res Ther 2012; 8(3): 348-54.
[http://dx.doi.org/10.4103/0973-1482.103511] [PMID: 23174713]

[215] Thomas LA, Dekker L, Kallumadil M, *et al.* Carboxylic acid-stabilised iron oxide nanoparticles for use in magnetic hyperthermia. J Mater Chem 2009; 19(36): 6529-35.
[http://dx.doi.org/10.1039/b908187a]

[216] Guardia P, Di Corato R, Lartigue L, *et al.* Water-soluble iron oxide nanocubes with high values of specific absorption rate for cancer cell hyperthermia treatment. ACS Nano 2012; 6(4): 3080-91.
[http://dx.doi.org/10.1021/nn2048137] [PMID: 22494015]

[217] Neshasteh-Riz A, Rahdani R, Mostaar A. Evaluation of the combined effects of hyperthermia, cobalt-60 gamma rays and IUdR on cultured glioblastoma spheroid cells and dosimetry using TLD-100. Cell J 2014; 16(3): 335-42.
[PMID: 24611138]

[218] Huang YY, Sharma SK, Yin R, Agrawal T, Chiang LY, Hamblin MR. Functionalized fullerenes in photodynamic therapy. J Biomed Nanotechnol 2014; 10(9): 1918-36.
[http://dx.doi.org/10.1166/jbn.2014.1963] [PMID: 25544837]

[219] Santos T, Fang X, Chen MT, *et al.* Sequential administration of carbon nanotubes and near-infrared radiation for the treatment of gliomas. Front Oncol 2014; 4: 180.
[http://dx.doi.org/10.3389/fonc.2014.00180] [PMID: 25077069]

[220] Jaymand M, Davatgaran Taghipour Y, Rezaei A, *et al.* Radiolabeled carbon-based nanostructures: New radiopharmaceuticals for cancer therapy? Coord Chem Rev 2021; 440: 213974.
[http://dx.doi.org/10.1016/j.ccr.2021.213974]

[221] Gao J, Xu B. Applications of nanomaterials inside cells. Nano Today 2009; 4(1): 37-51.
[http://dx.doi.org/10.1016/j.nantod.2008.10.009]

[222] Dong H, Ding L, Yan F, Ji H, Ju H. The use of polyethylenimine-grafted graphene nanoribbon for cellular delivery of locked nucleic acid modified molecular beacon for recognition of microRNA. Biomaterials 2011; 32(15): 3875-82.
[http://dx.doi.org/10.1016/j.biomaterials.2011.02.001] [PMID: 21354613]

[223] Hu L, Sun Y, Li S, *et al.* Multifunctional carbon dots with high quantum yield for imaging and gene delivery. Carbon 2014; 67: 508-13.
[http://dx.doi.org/10.1016/j.carbon.2013.10.023]

[224] Zhou J, Deng W, Wang Y, *et al.* Cationic carbon quantum dots derived from alginate for gene delivery: One-step synthesis and cellular uptake. Acta Biomater 2016; 42: 209-19.
[http://dx.doi.org/10.1016/j.actbio.2016.06.021] [PMID: 27321673]

[225] Wang L, Shi J, Zhang H, *et al.* Synergistic anticancer effect of RNAi and photothermal therapy mediated by functionalized single-walled carbon nanotubes. Biomaterials 2013; 34(1): 262-74.
[http://dx.doi.org/10.1016/j.biomaterials.2012.09.037] [PMID: 23046752]

[226] Mullick Chowdhury S, Zafar S, Tellez V, Sitharaman B. Graphene nanoribbon-based platform for highly efficacious nuclear gene delivery. ACS Biomater Sci Eng 2016; 2(5): 798-808.
[http://dx.doi.org/10.1021/acsbiomaterials.5b00562] [PMID: 33440577]

[227]   Martín R, Álvaro M, Herance JR, García H. Fenton-treated functionalized diamond nanoparticles as gene delivery system. ACS Nano 2010; 4(1): 65-74.
[http://dx.doi.org/10.1021/nn901616c] [PMID: 20047335]

[228]   Sitharaman B, Zakharian TY, Saraf A, *et al.* Water-soluble fullerene (C60) derivatives as nonviral gene-delivery vectors. Mol Pharm 2008; 5(4): 567-78.
[http://dx.doi.org/10.1021/mp700106w] [PMID: 18505267]

[229]   Bhirde AA, Patel V, Gavard J, *et al.* Targeted killing of cancer cells *in vivo* and *in vitro* with EGF-directed carbon nanotube-based drug delivery. ACS Nano 2009; 3(2): 307-16.
[http://dx.doi.org/10.1021/nn800551s] [PMID: 19236065]

[230]   Wu W, Li R, Bian X, *et al.* Covalently combining carbon nanotubes with anticancer agent: Preparation and antitumor activity. ACS Nano 2009; 3(9): 2740-50.
[http://dx.doi.org/10.1021/nn9005686] [PMID: 19702292]

[231]   Karchemski F, Zucker D, Barenholz Y, Regev O. Carbon nanotubes-liposomes conjugate as a platform for drug delivery into cells. J Control Release 2012; 160(2): 339-45.
[http://dx.doi.org/10.1016/j.jconrel.2011.12.037] [PMID: 22245689]

[232]   Ren J, Shen S, Wang D, *et al.* The targeted delivery of anticancer drugs to brain glioma by PEGylated oxidized multi-walled carbon nanotubes modified with angiopep-2. Biomaterials 2012; 33(11): 3324-33.
[http://dx.doi.org/10.1016/j.biomaterials.2012.01.025] [PMID: 22281423]

[233]   Ajima K, Murakami T, Mizoguchi Y, *et al.* Enhancement of *in vivo* anticancer effects of cisplatin by incorporation inside single-wall carbon nanohorns. ACS Nano 2008; 2(10): 2057-64.
[http://dx.doi.org/10.1021/nn800395t] [PMID: 19206452]

[234]   Zhang W, Guo Z, Huang D, Liu Z, Guo X, Zhong H. Synergistic effect of chemo-photothermal therapy using PEGylated graphene oxide. Biomaterials 2011; 32(33): 8555-61.
[http://dx.doi.org/10.1016/j.biomaterials.2011.07.071] [PMID: 21839507]

[235]   Kim H, Lee D, Kim J, Kim T, Kim WJ. Photothermally triggered cytosolic drug delivery *via* endosome disruption using a functionalized reduced graphene oxide. ACS Nano 2013; 7(8): 6735-46.
[http://dx.doi.org/10.1021/nn403096s] [PMID: 23829596]

[236]   Kolosnjaj-Tabi J, Hartman KB, Boudjemaa S, *et al.* *In vivo* behavior of large doses of ultrashort and full-length single-walled carbon nanotubes after oral and intraperitoneal administration to Swiss mice. ACS Nano 2010; 4(3): 1481-92.
[http://dx.doi.org/10.1021/nn901573w] [PMID: 20175510]

[237]   Li X, Liu H, Niu X, *et al.* The use of carbon nanotubes to induce osteogenic differentiation of human adipose-derived MSCs *in vitro* and ectopic bone formation *in vivo*. Biomaterials 2012; 33(19): 4818-27.
[http://dx.doi.org/10.1016/j.biomaterials.2012.03.045] [PMID: 22483242]

[238]   Shin SR, Jung SM, Zalabany M, *et al.* Carbon-nanotube-embedded hydrogel sheets for engineering cardiac constructs and bioactuators. ACS Nano 2013; 7(3): 2369-80.
[http://dx.doi.org/10.1021/nn305559j] [PMID: 23363247]

[239]   Lovat V, Pantarotto D, Lagostena L, *et al.* Carbon nanotube substrates boost neuronal electrical signaling. Nano Lett 2005; 5(6): 1107-10.
[http://dx.doi.org/10.1021/nl050637m] [PMID: 15943451]

[240]   Kalbacova M, Broz A, Kong J, Kalbac M. Graphene substrates promote adherence of human osteoblasts and mesenchymal stromal cells. Carbon 2010; 48(15): 4323-9.
[http://dx.doi.org/10.1016/j.carbon.2010.07.045]

[241]   Lee J.H, Shin YC, Hwang D.G, Kim J.S, Jin O.S. Graphene oxide-decorated PLGA/collagen hybrid fiber sheets for application to tissue engineering scaffolds. Biomater Res 2014; 18(1): 18-24.
[http://dx.doi.org/10.1186/s40824-023-00357-y] [PMID: 36855173]

[242]   Park SY, Choi DS, Jin HJ, *et al.* Polarization-controlled differentiation of human neural stem cells using synergistic cues from the patterns of carbon nanotube monolayer coating. ACS Nano 2011; 5(6): 4704-11.
[http://dx.doi.org/10.1021/nn2006128] [PMID: 21568294]

[243]   Song J, Jia X, Minami K, *et al.* Large-area aligned fullerene nanocrystal scaffolds as culture substrates for enhancing mesenchymal stem cell self-renewal and multipotency. ACS Appl Nano Mater 2020; 3(7): 6497-506.
[http://dx.doi.org/10.1021/acsanm.0c00973]

[244]   Abarrategi A, Gutiérrez MC, Moreno-Vicente C, *et al.* Multiwall carbon nanotube scaffolds for tissue engineering purposes. Biomaterials 2008; 29(1): 94-102.
[http://dx.doi.org/10.1016/j.biomaterials.2007.09.021] [PMID: 17928048]

[245]   Shi X, Sitharaman B, Pham QP, *et al.* Fabrication of porous ultra-short single-walled carbon nanotube nanocomposite scaffolds for bone tissue engineering. Biomaterials 2007; 28(28): 4078-90.
[http://dx.doi.org/10.1016/j.biomaterials.2007.05.033] [PMID: 17576009]

[246]   Shao S, Zhou S, Li L, *et al.* Osteoblast function on electrically conductive electrospun PLA/MWCNTs nanofibers. Biomaterials 2011; 32(11): 2821-33.
[http://dx.doi.org/10.1016/j.biomaterials.2011.01.051] [PMID: 21292320]

[247]   Liu Z, Cai W, He L, *et al. In vivo* biodistribution and highly efficient tumour targeting of carbon nanotubes in mice. Nat Nanotechnol 2007; 2(1): 47-52.
[http://dx.doi.org/10.1038/nnano.2006.170] [PMID: 18654207]

[248]   Liu KK, Wang CC, Cheng CL, Chao JI. Endocytic carboxylated nanodiamond for the labeling and tracking of cell division and differentiation in cancer and stem cells. Biomaterials 2009; 30(26): 4249-59.
[http://dx.doi.org/10.1016/j.biomaterials.2009.04.056] [PMID: 19500835]

[249]   Heller DA, Jin H, Martinez BM, *et al.* Multimodal optical sensing and analyte specificity using single-walled carbon nanotubes. Nat Nanotechnol 2009; 4(2): 114-20.
[http://dx.doi.org/10.1038/nnano.2008.369] [PMID: 19197314]

[250]   Kalluru P, Vankayala R, Chiang CS, Hwang KC. Nano-graphene oxide-mediated *In vivo* fluorescence imaging and bimodal photodynamic and photothermal destruction of tumors. Biomaterials 2016; 95: 1-10.
[http://dx.doi.org/10.1016/j.biomaterials.2016.04.006] [PMID: 27108401]

[251]   Miao W, Shim G, Kim G, *et al.* Image-guided synergistic photothermal therapy using photoresponsive imaging agent-loaded graphene-based nanosheets. J Control Release 2015; 211: 28-36.
[http://dx.doi.org/10.1016/j.jconrel.2015.05.280] [PMID: 26003041]

[252]   Chen L, Zhong X, Yi X, *et al.* Radionuclide 131I labeled reduced graphene oxide for nuclear imaging guided combined radio- and photothermal therapy of cancer. Biomaterials 2015; 66: 21-8.
[http://dx.doi.org/10.1016/j.biomaterials.2015.06.043] [PMID: 26188609]

[253]   Guan M, Ge J, Wu J, *et al.* Fullerene/photosensitizer nanovesicles as highly efficient and clearable phototheranostics with enhanced tumor accumulation for cancer therapy. Biomaterials 2016; 103: 75-85.
[http://dx.doi.org/10.1016/j.biomaterials.2016.06.023] [PMID: 27376559]

[254]   Shi S, Yang K, Hong H, *et al.* Tumor vasculature targeting and imaging in living mice with reduced graphene oxide. Biomaterials 2013; 34(12): 3002-9.
[http://dx.doi.org/10.1016/j.biomaterials.2013.01.047] [PMID: 23374706]

[255]   Liu Y, Chen C, Qian P, *et al.* Gd-metallofullerenol nanomaterial as non-toxic breast cancer stem cell-specific inhibitor. Nat Commun 2015; 6(1): 5988.
[http://dx.doi.org/10.1038/ncomms6988] [PMID: 25612916]

[256]   Meng H, Xing G, Sun B, *et al.* Potent angiogenesis inhibition by the particulate form of fullerene

derivatives. ACS Nano 2010; 4(5): 2773-83.
[http://dx.doi.org/10.1021/nn100448z] [PMID: 20429577]

[257]   Jiao F, Liu Y, Qu Y, *et al.* Studies on anti-tumor and antimetastatic activities of fullerenol in a mouse breast cancer model. Carbon 2010; 48(8): 2231-43.
[http://dx.doi.org/10.1016/j.carbon.2010.02.032]

[258]   Zhou Y, Deng R, Zhen M, *et al.* Amino acid functionalized gadofullerene nanoparticles with superior antitumor activity *via* destruction of tumor vasculature *in vivo*. Biomaterials 2017; 133: 107-18.
[http://dx.doi.org/10.1016/j.biomaterials.2017.04.025] [PMID: 28433934]

[259]   Yao H, Zhang Y, Sun L, Liu Y. The effect of hyaluronic acid functionalized carbon nanotubes loaded with salinomycin on gastric cancer stem cells. Biomaterials 2014; 35(33): 9208-23.
[http://dx.doi.org/10.1016/j.biomaterials.2014.07.033] [PMID: 25115788]

[260]   Qin Y, Chen J, Bi Y, *et al.* Near-infrared light remote-controlled intracellular anti-cancer drug delivery using thermo/pH sensitive nanovehicle. Acta Biomater 2015; 17: 201-9.
[http://dx.doi.org/10.1016/j.actbio.2015.01.026] [PMID: 25644449]

[261]   Lai PX, Chen CW, Wei SC, *et al.* Ultrastrong trapping of VEGF by graphene oxide: Anti-angiogenesis application. Biomaterials 2016; 109: 12-22.
[http://dx.doi.org/10.1016/j.biomaterials.2016.09.005] [PMID: 27639528]

[262]   Ma W, Cheetham AG, Cui H. Building nanostructures with drugs. Nano Today 2016; 11(1): 13-30.
[http://dx.doi.org/10.1016/j.nantod.2015.11.003] [PMID: 27066106]

[263]   DeWitt MR, Pekkanen AM, Robertson J, Rylander CG, Nichole Rylander M. Influence of hyperthermia on efficacy and uptake of carbon nanohorn-cisplatin conjugates. J Biomech Eng 2014; 136(2): 021003.
[http://dx.doi.org/10.1115/1.4026318] [PMID: 24763615]

[264]   Liu Z, Tabakman S, Welsher K, Dai H. Carbon nanotubes in biology and medicine: *In vitro* and *in vivo* detection, imaging and drug delivery. Nano Res 2009; 2(2): 85-120.
[http://dx.doi.org/10.1007/s12274-009-9009-8] [PMID: 20174481]

[265]   Tay ZW, Chandrasekharan P, Chiu-Lam A, *et al.* Magnetic particle imaging-guided heating *in vivo* using gradient fields for arbitrary localization of magnetic hyperthermia therapy. ACS Nano 2018; 12(4): 3699-713.
[http://dx.doi.org/10.1021/acsnano.8b00893] [PMID: 29570277]

[266]   Chen D, Wang H, Dong L, *et al.* The fluorescent bioprobe with aggregation-induced emission features for monitoring to carbon dioxide generation rate in single living cell and early identification of cancer cells. Biomaterials 2016; 103: 67-74.
[http://dx.doi.org/10.1016/j.biomaterials.2016.06.055] [PMID: 27372422]

[267]   Lee PC, Peng CL, Shieh MJ. Combining the single-walled carbon nanotubes with low voltage electrical stimulation to improve accumulation of nanomedicines in tumor for effective cancer therapy. J Control Release 2016; 225: 140-51.
[http://dx.doi.org/10.1016/j.jconrel.2016.01.038] [PMID: 26812005]

[268]   Miao W, Shim G, Lee S, Lee S, Choe YS, Oh YK. Safety and tumor tissue accumulation of pegylated graphene oxide nanosheets for co-delivery of anticancer drug and photosensitizer. Biomaterials 2013; 34(13): 3402-10.
[http://dx.doi.org/10.1016/j.biomaterials.2013.01.010] [PMID: 23380350]

[269]   Bussy C, Ali-Boucetta H, Kostarelos K. Safety considerations for graphene: Lessons learnt from carbon nanotubes. Acc Chem Res 2013; 46(3): 692-701.
[http://dx.doi.org/10.1021/ar300199e] [PMID: 23163827]

CHAPTER 10

Advances in Polymer/Ceramic Composites for Bone Tissue Engineering Applications

Luciano Benedini[1,3,*] and **Paula Messina**[1,2]

[1] *INQUISUR-CONICET, Universidad Nacional del Sur, Bahía Blanca, Argentina*

[2] *Department of Chemistry, Universidad Nacional del Sur, Bahía Blanca, Argentina*

[3] *Department of Biology, Biochemistry and Pharmacy, Universidad Nacional del Sur, Bahía Blanca, Argentina*

Abstract: Tissue engineering and regenerative medicine have accomplished enormous progress in the last few years. The application of recently designed nano-textured surface characteristics has shown increased enhancement in bone tissue regeneration. The development of materials that fulfill the exact requirements of bone tissue is still under investigation. However, we are approaching this aim. Composite materials are some of those materials under consideration, and they have emerged as a consequence of the logical unraveling of bone composition. Principal components of bone tissue are inorganic and organic matrices and water, in other words, ceramics and polymers. Accordingly, the design of these materials by combining different types of ceramics and polymers has opened a wide range of possibilities for bone regeneration treatments. Not all polymers nor all ceramics can be used for this purpose. Materials must gather particular properties to be applied in bone tissue engineering. Both types have to be safe, which means biocompatible and non-toxic. They, additionally, should have efficient surface behavior, bioactivity, and suitable mechanical properties. Sometimes, composites could behave as *in situ* drug delivery systems. Composites are engineering materials formed by two or more components, each bringing a unique physical property, and generating synergism. For these reasons, in this work, we will discuss features of host tissue, concepts such as bioactivity, osteoconductivity, and osteoinductivity, and the most significant polymers and ceramics used for developing composed materials. Finally, we focus on examples of composite materials based on these components applied for bone tissue regeneration.

Keywords: Alginate, Bioactive glass, Bioactivity, Bioadhesive, Biocompatible, Biodegradability, Bone pathologies, Bone tissue engineering, Calcium phosphate, Collagen, Compact and spongy bone, Gelatin, Hyaluronic acid, Hydroxyapatite, Non-immunogenic, Non-toxic, Polycaprolactone, Polymer/ceramic composites, Tunable properties.

* **Corresponding author Luciano Benedini:** INQUISUR-CONICET, Universidad Nacional del Sur, Bahía Blanca, Argentina; E-mail: lbenedini@uns.edu.ar

Saeid Kargozar and Francesco Baino (Eds.)

INTRODUCTION

Scientists have worked, for years, using different materials to improve the life quality of people with bone pathologies. These pathologies can be produced by illnesses, accidents, or attributes of aging of human beings. It would be desirable that the properties of materials can be tunable for the different requirements considering the type of pathology, the type of bone, that means long, short, or flat, and the features of the tissue sponge or compact [1].

Autografts have been frequently used strategies for addressing bone defect treatments. It is referred to transplant from one part of the body to another zone of the same patient. Autograft is an optimal material because it provides an osteoconductive surface and contains cells that contribute to the osteointegration process. However, autografts are not osteoinductive in a not orthotopic site because they are reabsorbed. In the orthotopic zone, they also exhibit reabsorption, but their osteogenic properties are enough for bone regeneration. A limited quantity of bone is obtained, with pain or loss of sensitivity in the donor site, and the risk of infection are some disadvantages of autografts. Consequently, allografts (same species), xenografts (different species), and later synthetic bone substitutes have become alternative strategies to overcome these situations [2, 3]. In this work, we will focus on this last strategy.

When we talk about bone pathologies, we immediately think of the bones of the legs or arms, which is correct. However, some particular treatment features are shown in bones containing teeth as short dimension defects and a low quantity of materials applied. Therefore, these materials can be applied for the potential treatment of moderate and severe periodontal disease and require a special section [4]. It is crucial to define the strategic approach for addressing these problems. Many different materials, such as implants based on metals or smart gels, have been proposed for treating bone pathologies [5]. Among those materials, composite materials have been widely spread during the last years [6]. The treatment of several bone-related disorders, diseases, or ailments has been addressed using biodegradable polymer-ceramic composites materials. Therefore, materials used for bone reparation or as filler materials have evolved from inert materials to those that strongly interact with the tissue, and thus, they achieve tissue requirements [7, 8].

Bone is a natural composite material formed by 55-70% (w/w) of inorganic components, 20–30% (w/w) matrix, and 10–20% (w/w) water. The main inorganic mineral constituent of bone is a substituted calcium phosphate with similar composition and structure to hydroxyapatite [9]. The organic component is formed by highly aligned triple helix type I collagen fibrils. The inorganic component provides mechanical properties to the bone, and the organic one offers flexibility [10]. By observing this natural structure, it is possible to propose the development of materials based on organic (or blends) and inorganic components (or combinations) for application in bone tissue pathologies. Consequently, for obtaining the materials and tailoring their properties, tissue engineering has emerged as a new discipline combining biological and engineering principles for creating a new organ or repairing and promoting the regeneration of damaged tissue. Today, regenerative medicine is applied in cardiovascular, nervous, musculoskeletal, and orthopedic therapies [11].

The first section of this work describes the structure of bones, functions, and formation. The second section addresses bone tissue as a nanostructured material. This view depicts principal architectures, compositions, and sizes that must be considered for designing bone tissue implants or devices. The third section shows an overview of the main types of biomaterials, their features, and concepts such as bioactivity, osteoconductivity, and osteoinductivity. The following section deals with groups of materials such as polymers and ceramics used for developing devices for bone tissue applications. Finally, the remarkable composite materials used for bone tissue engineering are described.

Bone Structure and Mechanical Properties, Functions and Formation

Structure and Mechanical Properties

Two types of bone are recognized: spongy, trabecular, or cancellous, and compact, cortical, or dense. This last type is mainly limited to the external shell of the bone or cortex. Cortical bone is composed of osteons. These are a group of cylindrical structures constituted by 4 to 20 concentric lamellas oriented following the axis of the bone named Haversian systems. The unit of this system shows a transversal section of 250 µm and encloses the center of the Haversian canal that connects with the narrow cavity. Along this canal passes the neurovascular system [12]. Osteocytes, the living bone cells, are disposed circumferentially around the lamellas in specific places named lacunae. Each bone type (wet human bones) confers different mechanical properties. Three relevant mechanical parameters must be considered, compressive strength, tensile strength, and Young's modulus. The compressive strength in the compact bone of the femur, tibia, and radius varies from 167 to 115 MPa, and for the spongy bone in vertebrae 8.4. The tensile strength for those large bones is from 120 to 150 MPa and 3.7 for vertebrae. Finally, the Young's modulus of compact bone in large bones is from 17 to 19 GPa [10].

Functions

Bones contain two different types of structures spongy and compact bone. Their distribution and proportions depend on functions and age. The physical forces that act on bone determine its architecture and vascularization. Bones provide support, locomotion, and protect the organs. They provide a suitable place for the hematopoietic tissue, the bone marrow. They have a crucial role in metabolism by the maintenance of mineral homeostasis because they act as reservoirs of phosphate, calcium, and other ions [12]. Bones are constantly remodeled, and their resistance is associated with stress. This remodeling is associated with the surrounding mechanical stimuli that generate an adaptation of the bone structure by growth and resorption processes [13]. The periosteum is an outer shell. It covers the bone and participates in the reparation and growth of bones because it contains vessels and nerves.

Formation

Osteogenesis or bone formation gathers a series of interrelated processes with simultaneous occurrence such as cell migration, cell differentiation, modulation, synthesis and secretion of molecules, extracellular mineralization, and resorption. All these processes are involved in bone remodeling. There are two types of ossification, intramembranous and endochondral. These types of osteogeneses refer to the environment where the bone is produced. The endochondral ossification is responsible for long bones formation, and the bone is formed based on preexisting hyaline cartilage. Here, mesenchymal cells differentiate into chondrocytes, and after the development of cartilage, the mineralization is facilitated. The intramembranous type does not involve cartilage and is responsible for forming flat bones such as cranium bones. This last type of bone formation is carried out by osteoblasts which are the bone-forming cells in mesenchyme areas richly vascularized where they are differentiated. These types of bone formation yield cancellous, spongy, or trabecular, and dense, cortical, or compact bones. The second type of bone is produced from the first [12].

Nanostructuration of Tissues: The Bone

At the lower hierarchical level, bone tissue can be defined as a composite material constituted by an inorganic (mineral) and organic portion. Inorganic part is mainly composed of hydroxyapatite (calcium phosphate), and the organic portion is mainly formed by collagen. Consequently, it is considered that bone is a nanostructured array of hydroxyapatite crystals embedded within the intrafibrillar collagen matrix. Due to the complex nature of both phases, the bone tissue cannot be classified as a specific type of composite. The biological hydroxyapatite shows calcium deficiency compared with the synthetic apatite. Biological apatite is carbonate substituted between 3.2 to 5.8%. However, the mechanical properties of the synthetic hydroxyapatite ($Ca_{10}(PO_4)_6(OH)_2$) crystals have been considered suitable to be applied in tissue engineering [1, 14, 15]. The basic building block of bone tissue is composed of mineralized collagen fibrils of 200 nm to a few microns' lengths and 80–100 nm thicknesses. This protein shows small gap zones occupied by plate-shaped hydroxyapatite nanocrystals distributed in the long axis of collagen fibrils [16]. These well-arrayed nanocrystals of 25–50 nm in length reinforce the polymer matrix [10]. Finally, we can define bone tissue as a composite material formed by a polymer-ceramic blend showing different structures simultaneously.

Bioactive Materials and Bioactivity

Biomaterials or bioactive materials are used for replacing or repairing soft and hard tissues. Their investigation emerges in the 60s and outsets with inert materials. The application of a foreign body generates a reaction. To avoid this response, the application of inert implants was proposed. However, no materials implanted are completely inert. Thus, these materials keep isolated from the living tissues by the fibrous tissue with a non-adherent capability. This condition leads to movements in the implant-tissue interface failing the implant and probably a remotion need [17]. The existence of fibrous tissue, or a capsule surrounding the implant, could induce no chemically or biologically bonded [14]. For overcoming this problem, biological fixation and bioactive fixation were considered. The first approach focuses on the development of a rough surface and pores of 150 micrometers (or greater) to promote tissue ingrowth and guarantee its vascularization (angiogenesis). However, the apparition of interfacial movements and damage to the tissue occurs, and the material should remove. After that, the second approach, bioactive fixation, emerged. The evolution yields in second-generation materials: biomaterials or bioactive materials must lead to a strong mechanical interaction between both surfaces (tissue and material) [14]. The change from inert to active has brought a paradigm shift. Therefore, a bioactive material boosts osteogenesis by creating a suitable environment for bone growth. In this process, an interface between living and non-living material is developed [18]. There are different bioactive materials such as bioglass®, synthetic hydroxyapatite, bioactive composite materials, and bioactive coating materials [19]. Depending on the bonding mechanism, the time dependence for establishing the bond, the strength of the bond, and its thickness are the main features characterizing the behavior of implanted materials. In this context, the level of bioactivity can be associated with the rate of interfacial bond generation. Resorbable materials have been applied for designing other types of implants. They are degraded and replaced by normal tissue gradually. These implants are formed by tricalcium phosphate ceramics and, polymers, and others. Remarkably, subproducts obtained by material degradation cannot be toxic. Their amounts released should be handled by cells of the host tissue. The selection of these materials deals with the repair capability of the tissue [20].

Materials that promote osteoblasts' adhesion and proliferation, such as bioactive materials, are considered osteoconductive. Osteoconductive ceramics can bind to soft and hard tissues and can be applied for managing periodontal defects and tympanic ossicle substitution. Different glasses and ceramics have this property of improving bone regeneration. These materials are osteoconductive and osteoinductive because they promote the release of Si^{4+} and Ca^{2+} ions to produce the formation of bone by stimulating the response of the genes in cells. This response is intracellular conversely to the extracellular response associated with osteoconduction. Biomaterials are used in tissue engineering because they are designed to recover the biological functions of damaged tissues by promoting cell activation and regeneration. Therefore, bioactivity is a crucial feature of material used as a bone substitute or bone graft. This property shows the capability of a material to interact with tissues of the organism and return their functionality [21].

Bioactive Materials Used for Composite Development

Three materials have been described in this section as inorganic components of the scaffolds: Hydroxyapatite, tricalcium phosphate, and bioactive glass.

Hydroxyapatite (HA)

Among synthetic materials applied to bone grafting are calcium phosphate ceramics, in particular, hydroxyapatite, β-tricalcium phosphate, and their blends [8, 22, 23]. Hydroxyapatite is a class of calcium phosphates that display suitable properties and show a Ca/P ratio of HA is 1.67 [10]. Synthetic hydroxyapatite is a calcium phosphate bioceramic widely used for the renewal of hard tissues augmenting bone in deficient regions caused by different pathologies, in periodontal disease, in bone repair as a filler in defects produced by infections and tumors, and as a coating of metallic implants [24]. Its applications are due to the suitable biocompatibility, biodegradability, and osteoconductivity [25]. Hydroxyapatite allows chemical interactions with the bone tissue that improves the recuperation of the organ because it resembles the inorganic matrix of bones. Regular hydroxyapatite crystals exhibit poor mechanical properties, and for this reason, they cannot be used for filling broad defects. Nanoscale hydroxyapatite or nanocrystalline hydroxyapatite has overcome this problem because of the superior surface area. This fact leads to sinterability, enhanced densification, and improved fracture toughness [1, 26]. Nanosized hydroxyapatite is applied for improving the behavior of calcium phosphate cement because it increases the setting time and the transfiguration (to apatite during the self-hardening) of cement. Therefore, it promotes the nucleation of new and smaller biogenic hydroxyapatite on its surface[1]. Hydroxyapatite can be synthesized by hydrothermal procedures [25], and calcium ions can substitute for magnesium ions in their lattice and thus, improve their antibacterial properties [27]. Nanohydroxyapatite has been successfully loaded with antibiotics and anti-inflammatory drugs for osteomyelitis treatment. Benedini *et al.* [26] have demonstrated a suitable release of ciprofloxacin in a concentration above minimal inhibitory for three different types of bacteria strains. Hydroxyapatite has been combined with particular polymers for improving its properties. Composite materials based on this ceramic and alginate loaded and not loaded with drugs have been made by Benedini *et al.* [6, 28]. Gelatin [29], chitosan [30], collagen [31], polyacrylic acid [32], and other polymers (or blends) were used for developing composite materials using hydroxyapatite as a ceramic matrix. Benedini *et al.* [32] have shown that chemical modifications of chemical groups exposed on the hydroxyapatite surface yield tailored properties of nanoparticles.

β-tricalcium Phosphate (β-TCP)

β tricalcium phosphate, $\beta\text{-}Ca_3(PO_4)_2$, (β-TCP) is one the most diffused synthetic ceramics proposed as the bone graft substitute [23] because apatite calcium phosphate mineral is the main component of bones. The Ca/P ratio of β-TCP is 1.5 [10]. Hydroxyapatite shows more crystallinity than natural bone, and this feature can be associated with more stability up to a certain point of fragility or an increased risk of fracture in surrounding hydroxyapatite implantation. β tricalcium phosphate is replaced faster than hydroxyapatite by the new bone. It shows thermal stability such as hydroxyapatite and can be degraded in acid conditions. This fact is governed by osteoclast cells *in vivo*. β-TCP can be obtained by the precipitation method, thermal conversion, and solid-state reaction. Tricalcium phosphate can be combined with alginate and gelatin [33], polycaprolactone [34], and hyaluronic acid [35] for obtaining scaffolds for bone tissue regeneration.

Bioactive Glass

Bioactive glasses are one of the most widely applied materials for bone tissue engineering due to their ability to favorably interact with bone (showing the capability to bind to hard and soft tissues without rejection) and their angiogenic and osteogenic properties. The latter is due to these glasses precipitating biogenic hydroxyapatite in natural or simulated physiological conditions [36]. Accordingly, the adsorption of proteins onto the surface of implanted materials is improved, and consequently, the integration with the surrounding bone [3]. Bioactive glasses provide highly reactive surfaces and, for this reason, have partially replaced non-active materials previously used in bone tissue regeneration as devices based on metallic and plastic materials. They are carried out by melt or sol-gel techniques [37].

Bioactive glasses have demonstrated antimicrobial properties due to the release of phosphorous, calcium, and sodium ions. Through this mechanism, the adhesion and bacterial proliferation are reduced due to the released ions contributing to the increase of the osmotic pressure and pH at the implantation site. This mechanism is based on physicochemical behavior and cannot be associated with bacterial resistance. The increase in pH additionally contributes to the development of biogenic hydroxyapatite and cell proliferation [3]. Commercially trademarked Bioglass® 45s5 shows the composition: 45% SiO_2, 24.5% Na_2O, 24.5% CaO, and 6% P_2O_5 [38]. These types of glass were used for bone defects, craniofacial surgeries, and restorative dentistry. When it was applied for periodontal disease, it has kept unchanged the coagulation rates [39]. Additionally, it has demonstrated *in vitro* studies that the phosphate content of the glass plays an important role in osteoblast response [40].

Bioactive glasses have been applied alone and combined with polymers such as gelatin, alginate, collagen, chitosan, and cellulose derivatives [41] for carrying out composite materials with improved mechanical properties. The polymer type can influence the adhesion to the target site.

Polymers

The application of polymers in bone tissue engineering is due to different causes. The first one is the resorption of bone in the implantation place. When a ceramic material or metallic is implanted in a bone, it suffers from resorption because of the differential mechanical properties between the implanted and host material. For example, aseptic loosening of the prosthesis in total hip replacement is a common problem caused by the state of stress and strain in the femoral cortex [42]. Another problem related to these types of implants is the degradability in biological conditions (biodegradability). For these two disadvantages, polymers have been introduced into clinical use. Polymers are easy to manufacture, at low costs, can be tailored to complex shapes, can be blended with other polymers and ceramic to obtain the sought mechanical properties, and are biocompatible. It has been demonstrated that the tensile strength and elastic modulus make possible their application in bone implants [43]. In this section, the most remarkable polymers used jointly with inorganic matrices have been described.

Hyaluronic Acid

Hyaluronan is a linear, anionic, and unbranched polysaccharide formed by alternating units of D-glucuronic acid and N-acetyl-D-glucosamine with a molecular weight of ~ 104 kDa. It is a very flexible molecule, and the ionization is affected by the pH of the medium (because its pka is ~ 3) and the ionic strength [44]. Mechanical parameters such as elastic modulus, shear modulus, viscosity, and viscoelasticity depend on the concentration of this acid and the crosslinking degree of the formed hydrogels [45]. Hyaluronic acid is a constitutive portion of the extracellular matrix of joints, skin, and eyes. For this reason, it is a biocompatible, biodegradable, non-immunogenic, non-inflammatory, and consequently, non-toxic molecule, which plays an essential role in many physiological processes in humans [46].

Hyaluronic acid is a natural polymer widely used in combination with ceramic materials for the fabrication of organic-inorganic composites for biomedical applications as hyaluronan acid promotes biomineralization [47]. Among the inorganic materials used for this aim, we mention hydroxyapatite, biocements based on calcium-phosphate, bioceramics based on silica (SiO_2), and glasses. Each component of the composite material provides singular features generating a synergism and improving the features. In this context, the combination of ceramics-hyaluronan acid has yielded several formulations such as gels, scaffolds, films, and coating materials. Additionally, these formulations have been proposed as a structure for drug delivery [46].

Gelatin

Gelatins are natural helical proteins derived from collagen-containing glycine, proline, hydroxyproline, and alanine as the main amino acids [45]. The amino acid content of gelatin confers rheological and physical-chemical features that must be considered in the design and development of devices. For example, proline and hydroxyproline stabilize the molecule. Consequently, a lower proportion of these amino acids improves its flexibility and facilitates the reorganization of the network. Gelatin is a biodegradable, biocompatible, non-immunogenic, and non-toxic polymer. It is inexpensive and shows high water solubility. This protein has been applied in different pharmaceutical formulations, from oral capsules to scaffolds combined with hydroxyapatite proposed for tissue regeneration [20, 48]. Nanostructured scaffolds were designed using hydroxyapatite and gelatin as the main components for bone repair [49]. In another work, biomimetic fiber mesh scaffolds based on gelatin and hydroxyapatite nano-rods with bone reparation abilities were obtained [29].

Collagen

Collagen represents 30% of all body proteins and is the main structural protein of hard and soft tissues. It plays a crucial role in maintaining the biological and structural integrity of the extracellular matrix and provides physical support to tissues [50]. There are at least 13 types of collagens classified by the length of the chain. Collagen is formed by more than 1000 amino acids that develop a unique triple-helix sequence. The average molecular weight of the helix is 300 kDa with 300 nm length and 1.5 nm diameter [45]. The Young's modulus of dry fibrils of type I collagen is 5 ± 2 GPa. This value decreases when fibrils are immersed in phosphate-buffered saline to 0.2 GPa.

Collagen can be obtained from sources and is widely used in tissue engineering due to its excellent properties, such as low immunogenicity, porous structure, good permeability, biocompatibility, and biodegradability. However, the poor mechanical property of collagen scaffolds limits their applications [51]. For this reason, it has been combined with other polymers and inorganic materials for performing scaffolds that mimic natural extracellular matrix and can be cross-linked by chemical or physical methods. In this context, collagen-hydroxyapatite scaffolds have been reported by Chen *et al.* [52]. The amount of collagen regulates the porosity of the material, compressive modulus, and cellular interaction. In another work, this type of composite is used as a coating material [53].

Alginates

Alginates are algae-derivative polysaccharides. Strictly, they are linear anionic hetero-polysaccharides. They are freely soluble in water and can develop hydrogels. Additionally, alginates are considered allies in many pharmaceutical formulations because of their biodegradability, bioadhesive capability, low toxicity, and low immunogenic risk [45]. Carboxylate groups confer a negative charge to this molecule. This property has a crucial role in biogenic hydroxyapatite formation under certain conditions (*in vivo* and *in vitro*) [6] because they work as nucleation centers favoring osteointegration into the host tissue. This charge participates in electrostatic interactions with cations, mainly divalent cations, and through hydrogen bonds. The relation of this polymer with others and inorganic matrixes for developing composite materials is mediated by these interactions. Therefore, both hydrogels and composite materials can be considered tridimensional arrangements. Divalent ions, such as calcium, reinforce the alginate hydrogel structure. Highly soluble calcium salts, $CaCl_2$, generate rapid gelation, and aggregates appear. However low solubility salts, such as $CaCO_3$, produce smooth films. Alginates can be combined with hydroxyapatite to develop composite materials. The concentration of this ceramic can modify the mechanical properties of the composite [28]. Benedini *et al.* [28] have demonstrated that composite materials based on alginates and synthetic nanohydroxyapatite show pH-dependent behavior, and this property participates in mesogens' development and the release of drugs loaded into them [6, 28].

Polycaprolactone

Polycaprolactone is an aliphatic polyester formed by a homopolymer of caprolactone units with a melting point of 60 °C and Mw between 80000 and 150000 Da. This polymer is biodegradable (~24 months), biocompatible, and safe when applied in injectable formulations or implantable drug delivery systems. In aqueous solutions, it shows hydrolytic degradation, yielding nontoxic hydroxycarboxylic acids. It is included in many formulations such as rods, microspheres, films, beads, and capsules, and in liquid formulations, it has been used as controlled-release systems [44].

Because of these properties, polycaprolactone has been proposed for bone and cartilage repair [54]. Porous scaffolds carried out by the selective laser sintering based on polycaprolactone have been used for delivering bone morphogenetic protein-7 (BMP-7) transduced fibroblasts. Additionally, it has been blended with other polymers [55] and ceramics such as hydroxyapatite [56] for developing composite materials applied to bone tissue engineering.

Table 1. Biomaterials for bone tissue regeneration.

Biomaterials	Properties
Natural Polymers	
Collagen	• Biodegradable, it forms composite materials with ceramics, and coating
Gelatin	• Biodegradable, it forms composite materials with ceramics
Alginate	• Biodegradability, bioadhesive, and pH-sensitive. Its solubility depending on pKa. It can forms hydrogels. It forms composite materials with ceramics
Hyaluronic acid	• Biodegradability, and pH-sensitive. It forms composite materials with ceramics
Synthetic Polymers	
Polycaprolactone	• Bioabsorbable and biodegradable. Thermal stability. It forms composite materials with ceramics
Inorganic Ceramics	
Hydroxyapatite	• It can generate biogenic hydroxyapatite, intermediate mechanical properties, osteoconductivity, and it forms composite materials with polymers
Bioactive Glass	• It precipitates biogenic hydroxyapatite and it forms composite materials with polymers
Tricalcium Phosphate	• Biodegradable. It precipitates biogenic hydroxyapatite. It forms composite materials with polymers

Composite Devices used in Bone Tissue Engineering

Tissue engineering, and principally, bone tissue engineering, emerges to overcome the hindrances of current treatments based on homologous and heterologous bone grafting inserts. These processes have been conceived as a combination of different strategies and materials that involve natural or synthetic polymers, ceramics, and active molecules for restoring and repairing functions of the injured tissue. For this aim, scaffolds and composite materials should gather certain features. The most important properties are: physicochemical and mechanical. Additionally, they must promote the correct cell attachment generating the proliferation and repairing of the tissue and consequently recovering its function [57]. In this section, different composite materials and scaffolds based on materials previously described are addressed. In Table **1**, biomaterials for bone tissue regeneration are summarized.

Hydroxyapatite is one of the most relevant ceramics used for bone tissue engineering. It has been combined with other ceramics and polymers. This strategy improves its mechanical properties or contributes to additional properties, such as elasticity and bio adhesion, to composites. In this sense, blends of hydroxyapatite and β-tricalcium phosphate have been combined with different polymers by applying diverse techniques. Huang *et al.* [58] have developed scaffolds based on the mentioned ceramics and polycaprolactone. The authors have designed composites with 10% polycaprolactone and 20% each ceramic using the 3D printed technique. That is a screw-assisted additive manufacturing system by applying a 0 to 90° lay-down pattern and using a melting temperature of 90°C. The behavior of each composite (polycaprolactone-hydroxyapatite and polycaprolactone- β-tricalcium phosphate) has been compared with a scaffold based on the polymer. All systems have shown a porosity with a porous size of around 300 μm, but other systems made with hydroxyapatite have shown smoother surfaces. The authors have demonstrated that the working temperature does not alter the properties, and the addition of ceramics does not influence the hydrophilic properties of the polymer. During 14 days, scaffolds made with hydroxyapatite exhibit higher proliferation cell rates than the other structures. After mechanical assays (compression tests), the authors suggest that a blend formed by both ceramics in defined proportions and polycaprolactone could improve the mechanical properties of scaffolds. This conclusion is based on hydroxyapatite which shows brittleness and low mechanical stability [59]. Through the cryo-printing technique, another type of scaffold based on polycaprolactone and hydroxyapatite was developed [56]. In this work, with both components, glacial acetic acid, a stable slurry was developed and then scaffolds with 20%, 40%, and 60% hydroxyapatite were carried out. Morphological features were evaluated by SEM, crystallinity by XRD, and surface functional groups by EDX. The tridimensional lattice of scaffolds was made up of orthogonal filaments layer by layer of ~200 μm diameter. This distribution forms interconnected pores of ~240 μm and channels. While the content of hydroxyapatite increases in the scaffolds, the compressive modulus increases. The compression yield and tensile break strain are reduced while the hydroxyapatite content increases. *In vitro* biomineralization test has demonstrated that biogenic hydroxyapatite deposits and protein adsorption onto the materials increases when the ratio of hydroxyapatite increases in the composite material . Finally, the authors have shown through biocompatibility *in vitro* assay an adequate cell adhesion, proliferation, and osteogenic differentiation.

Other types of composite materials have been carried out using hydroxyapatite and collagen. Chen *et al.* [52] have shown the relevance of the proportion of collagen in these scaffolds. The authors reported that the osteogenic differentiation capability of mesenchymal stem cells seeded in the composite material increased when collagen concentration was high (0.70% w/w). However, a suitable proliferation of cell and viability parameters are reported with the use of 0.35% w/w of collagen. In another work [60], tridimensional scaffolds have been developed based on collagen and hydroxyapatite with different stiffness. Three proportions of collagen (0.35, 0.5, and 0.7%w/w) and fixed concentrations of hydroxyapatite (22% w/w) were used for carrying out solutions where the porcine decellularized cancellous bone, was soaked. In this sense, after drying, the solution forms a coating.

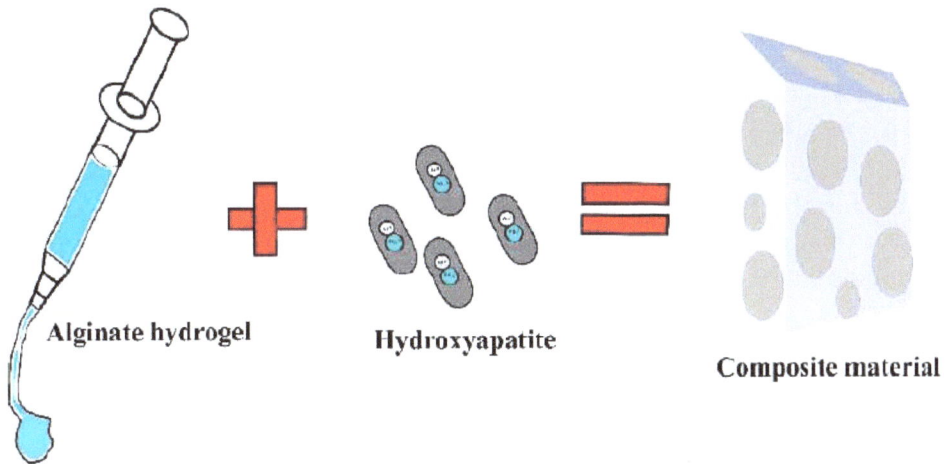

Alginate hydrogel   **Hydroxyapatite**   **Composite material**

Fig. (1). A scheme of alginate matrix reinforced with hydroxyapatite nanoparticles.

Alginate is another polymer that yields composite materials when combined with hydroxyapatite. Benedini *et al.* [28] have reported the application of hydroxyapatite as a calcium source for reinforcing the structure of the polymer for potential application in bone tissue engineering. In this work, different concentrations of hydroxyapatite and $CaCl_2$ were used for gelation. Composites with ratios 2:1, hydroxyapatite:$CaCl_2$, have shown improved behavior in swelling tests when pH values vary in bioactive assays. Alginate 2% in the aqueous solution, reinforced by different sources of calcium, shows good biogenic properties of composites and the capability for developing liquid crystals when pH conditions vary. A scheme of an alginate matrix reinforced with hydroxyapatite nanoparticles is shown in Fig. (**1**). In another study, Benedini *et al.* [6] developed the same composite materials but added antibiotic properties. In this work, two materials were used. In the first material, ciprofloxacin (an antibiotic fluoroquinolone) was loaded onto hydroxyapatite nanoparticles by adsorption method [26] and then incorporated into the alginate solution. In the second material, ciprofloxacin was loaded into an alginate solution, and then nanoparticles were incorporated. Thus, composites formed by loaded nanoparticles have shown improved cell viability, activity against *Staphylococcus aureus* and *Pseudomonas aeruginosa,* and differential behavior against pH variations. Both types of composites have shown suitable activity against *Escherichia coli*. The bioactivity assays have shown no difference between the composites. Finally, both the materials displayed satisfactory cell viability and unmodified cell morphology after 72 h incubation.

These systems based on alginate and nanoparticles of hydroxyapatite have inspired membranes that resemble bone periosteum [61]. In this work, the authors have proposed two-sided membranes where fibroblasts have grown on one side and osteoblast on the other. These membranes have been carried out with different ratios of the ceramic and show different properties; for example, those with the highest amount of nanohydroxyapatite significantly induced osteoblast differentiation, and their degradation is increased. This last fact could be due to a certain instability produced by the crystalline phase. Young's modulus is an important parameter studied in this work, and it demonstrates that for these membranes, this modulus is comparable with that of cranial bone (5-18 GPa). This result is very encouraging because a similar modulus can guarantee the success of the implant.

Composites based on hydroxyapatite and whey protein isolate were designed for bone regeneration treatments [62]. Whey protein was proposed for increasing the strength of the composite without a loss of bioactivity. In this work, composites were carried out through the gelation of gel formed by 40% whey protein and three concentrations of two different synthetic hydroxyapatite (5, 10, and 15%). These hydroxyapatites have shown differences in crystallinity degree observed by X-ray diffraction. With scanning electron microscopy and energy-dispersive X-ray spectroscopy, after 15 days of incubation in simulated body fluid, the strongest biomineralization was reached by a composite formed by 15% of the ceramic. The hydroxyapatite concentration influences the swelling behavior. At higher content of the ceramic, the swelling is lower. Additionally, whey proteins promote monocyte migration and activation in the implantation site and encourage tissue regeneration.

Gelatin, as a collagen derivative, is a widely used polymer for developing composite materials. In this context, biomimetic fiber mesh scaffolds have been carried out with this polymer and hydroxyapatite nanorods. In this work, the intrinsic skills of the composite material have been tuned to attain bone reparation abilities [30]. The composites were prepared using 0.8 g/ml of gelatin added with 0.85 mg/ml of hydroxyapatite nanoparticles previously synthesized. For the reinforcement of the material, two concentrations of the crosslinker agent, tannic acid (12.4 and 33 mg per gram of gelatin), were added, and consequently, two materials were obtained. Increased concentration of tannic acid increases the ultimate strain values and yield strength. However, ultimate strength was lower for the composite with high tannic acid content. The bioactivity assay has demonstrated that the increased biogenic hydroxyapatite deposits reach a peak at 14 days, and after that, the material starts to degrade. About 15% of the scaffold degrades in 21 days. Therefore, the scaffold developed system with 12.4 mg of tannic acid has shown the best mechanical properties, thermal stability, open porosity volume fraction, and structural permeability to indirectly evaluate their potential use in bone tissue repair. Factors that influence the interaction between gelatin and hydroxyapatite in crosslinked scaffolds have been studied by Zhang *et al.* [63]. Two types of scaffolds with gelatin, hydroxyapatite, and the crosslinker, glutaraldehyde, were carried out: gel and solid scaffolds. After incubation in SBF, gel and solid scaffolds have shown an increase in Young's modulus compared with the same scaffolds provided with commercial apatite. The authors demonstrate that this improvement in mechanical features is due to carbonate hydroxyapatite deposits such as the natural bone remodeling mechanism. Additionally, the growth of biogenic apatite inside the gel was more than in a solid scaffold due to channels that deliver water and ions inside the gel.

Reiter *et al.* [64] have carried out different 3D sponge-like scaffolds based on 45s5 bioactive glass by foam application technique. In this work, gelatin was used as a coating agent and reservoir of icariin. Icariin is a flavonoid used for treating fractures, bone and joint diseases. Once prepared, scaffolds were soaked into different gelatin solutions (with icariin dissolved). The relative efficiencies of crosslinking agents, such as caffeic acid and N-(3-Dimethylaminopropyl)-N′-ethylcarbodiimide hydrochloride (EDC)/N-hydroxysuccinimide (NHS), were investigated. These agents modify the mechanical properties, bioactive behavior, and icariin release capability of the diverse scaffold types. The compression assay has shown that the compressive strength and the work of fracture (area under the curve) have increased drastically in gelatin-coated scaffolds compared to uncovered ones. The range of compressive strength was 1.8–2.9 MPa, independently of the crosslinking agent used. For all these systems, icariin has shown a burst release during the first 6 h, and then a controlled release was observed. However, scaffolds crosslinked with caffeic acid have shown higher release concentration than scaffolds built with the other agent and during more time (900 h). In this context, Hum *et al.* have developed scaffolds based on 45s5 bioactive glass coated with collagen membranes for potential application in bone tissue engineering [65].

For application in scaffold development, collagen can be purchased or isolated from rat tails and other sources such as pigs or bovine skins. However, Dhinasekaran *et al.* [66]. have obtained collagen from fish scales. In this work, the authors have used bioactive glass (45s5) to carry out hollow and solid fibers by electrospinning technique, and then collagen sheets were layered onto them. First, collagen is obtained from fish scales and dried. Second, the hollow bioactive glass fibers and solid bioactive glass fibers are obtained. Finally, both bioactive glasses are cast onto collagen fibers, and materials are obtained. Mechanical tests have shown that these scaffolds have mechanical properties ranging from cortical bone. The membranes have shown high bioactivity and wide hemocompatibility. The cell viability was tested with fibroblast cell lines. This assay has shown a random cell growth onto solid fibers' bioactive glass-collagen membranes, but hollow ones have shown a particular alignment. Due to this fact, these last membranes can be used for promoting the directional growth of cells.

Bioactive glass doped with strontium combined with collagen has been used for developing scaffolds for evaluating their bone regeneration effects [67]. Strontium has been included in the materials because it has demonstrated antibiotic properties against *Escherichia Coli* and *Porphyromonas gingivalis* [68] and an enhancement of femoral bone restoration [69]. In this work, scaffolds were prepared by physically mixing SrO and bioactive glass in ethanol, and then the sample was milled. Then, collagen solution was added to the doped bioglass in a ratio collagen:bioglass 1:2. Finally, the polymer was crosslinked with EDC/NHS, washed, and freeze-dried. The authors have demonstrated suitable cell viability and proliferation of human stem cells. Sr has shown a relevant role in cell differentiation and the doped scaffolds have promoted protein synthesis crucial for bone health. An *in vivo* study has shown that Sr-doped scaffolds have low degradation rates according to bone regeneration rates. Their water absorption capability displays the possibility for the transportation of cells and nutrients.

Bioactive glass can be combined with hyaluronic acid for developing scaffolds applied to bone tissue regeneration. In this sense, injectable biocomposites based on sol/gel-derived bioactive glass nanoparticles and hyaluronic acid solution have been carried out with this aim [70]. The authors have produced two types of active bioglasses by sol-gel processes, and the resultant products were characterized by x-RD to know the phase composition, DSC for thermal behavior, N2 adsorption-desorption isotherms, and their morphology by TEM. The obtained nanoparticles were 300 and 50 nm long, respectively. The composites were performed with a 3% hyaluronic acid solution and a ratio of solid-liquid (bioglass-hyaluronic solution) of 0.25 and 0.33 mg/ml. Both composites have shown good injectability, but composites with lower-size nanoparticles are more bioactive than the other. This composite exhibits high viscosity, and both behave as non-thixotropic fluids.

Mentioned scaffolds have shown many advantages compared to first-generation materials. Biocompatibility with cell lines or *in vivo*, bioactivity, osteoconduction, low reabsorption rates, low toxicity, and low rates of immunogenic reactions are some advantages of these materials. However, some mechanical properties have not been reached to be applied and commercialized freely. With constant advances in this field, the challenges will be overcome in the following decades, and in this way, the life quality of human beings will be improved.

CONCLUDING REMARKS

Ceramic/polymer composites for bone tissue regeneration emerged some decades ago as a promising strategy in response to different requirements imposed by bone diseases. Additionally, they have been proposed to overcome the limitations of autografts, xenografts, and allografts. Ceramics provide strength, osteoconductivity, and osteoinductivity to the synthesized material, and polymers contribute to flexibility, adaptability, and channels for the transportation of water and nutrients. We are still seeking best biologically active systems for bone regeneration. However, the materials based on hyaluronic acid, alginates, polycaprolactone, collagen, and gelatin, combined with ceramics such as hydroxyapatite, tricalcium phosphate, and bioactive glass, gather many of the relevant features for facing the challenges of bone tissue regeneration.

ACKNOWLEDGEMENTS

The authors acknowledge Universidad Nacional del Sur (UNS-PGI 24/Q092), Concejo Nacional de Investigaciones Científicas y Técnicas de la República Argentina (CONICET, PIP – 11220130100100CO) and Agencia Nacional de Promoción Científica y Tecnológica (ANPCyT, PICT 201-0126), Argentina. LB and PM are researchers of CONICET.

REFERENCES

[1]   Messina PV, D'Elía NL, Benedini LAJNfNT. Bone tissue regenerative medicine *via* bioactive nanomaterials. 2017; 769-92.
[http://dx.doi.org/10.1016/B978-0-323-46142-9.00028-1]

[2]   Perry CRJCO, Research R. Bone repair techniques, bone graft, and bone graft substitutes. 1999; 71-86.
[http://dx.doi.org/10.1097/00003086-199903000-00010]

[3]   Boyan BD, Cohen DJ, Schwartz Z. Schwartz, 7.17 Bone tissue grafting and tissue engineering concepts Comprehensive Biomaterials II 2017; 298-313.

[4]   Joshi D, Garg T, Goyal AK, Rath G. Advanced drug delivery approaches against periodontitis. Drug Deliv 2016; 23(2): 363-77.
[http://dx.doi.org/10.3109/10717544.2014.935531] [PMID: 25005586]

[5]   Lienemann PS, Vallmajo-Martin Q, Papageorgiou P. *et al.* Smart hydrogels for the augmentation of bone regeneration by endogenous mesenchymal progenitor cell recruitment. 2020. 7(7): 1903395.
[http://dx.doi.org/10.1002/advs.201903395]

[6]   Benedini L, Laiuppa J, Santillán G, Baldini M, Messina P. *et al.* Antibacterial alginate/nano-hydroxyapatite composites for bone tissue engineering: Assessment of their bioactivity, biocompatibility, and antibacterial activity. 2020. 115: 111101.

[7]   Messina P, Luciano B, Placente D. Tomorrow's healthcare by nano-sized approaches: A bold future for medicine. CRC Press 2020.
[http://dx.doi.org/10.1201/9780429400360]

[8]   Alizadeh-Osgouei M, Li Y, Wen CJBm. A comprehensive review of biodegradable synthetic polymer-ceramic composites and their manufacture for biomedical applications. 2019. 4: 22-36.
[http://dx.doi.org/10.1016/j.bioactmat.2018.11.003]

[9]   Park J, Lakes RS. Biomaterials: an introduction. Springer Science & Business Media 2007.

[10]   Victor SP, Muthu JJJoM. Polymer ceramic composite materials for orthopedic applications—relevance and need for mechanical match and bone regeneration. 2014. 2(1): 1-10.

[11]   Messina PVB. L. A.; Placente, D., Nanotechnology application in tissue regeneration and regenerative medicine: Smart tools in modern healthcare.Tomorrow's Healthcare by Nano-sized Approaches. Boca Raton: CRC Press 2020; pp. 253-75.
[http://dx.doi.org/10.1201/9780429400360-10]

[12]   Ovalle WK, Nahirney PC. Netter's essential histology e-book: With correlated histopathology. Elsevier Health Sciences 2020.

[13]   Carter D, Van der Meulen M, Beaupre GJB. Mechanical factors in bone growth and development. 1996. 18(1): S5-S10.
[http://dx.doi.org/10.1016/8756-3282(95)00373-8]

[14]   Cao W, Hench LLJCi. Bioact Mater 1996; 22(6): 493-507.

[15]   Meyers MA, Chen P.Y, Lin AYM, Seki Y. Biological materials: Structure and mechanical properties. 2008. 53(1): 1-206

[16]   Landis WJ, Hodgens KJ, Arena J, Song MJ, McEwen BF. Structural relations between collagen and mineral in bone as determined by high voltage electron microscopic tomography. 1996. 33(2): 192-202.
[http://dx.doi.org/10.1002/(SICI)1097-0029(19960201)33:2<192::AID-JEMT9>3.0.CO;2-V]

[17]   Salinas AJ, Vallet-Regí MJRa. Bioactive ceramics: from bone grafts to tissue engineering. 2013. 3(28): 11116-11131
[http://dx.doi.org/10.1039/c3ra00166k]

[18]   Takadama H, Kokubo T. *In vitro* evaluation of bone bioactivity. Bioceramics and their clinical

applications. Elsevier 2008; pp. 165-82.
[http://dx.doi.org/10.1533/9781845694227.1.165]

[19] Zhao X. Introduction to bioactive materials in medicine.Bioactive Materials in Medicine. Elsevier 2011; pp. 1-13.

[20] Benedini LA, Messina PV. Nanodevices for facing new challenges of medical treatments: Stimuli-responsive. Drug Deliv Syst 2021.

[21] Kokubo T, Takadama HJB. How useful is SBF in predicting *in vivo* bone bioactivity? 2006. 27(15): 2907-2915.
[http://dx.doi.org/10.1016/j.biomaterials.2006.01.017]

[22] Ginebra MP, Espanol M, Maazouz Y, Bergez V, Pastorino D. Bioceramics and bone healing. 2018. 3(5): 173-183.
[http://dx.doi.org/10.1302/2058-5241.3.170056]

[23] Bohner MJI. Calcium orthophosphates in medicine: from ceramics to calcium phosphate cements. 2000. 31: D37-D47.
[http://dx.doi.org/10.1016/S0020-1383(00)80022-4]

[24] Dorozhkin SVJM. Nanodimensional and nanocrystalline apatites and other calcium orthophosphates in biomedical engineering, biology and medicine. 2009. 2(4): 1975-2045.
[http://dx.doi.org/10.3390/ma2041975]

[25] D'Elía NL, Gravina AN, Ruso JM, Laiuppa JA, Santillán GE, Messina PV. Manipulating the bioactivity of hydroxyapatite nano-rods structured networks: effects on mineral coating morphology and growth kinetic. 2013. 1830(11): 5014-5026.
[http://dx.doi.org/10.1016/j.bbagen.2013.07.020]

[26] Benedini L, Placente D, Ruso J, Messina P. Adsorption/desorption study of antibiotic and anti-inflammatory drugs onto bioactive hydroxyapatite nano-rods. 2019. 99: 180-190.

[27] Andrés NC, Sieben JM, Baldini M, Rodríguez CH, Famiglietti Á, Messina PV. Electroactive Mg2+-hydroxyapatite nanostructured networks against drug-resistant bone infection strains. 2018. 10(23): 19534-19544.

[28] Benedini L, Placente D, Pieroni O, Messina P. Assessment of synergistic interactions on self-assembled sodium alginate/nano-hydroxyapatite composites: to the conception of new bone tissue dressings. 2017. 295: 2109-2121.

[29] Sartuqui J, Gravina AN, Rial R. *et al.* Biomimetic fiber mesh scaffolds based on gelatin and hydroxyapatite nano-rods: Designing intrinsic skills to attain bone reparation abilities. 2016. 145: 382-391.

[30] Gritsch L, Maqbool M, Mouriño V. *et al.* Chitosan/hydroxyapatite composite bone tissue engineering scaffolds with dual and decoupled therapeutic ion delivery: Copper and strontium. 2019. 7(40): 6109-6124.

[31] Yu L, Rowe DW, Perera IP. *et al.* Intrafibrillar mineralized collagen–hydroxyapatite-based scaffolds for bone regeneration. 2020. 12(16): 18235-18249.
[http://dx.doi.org/10.1021/acsami.0c00275]

[32] Benedini LA, Moglie Y, Ruso JM, Nardi S, Messina PV. Hydroxyapatite Nanoparticle Mesogens: Morphogenesis of pH-Sensitive Macromolecular. Liq Cryst 2021; 21(4): 2154-66.

[33] Eslaminejad MB, Mirzadeh H, Mohamadi Y, Nickmahzar A. Bone differentiation of marrow-derived mesenchymal stem cells using β-tricalcium phosphate–alginate–gelatin hybrid scaffolds. 2007. 1(6): 417-424.
[http://dx.doi.org/10.1002/term.49]

[34] Erisken C, Kalyon DM, Wang HJB. Functionally graded electrospun polycaprolactone and β-tricalcium phosphate nanocomposites for tissue engineering applications. 2008. 29(30): 4065-4073.

[http://dx.doi.org/10.1016/j.biomaterials.2008.06.022]

[35] Han SH, Jung SH, Lee JHJJoBS. Preparation of beta-tricalcium phosphate microsphere-hyaluronic acid-based powder gel composite as a carrier for rhBMP-2 injection and evaluation using long bone segmental defect model. 2019. 30(8): 679-693.

[36] Hench LL, Polak JMJS. Third-generation biomedical materials. 2002. 295(5557): 1014-1017.
[http://dx.doi.org/10.1126/science.1067404]

[37] Tahriri M, Bader R, Yao W. *et al.* Bioactive glasses and calcium phosphates. 2017: 7-24.
[http://dx.doi.org/10.1016/B978-0-08-100961-1.00002-5]

[38] Hench LL, Jones JRJFib. Bioactive glasses: frontiers and challenges. 2015. 3: 194.

[39] Profeta AC, Prucher GMJDmj. Bioactive-glass in periodontal surgery and implant dentistry 2015.
[http://dx.doi.org/10.4012/dmj.2014-233]

[40] Lossdörfer S, Schwartz Z, Lohmann CH, Greenspan DC, Ranly DM, Boyan BD. Osteoblast response to bioactive glasses *in vitro* correlates with inorganic phosphate content. Biomaterials. 2004 Jun; 25(13): 2547-55.
[http://dx.doi.org/10.1016/j.biomaterials.2003.09.094]

[41] Sergi R, Bellucci D, Cannillo VJM. A review of bioactive glass/natural polymer composites: State of the art. 2020. 13(23): 5560.

[42] Wang MJB. Developing bioactive composite materials for tissue replacement 2003.
[http://dx.doi.org/10.1016/S0142-9612(03)00037-1]

[43] Pielichowska K, Blazewicz SJBl. Bioactive polymer/hydroxyapatite (nano) composites for bone tissue regeneration. 2010: 97-207.

[44] Sheskey RCRPJ, Quinn ME. Handbook of pharmaceutical excipients. 2009.

[45] Ruso JM, Messina PV. Application of natural, semi-synthetic, and synthetic biopolymers used in drug delivery systems design.Biopolymers for medical applications. CRC Press 2017; pp. 46-73.
[http://dx.doi.org/10.1201/9781315368863]

[46] Sikkema R, Keohan B, Zhitomirsky IJM. Hyaluronic-Acid-Based Organic-Inorganic Composites for Biomedical Applications 2021; 14(17): 4982.

[47] Chen ZH, Ren XL, Zhou HH. *et al.* The role of hyaluronic acid in biomineralization. 2012.
[http://dx.doi.org/10.1007/s11706-012-0182-4]

[48] Benedini LJCP, Science P. Advanced protein drugs and formulations 2022.
[http://dx.doi.org/10.2174/1389203722666211210115040]

[49] Azami M, Tavakol S, Samadikuchaksaraei A. *et al.* A porous hydroxyapatite/gelatin nanocomposite scaffold for bone tissue repair: *in vitro* and *in vivo* evaluation. 2012. 23(18): 2353-2368.

[50] Gelse K, Pöschl E, Aigner TJAddr. Collagens—structure, function, and biosynthesis. 2003. 55(12): 1531-1546.

[51] Dong C, Lv YJP. Application of collagen scaffold in tissue engineering: recent advances and new perspectives 2016.
[http://dx.doi.org/10.3390/polym8020042]

[52] Chen G, Lv Y, Guo P. *et al.* Matrix mechanics and fluid shear stress control stem cells fate in three dimensional microenvironment. Curr Stem Cell Res Ther. 2013 Jul; 8(4): 313-23
[http://dx.doi.org/10.2174/1574888X11308040007]

[53] Chen G, Lv Y, Dong C, Yang L. Effect of internal structure of collagen/hydroxyapatite scaffold on the osteogenic differentiation of mesenchymal stem cells. 2015. 10(2): 99-108.

[54] Williams JM, Adewunmi A, Schek RM. *et al.* Bone tissue engineering using polycaprolactone scaffolds fabricated *via* selective laser sintering. 2005.

[http://dx.doi.org/10.1016/j.biomaterials.2004.11.057]

[55]   Cai Q, Bei J, Wang SJJoBS. Polymer Edition, Synthesis and degradation of a tri-component copolymer derived from glycolide, L-lactide, and ε-caprolactone. 2000. 11(3): 273-288.

[56]   Li Y, Yu Z, Ai F. *et al.* Characterization and evaluation of polycaprolactone/hydroxyapatite composite scaffolds with extra surface morphology by cryogenic printing for bone tissue engineering. 2021. 205: 109712.

[57]   Murugan S, Parcha SRJJoMSMiM. Parcha, Fabrication techniques involved in developing the composite scaffolds PCL/HA nanoparticles for bone tissue engineering applications. 2021. 32(8): 93.

[58]   Huang B, Caetano G, Vyas C. *et al.* Polymer-ceramic composite scaffolds: The effect of hydroxyapatite and β-tri-calcium phosphate. 2018. 11(1): 129.

[59]   Akpan E, Dauda M, Kuburi LS, Obada DO, Dodoo-Arhin D. A comparative study of the mechanical integrity of natural hydroxyapatite scaffolds prepared from two biogenic sources using a low compaction pressure method. 2020.
[http://dx.doi.org/10.1016/j.rinp.2020.103051]

[60]   Chen G, Dong C, Yang L, Lv Y. 3D scaffolds with different stiffness but the same microstructure for bone tissue engineering. 2015. 7(29): 15790-15802.

[61]   D'Elía NL, Rial Silva R, Sartuqui J, *et al.* Development and characterisation of bilayered periosteum-inspired composite membranes based on sodium alginate-hydroxyapatite nanoparticles. J Colloid Interface Sci. 2020 Jul 15;572: 408-420.
[http://dx.doi.org/10.1016/j.jcis.2020.03.086]

[62]   Słota D, Głąb M, Tyliszczak B, *et al.* Composites based on hydroxyapatite and whey protein isolate for applications in bone regeneration. Materials (Basel). 2021 Apr 29; 14(9): 2317.

[63]   Zhang Z, Li K, Zhou W. *et al.* Factors influencing the interactions in gelatin/hydroxyapatite hybrid materials. 2020. 8: 489.

[64]   Reiter T, Panick T, Schuhladen K, Roether JA, Hum J, Boccaccini AR. Bioactive glass based scaffolds coated with gelatin for the sustained release of icariin. 2019.
[http://dx.doi.org/10.1016/j.bioactmat.2018.10.001]

[65]   Hum J, Boccaccini ARJIjoms. Collagen as coating material for 45S5 bioactive glass-based scaffolds for bone tissue engineering 2018.
[http://dx.doi.org/10.3390/ijms19061807]

[66]   Dhinasekaran D, Vimalraj S, Rajendran AR, Saravanan S, Purushothaman B, Subramaniam B. Bio-inspired multifunctional collagen/electrospun bioactive glass membranes for bone tissue engineering applications. 2021. 126: 111856.

[67]   Mosaddad SA, Yazdanian M, Tebyanian H. *et al.* Fabrication and properties of developed collagen/strontium-doped Bioglass scaffolds for bone tissue engineering. 2020. 9(6): 14799-14817.

[68]   Kargozar S, Montazerian M, Fiume E, Baino F. Multiple and promising applications of strontium (Sr)-containing bioactive glasses in bone tissue engineering. 2019. 7: 161.

[69]   Wei L, Ke J, Prasadam I. A comparative study of Sr-incorporated mesoporous bioactive glass scaffolds for regeneration of osteopenic bone defects. Osteoporos Int. 2014 Aug; 25(8): 2089-96.
[http://dx.doi.org/10.1007/s00198-014-2735-0]

[70]   Sohrabi M, Hesaraki S, Kazemzadeh A, Alizadeh M. Development of injectable biocomposites from hyaluronic acid and bioactive glass nano-particles obtained from different sol-gel routes. Mater Sci Eng C Mater Biol Appl. 2013 Oct; 33(7): 3730-44.
[http://dx.doi.org/10.1016/j.msec.2013.05.005]

SUBJECT INDEX

A

Acid 38, 41, 42, 43, 70, 91, 203, 212, 231, 237, 238, 239, 240, 241, 242, 245, 246, 247
  amino 239, 240
  caffeic 245, 246
  glacial acetic 242
  hyaluronan 239
  hyaluronic 231, 237, 238, 239, 241, 247
  hydrochloride 91
  nitric 91
  phytic 38
  poly lactic (PLA) 41, 42, 43, 212
  polyacrylic 237
  pyrenebutanoic 203
  tannic 245
Activity 45, 65, 92, 94, 129, 132, 142, 145, 188, 192, 201, 244
  anti-inflammatory 92
  antioxidant 188
  electrophysiological 201
  fluorescent 192
  hydrolytic 129
  metabolic 142, 145
  mitochondrial 45
  osteoblastic 65
Adsorbing fibrinogen 159
Adsorption 65, 117, 190, 195, 237
  gas 190
Alanine transaminase 94
Alkoxides, organometallic 87
Amorphous calcium phosphates (ACP) 6, 37, 129, 132, 134, 152, 160
Angiogenesis 2, 3, 5, 14, 15, 38, 43, 83, 93, 150, 162
Anti-cancer drugs 194
Anti-inflammatory 6, 137, 237
  drugs 137, 237
  responses 6
Antibacterial 1, 2, 6, 19, 33, 46, 47, 58, 66, 73, 83, 89, 90, 117

activity 1, 2, 6, 19, 83, 89, 90
  effect 47, 58, 73, 117
Anticancer 2, 137, 196
  activity 2, 196
  drugs 137
Apatite 45, 47, 111, 135
  formation 45, 47
  growth 111
  nucleation 135
Apoptosis 15, 141, 209
Applications 13, 14, 17, 59, 112, 119, 122, 128, 150, 192
  musculoskeletal 119
  orthopedic 13, 17, 59, 112, 119, 122, 128, 150
  osseointegration 14
  therapeutic 192
Arthrodesis Application 12, 19

B

BCP-biphasic calcium phosphates 157
Bioabsorbable bioactive microparticles 159
Bioactive 8, 13, 95, 125, 126, 138, 235
  and bioresorbable ceramics 125
  borate glass fibers (BBGFs) 95
  ceramics 8, 13, 125, 126, 138
  fixation 235
Bioactive glass 58, 68
  properties 58
  synthesis methods 68
Biocompatibility 12, 84, 108
  and corrosion resistance 108
  assays 84
  properties 12
Biomaterials 36, 74, 86, 115, 132, 156
  polymeric 115, 132
  silicate-based 74
  third-generation 86, 156
  traditional metallic 36

www.ingramcontent.com/pod-product-compliance
Lightning Source LLC
Chambersburg PA
CBHW050820220326
41598CB00006B/268